ACKNOWLEDGEMENTS

The following companies and individuals have contributed to the publication of this manual through technical knowledge and information or through photos and drawings. Their contributions have greatly added to the production of this manual and very much appreciated by the authors and Hobar Publications.

Allen-Bradley Company, Milwaukee, Wisconsin 53204

Bear, William A., Photography Work, LaCrosse, Wisconsin 54601

Baldor Electric Company, Fort Smith, Arkansas 72902

Bodine Electric Company, Chicago, Illinois 60618

Delmar Publishers, Inc., Albany, New York 12205

General Electric Co., Ft. Wayne, Indiana 46804

Gould Inc., Electric Motor Division, St. Paul, Minnesota 55164

Grainger, W. W., Chicago, Illinois 60648

Lincoln Electric Co., Cleveland, Ohio 44117

Marathon Electric, Wausau, Wisconsin 54401

Reliance Electric, Cleveland, Ohio 44117

Square D. Co., Milwaukee, Wisconsin 53201

Twin City Industrial Motor Repair Inc., Components and Technical Review, St. Paul, Minnesota 55117

Westinghouse Electric Corporation, Buffalo, New York 14240

TABLE OF CONTENTS

ELECTRIC MOTORS
Principles, Controls, Service, and Maintenance

ELECTRIC MOTORS
Principles, Controls, Service, and Maintenance

HISTORY AND APPLICATION OF ELECTRIC MOTORS

DEVELOPMENT OF THE ELECTRIC MOTOR

The electric motor, as it appears today, has not had a great change in its physical appearance since its early development. The basic principles and rules of magnetism and electricity which made the early motors rotate are still practical and in practice today. As these principles were identified for this manual, individuals' names started appearing which sparked an interest for the development of a chronological listing of motor principles and people.

MAGNETISM

The ore of iron, Fe3 04, magnetite is a crystalline mineral, dark in color has a metallic luster, and is magnetic. The magnetic property of the lodestone, no doubt, prompted the name for the ore. Whether fact or fiction, references have been made to Chinese historians as early as 2637 B.C., about applications of metals with magnetic properties. The magnetic compass was a practical outgrowth resulting from experiments with magnetism. The actual time of development is unknown but written details were recorded in 1269 by an Italian, Petrus Peregrinus. In 1600, William Gilbert from England published his research findings on electricity and magnetism.

ELECTRICITY

The first machine to make electricity, in a crude form, was made by Ott von Guericke, a German in 1660. Materials serving as conductors were identified in 1729 by Stephen Gray in England. In 1734 Charles Du Fay of France announced that some electricity repels similar charges and that opposite kinds attract. Sir William Watson of England is credited to naming electrical charges as positive and negative. Benjamin Franklin, U.S., in June 1752 tossed his kite into the electrical ring of activities and proved that lightning is an electrical discharge. In 1748, Franklin had already made a machine with a wheel that was rotated by static electricity. Although the rotating wheel was an interesting event, he was not credited with inventing the electric motor.

INDUCTION

Electrical induction occurs when an object becomes electrified when it is approached by another body carrying an electrical charge. Franklin and others were experimenting with induction in 1775, but the recognized inventor of electro-magnetic induction was Michael Faraday in England in 1831. In 1800 Alessandro Volta from England, made public the Volta pile where an electrical charge was not dissipated but constantly renewed. This and further findings resulted in Volta's definition of electromotive force (EMF). The Voltaic cell led to the invention of the battery. In 1812 Hans Oerstad from Denmark was expressing the belief that magnetic fields were associated with electricity. In 1820, Andre Ampere from France, discovered that there were forces between two wires carrying electric current. Ampere defined the following electrical concepts: (1) the relationship between the direction of current flow and the deflection of a magnetic needle; (2) parallel conductors carrying current in the same direction attract each other; and (3) conductors carrying current in opposite directions are mutually repelled. Ampere was also influential in discovering that a wire helix carrying an electric current is capable of magnetizing a soft iron bar around which it is wound.

ELECTROMAGNET

The first developed electromagnet was in 1821 by Joseph Henry, from U.S. A practical electromagnet as known today, was not developed until 1830. In 1821, Faraday discovered that a conductor carrying a current would rotate about a magnetic pole and that a magnetized needle would rotate about a wire carrying an electric current. George Simon Ohm, Germany, in 1827 published his studies on the law that Volts equals Ampere times resistance ($V=IR$) which was later called Ohms Law. Faraday's 1831 contribution in magnetism and electricity was a machine called a unipolar generator. His experiments with this machine verified the process of electromagnetism which opened the way for conversion of mechanical energy into chemical energy. Faraday has been credited with the discovery of electromanetic induction and Henry with the discovery of self-induction. Self-induction is the property of an electrical circuit that causes a counter electromotive force when the circuit is made or broken.

The discovery of electromagnetism, electromagnetic induction and self-induction by Oerstad, Faraday and Henry, respectively, provided the discoveries necessary for the generation of electricity by mechanical methods

GENERATION OF ELECTRICITY

Generation of electricity had to be accomplished before the discovery of an official electric motor. The early generators were made by Hypolite Pixii of France and Dal Negro of Italy in 1832; Saxton of U.S. in 1833; Clark of England in 1834; and Page of U.S. in 1835. A Belgian professor of physics, M Nollet in 1850 produced a practical revision of previous machines. The Nollet machine recorded a 50 volt output and produced 53, 1/3 cycles per second. Another new machine, the dynamo, had its development credited to four men:(1) Moses G. Farmer of U.S.; (2) Alfred Varley and Charles Wheatstone of England; and (3) Werner Siemers of Germany. The next invention recognized in the development of the electric motor was the slotted ring armature. This was developed by Antonio Pacinotti from Italy in 1860.

DC MOTORS

Next came the process of inventions which would convert electrical energy into mechanical energy. The listing of inventions and persons involved with the electric motor is in Table 1-1.

Table 1-1. Electric Motor Development

NAME	DATE AND PLACE	PRODUCT
Sturegon	1832	General Type Motor
Henry		Walking Beam Motor
Davenport	1834, U.S.	Motor for Railway
Davidson	1838, Scotland	Electric Car
Jacobi	1839, U.S.	Motor for a Boat

At an industrial exposition in Vienna, Austria, in 1873 a technician was preparing an exhibit of Gramme Dynamo-electric machines when a mistake was made. The leads on the second machine were reversed and the first machine served as the dynamo or generator producing the electricity and the second machine utilized this energy to respond as a motor. These early electric motors were DC types.

ALTERNATING CURRENT

In 1882, Goulard and Gibbs, from England, were credited with work on transformers as known today. George Westinghouse in 1885 purchased the Goulard and Gibbs rights in the U.S. By 1886, the first alternating current (AC) generating plant was in operation in Buffalo, New York. The opposition to Westinghouse and AC was the Edison Company that was producing DC. Edison stressed that the danger from AC would come from the high voltage and the transmission of the electrical energy. The Edison concept lost and Westinghouse AC was accepted. The early AC generating plants produced 25 cycle current for motors and 50 or 60 cycle for lighting purposes. The early plants produced single-phase current and later expanded to two- and three-phase. Man and machine have settled on single- and three-phase power generation.

AC MOTORS

Recognized as early inventors of today's induction run electric motors were: (1) F.C. Bailey of England in 1879; (2) Galileo Ferraris of Turin, Italy, in 1885; and (3) Nikola Tesla, Hungary, in 1888. George Westinghouse recognized the potential of the Tesla induction motor and acquired both the invention and services of Mr. Tesla. The AC motor came about only a few years after the DC motor.

The principles of magnetism and electricity associated with electric motors may have been introduced to you in other classes or programs. The names associated with these principles may or may not be familiar. These important and historical events have influenced the development of the electric motor. The knowledge of how electricity is generated is essential in the understanding of how electric motors operate. A simple explanation is that an electric motor is, in fact, the opposite of a generator or alternator. Note the different motor styles in Figure 1-1.

Figure 1-1. Electric Motor Styles

ELECTRIC MOTORS TODAY

Motors, in their basic operation, are generators or alternators operating in reverse. If we visualize a simple closed loop version as this basic principle, we have a transmission of power shown in Figure 1-1. In this example diesel fuel, which has a potential for power because of its available British Thermal Units (BTU) is burned in the diesel engine and converted to rotating forces which drive the power take-off (PTO) turning the generator or alternator. The turning of the generator or alternator produces power, measured in watts (W), which when connected by conductor wires to the motor, will turn the motor. The motor will then drive a machine, which does useful work for us.

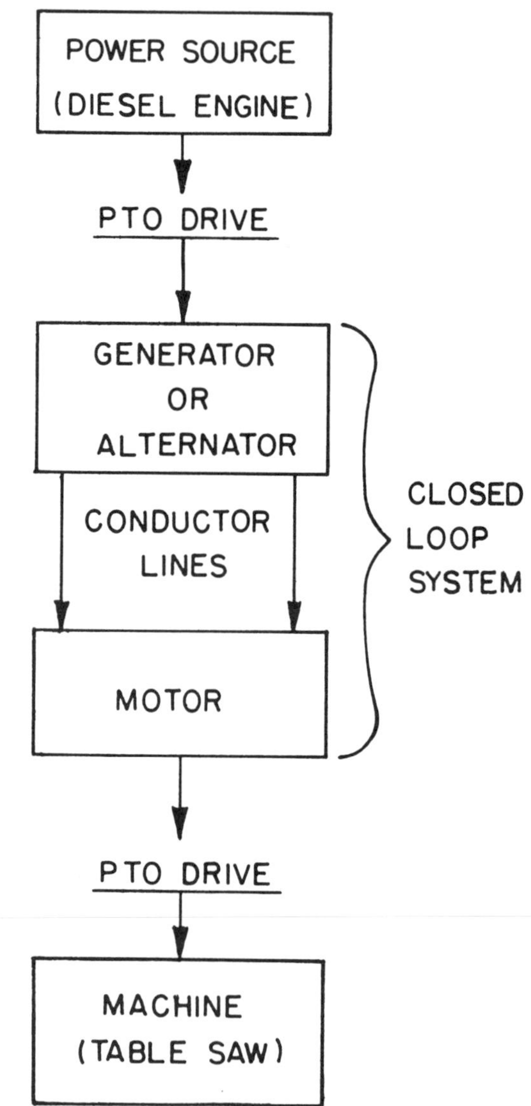

Fig. 1-2. Transmission of Power from Diesel Fuel for Sawing Wood

If you follow through the changes or conversion of energy from one form to another in Figure 1-2, there are four steps, namely:

(1) Diesel fuel to rotating PTO shaft through a diesel engine convertor.

(2) Rotating PTO shaft to watts (electric power) by the generator or alternator convertor and transmitted to conductor wires.

(3) Watts of electrical power on transmission conductor wires to a motor convertor and then rotating PTO drive.

(4) Rotating PTO shaft to a saw blade that cuts wood (as an example) or does other desired work because of the machine convertor.

In summary, there were four major energy convertors in this example; the diesel engine, the generator or alternator, the motor and the machine. Unfortunately, each time energy is converted from one form to another, there is a loss and sometimes this loss is large. In the previous example, the efficiency from the beginning to the end might be as follows: (1) diesel engine, 35 percent; (2) generator or alternator, 60 percent; (3) electric motor, 60 percent; and, (4) the machine 40 percent. Accepting these percentage efficiencies and starting with 1000 BTU's of fuel, there would be only 144 BTU's available for cutting of the wood because: 1000 BTU x 60 percent (alternator or generator efficiency) x 60 percent motor efficiency x 40 percent machine efficiency equals 144 BTU output. If the conductor wires are many miles long then another 10 percent might be lost.

From this line of reasoning, we can always assume that the use of electrical power is an inefficient system as we use it today. However the electrical system has the following advantages: (1) easy to control by switches; (2) usually non-polluting at the site where the motor and machine usage takes place; (3) motors on alternating current (AC) give us constant speeds; (4) the rotating forces produced by motors are very smooth; (5) motors need less maintenance than engine-style convertors; and (6) capital investment in motors is often less than engines per work unit.

These advantages are so highly favorable over other energy converting machines, that in the highly developed countries, the consumer is willing to sacrifice the lower efficiency for the convenience and other favorable attributes provided by electric motors. For example, consider that internal combustion engines must be refueled regularly and the multitude of moving parts and maintenance are a big problem; external combustion engines, steam engines as example, must have a fire pot, water boiler, cylinders and cranks designed to convert fuel to rotary motion. Therefore, electric motors are very convenient pieces of equipment when we think about what we would have to use instead.

The only other big disadvantage besides the general inefficiency of the whole system, is in vehicular movers, especially farm tractors, heavy industrial equipment, automotive and lawn garden equipment. To make a ridiculous point, driving an electric motor driven automobile from St. Louis to Chicago would take a 300 mile power cord if plugged into a St. Louis power plant. Electric-battery driven tractors and autos have for the most part proven unsatisfactory to date because the battery bank must be very large and heavy to store enough power to provide any logical range before recharging, and this weight reduces the vehicle's efficiency. Experimental vehicles with internal combustion engines, charging battery banks and electric motor drives, have likewise proven unsatisfactory because of the efficiency loss in energy conversion at both the generator and the motor.

It seems that only the railways have been able to effectively use electric motors on a large scale for mobile transportation: (1) either by lines overhead as in suburban railways or underground as in subways, and (2) with a power flow of diesel engines to generator to motor to final drive all on one locomotive. Electric motors as the prime power convertor on airplanes seems completely out of the question for obvious reasons just discussed and due to the airplanes extremely low-energy efficiency as a mover of materials and people.

Electric motor power can only be made more efficient by: (1) engineering designs that make the motor more efficient, for example, a motor's efficiency might be improved from 60 to 75 percent; and (2) using less expensive fuels or energies to turn generators and alternators, such as wind which is generally unreliable; water gravity or hydroelectric which is unreliable in droughts and in some countries, such as the U.S. where the major waterways are used for navigational purposes or are not near high population centers; and, solar where research and experimental designs have not made this practical for large loads. The point is that we will probably use coal, oil and nuclear energy fuels to power our electrical plants for years to come. Unfortunately all three of these fuels are unrenewable and cause high environmental pollution. Yet they are still the least expensive considering the alternatives.

In the U.S. we use many Megawatts of electricity each year. The electric motor accounts for about 60 percent of the total usage of electricity. Lighting and electric heating are the other two main uses of generated electricity. Electricity in its generated and transmitted form which is usable as we know it today, is one of the most important inventions. The major differences between highly developed countries of the world is the extent the countries produce and use electricity.

Basic industries, whether a manufacturer of road building equipment or tillage tools to increase the food and fiber for the worlds population, are almost impossible to operate without electricity and electric motors to run the specialized tools and machines required for routine manufacturing.

The food producing effort of any country is tied directly to the availability of electricity on the farms and in the industries that process the foodstuffs. We take electricity so much for granted, perhaps as example of milk production will explain the importance of electricity and electric motors. A hundred dairy cows can easily be milked by one man in two hours with only a 3/4 hp motor driving the vacuum pump. The fresh milk, untouched by human hands, is pumped through lines to a bulk tank, which holds the milk at the farm site. The lines can even be washed and flushed by power washer pumps that are electric motor driven. The milk bulk tank cools the milk quickly with a refrigeration unit powered by an electric motor. The truck which picks up the milk from the farm uses an electric motor to pump the milk from the bulk tank into the truck. At the processing center, the milk is unloaded and moved through pasteurization and homogenization units by pumps driven by electric motors. Packaging equipment such as bottling and carton containing equipment is powered by electric motors. Even the forklift in the milk processing plants shipping department which places crates of milk packages onto the delivery truck destined for local stores will most likely be electric motor driven. Now, imagine what the milk industry would be without electricity and electric motors. Start by picturing 10 people milking those 100 cows by hand and you will have some idea of the importance of electricity. Actually this was probably a poor illustration unless the reader has had the experience of milking cows by the arm strong method.

Consider the electrical motor usage in your home. As one author is writing this manual he hears the electric clothes dryer running. The drum is being turned by an electric motor. The swimming pool can be seen outside the patio door. Its recirculating pump is turned by a motor. A dehumidifying unit is near by with another motor. The furnace turned on and the forced warm air system is motor powered. The air conditioner is a large refrigeration system, motor powered as is the kitchen refrigerator and the freezer. On the kitchen counter there is a food blender and food mixer, both electric motor powered. Under the counter is a dish washer which is basically a cabinet with a pump powered by an electric motor to splash water. A garbage disposal under the kitchen sink is powered by a 3/4 hp motor. Water vapors are added to the air with a water wheel style humidifier operated by a motor with an infinite speed control range, controlled by a special humidistat capable of demanding slow to fast speeds. Besides most of the convenience and comfort devices previously mentioned and found in many homes in the U.S., Canada and other highly developed nations, we have yet to mention the numerous other motor driven machines that add to comfortable living. Among these are; can openers, exhaust fans, vacuum sweepers, hair dryers, record and tape players, and even electric tooth brushes. Dozens of toys for children are motor driven from both the household 120 volt circuits, or from dry cell batteries. In the home shop are found numerous motor driven portable tools and stationary machines which are motor driven, for example, portable electric hand saws and drills, power wood planers and metal cutting band saws. This author, with all these electrical servants, would not complain if his electric bill increased a few cents per day!!

Electric motor usage in industry and in agriculture is so great that it would be impossible to list and explain the application of but a few in this manual. Anyone who has toured a factory must be impressed with the work completed by motors such as turning lathes, driving presses and punches, the cooling of welding machines, cooling the drinking water for employees, pumping chemicals through lines and powering electric typewriters in the office. The almost unlimited list of motor applications to help us do our work easier, faster, safer, more efficiently, and in a more pleasant surrounding is largely due to the invention of generators, alternators and motors. All this started only about 150 years ago and today has taken much of the physical work out of our life.

****

CONSTRUCTING EXPERIMENTAL MOTORS

As a special project, construct experimental electric motors. Refer to the Appendix in this manual and the Appendix of the Instructor Manual for three plans for series, shunt or compound wound DC motors and two plans for synchronous AC motors.

DEFINITION OF TERMS

ALTERNATING CURRENT (AC): Current which reverses its direction of flow.

ALTERNATOR: Rotating machine that produces AC and is often called an AC generator. Converts mechanical energy into electrical energy.

BATTERY: Dry or wet cell using chemical energy to produce electrical energy.

CURRENT: Electrical energy flowing through a conductor.

DIRECT CURRENT (DC): Current which does not change its direction of flow.

ELECTROMAGNET: Magnet produced by winding conductors around a metal core and when the circuit is energized with electricity a magnet is produced.

GENERATOR: Rotating machine that produces electrical energy. Converts mechanical energy into electrical energy.

HYDRO-ELECTRIC: Machine using flowing water as the source of energy to produce electricity.

INDUCTION: Current produced by moving a coil in a magnetic field or by moving a magnetic field pass a coil.

KILOWATT (KW): Equivalent to 1000 watts.

MAGNETISM, NATURAL: Power of attraction of the iron ore called magnetite.

MEGAWATT: Equivalent to 1,000,000 watts.

MOTOR, ELECTRIC: Machine that uses electrical energy to convert it to rotary motion and mechanical energy.

WATT (W): Measurement of electrical energy. Measured by a wattmeter. W = Volts x Amperes.(See Definition of Terms, Unit III).

NOTES

CLASSROOM EXERCISE I-A

Development of the Electric Motor

1. The basic principles and rules of __________ and __________ which caused early motors to rotate are still in practice today.

2. On the time line shown, identify the important fact or date that the fact was discovered which influenced the development of the electric motor.

<u>Matching Items</u>

A. George Westinghouse
B. Benjamin Franklin
C. 1827
D. William Gilbert
E. 1888
F. Materials for conductors
G. Michael Faraday
H. 1660
I. 1883
J. Andre Ampere

1600, _____ Published findings on electricity and magnetism

Date: _____ First machine to move machinery

1729 -- Gray Identified _____

1752, U.S. inventor _____ flew his kite into electrical ring

1820, _____ discovered there are forces between wires carrying current

Date: _____ Ohm published studies on his law

1831, _____ invented electromagnetic induction

Approximate date _____ of early generator development

1885, _____ purchased rights to transformer

Date: _____ George Westinghouse purchased AC induction run motor invention and services of Tesla of Russia

NOTES

EXTERNAL FEATURES OF AN ELECTRIC MOTOR

The external features of electric motors are varied. Electric motors are often part of the equipment to which they may be permanently mounted, such as the portable electric drill or an electric carving knife. The external shape may not appear in the form that we commonly think electric motors should appear.

When someone mentions an "electric motor" we think of a cylinder shaped object laying on its side with closed ends, a power shaft sticking out one or both ends, and the motor body attached to a frame. This latter impression is proper, because most motors used for auxiliary drive purposes are shaped as described. Examples would be motors that belt drive a drill press or are power take off(PTO) connected to a water pump.

Regardless of the electric motor's shape, internal parts and theory of operation, they have much in common whether an integral part of a portable electric hand saw or the power unit on a large fan for a crop drying unit. This and the units to follow will explain the various types and styles of commonly used motors for home, farm and industry.

MOUNT AND BASE

For motors used as auxiliary power units, it is necessary to have the motor secured to drive a piece of equipment, such as the industrial belt-driven blower application shown in Figure 2-1.

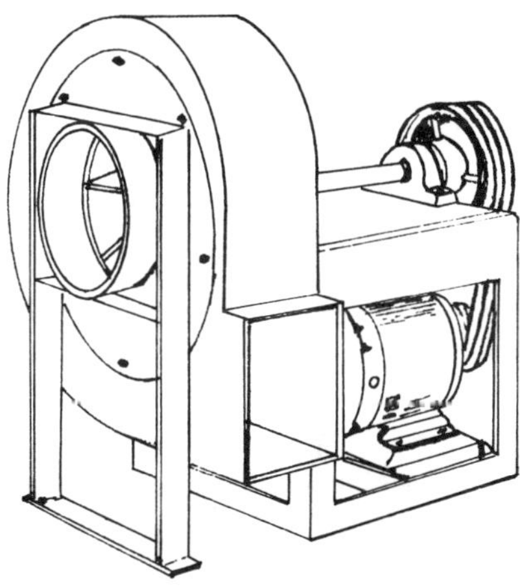

Fig. 2-1. Mounting of an Auxiliary Motor for Driving an Industrial Belt-Drive Blower

In this application the motor's stator frame is welded to the metal base. This style of mounting is called rigid mount and is illustrated in Figure 2-2.

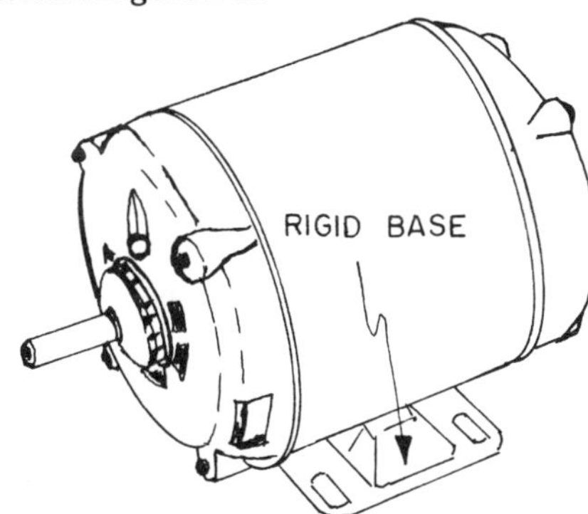

Fig. 2-2. Rigid Base Mounted to Motor Frame

Another commonly used rigid mount is shown in Figure 2-3. The motor's stator frame is a cast iron alloy as is the mounting base. Both are cast as one part making up the motor's main frame and the mount.

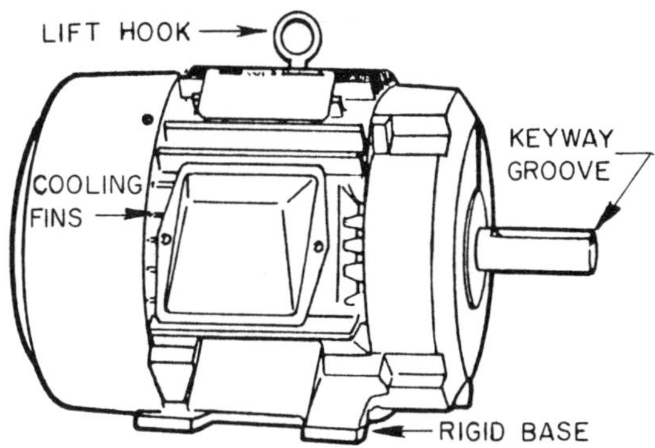

Fig. 2-3. Large Motor with the Base as a Part of the Motor Frame Casting

Generally smaller motors use the welded mild steel base system and larger motors use the casting system but there are many exceptions. The welded frame to base system would be less expensive to manufacture and is a reason it is used more frequently on smaller fractional horsepower (hp) motors. Motors which have the cooling fins for radiation of heat from the motor often have cast iron mount bases as found on larger integral hp motors.

A popular style of motor mounting for fractional hp motors is one shaped like a cradle called resilient mount system, and is shown in Figure 2-4. This system is sometimes called the cushion mount.

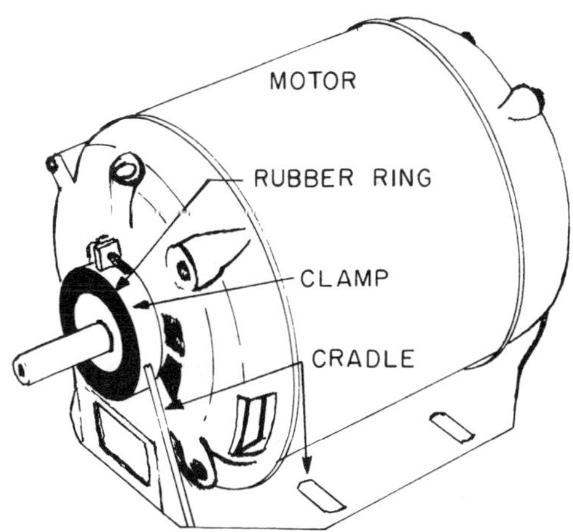

Fig. 2-4. Resilient Mounting System as a Cushion to Vibration

Regardless of its name, the intent is to minimize the amount of vibration caused by machinery being driven, the power shaft drivers (such as belts,couplers and chains) and/or the motor itself. Vibrations of the motor can place unnecessary wear upon motor bearings and the use of the resilient cushioned mount helps reduce this problem. Rubber rings on the mount help reduce shock caused in starting and stopping the motor. A very common application of the resilient mounted motor is on squirrel cage fans, Figure 2-5, similar to many forced warm air heating systems.

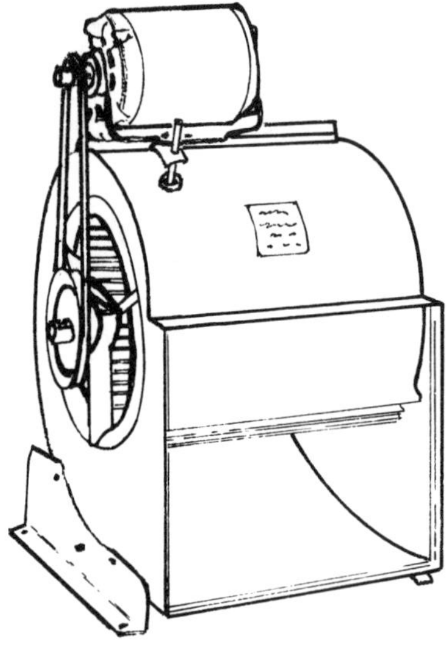

Fig. 2-5. A Resilient Mounted Motor Driving a Blower Unit

Motor bases are often slotted to permit sliding the motor back and forth to tension a belt, Figure 2-4. This adjusting system is the most common one used on fractional horsepower motors. It provides an opportunity to err slightly on belt replacements and to make adjustments after belts have stretched from normal use.

Electric motor bases can also be attached to special sliding rails as illustrated in Figure 2-6. Belts or chains would be tensioned and aligned by loosening the base mounting bolts. This system is common for larger motors.

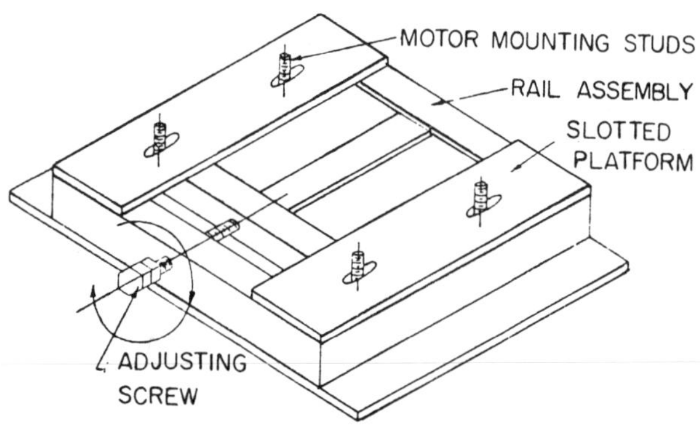

Fig. 2-6. Rail Adjusting System used with Large Motors

"Foot" mount methods, rigid base or resilient base, cannot meet the demands of all applications. Some motors will require "end" mounting. The two common end mountings are the "C-Face" and "D-Flange", note Figure 2-7. The "C-Face" and "D-Flange" motors are attached by their end shields. The "C-Face" is indicated in a NEMA (National Electrical Manufacturer's Association) frame number by the suffix letter "C". The motor's end shield has threaded bolt holes and the mounting bolts pass through mating holes on the connecting machine, note Figure 2-7, (a). The "D-Flange" mounting is a NEMA frame number with the suffix letter "D". In this case the end shield is called a flange and the "D-Flange" motor is used on machinery with tapped holes, using through bolts, where it would be inconvenient or impossible to use nuts, note Figure 2-7 (b). Other special motor mounts are available for definite purpose motors as noted in Figure 2-8. Pump motors (a) are identified by the suffix VP, P, PH; oil burner motors (b) by M or N; and sump pump motors (c) by suffix K. The fan or blower motor (d) can have through bolts from end to end permitting motor mounting from either end.

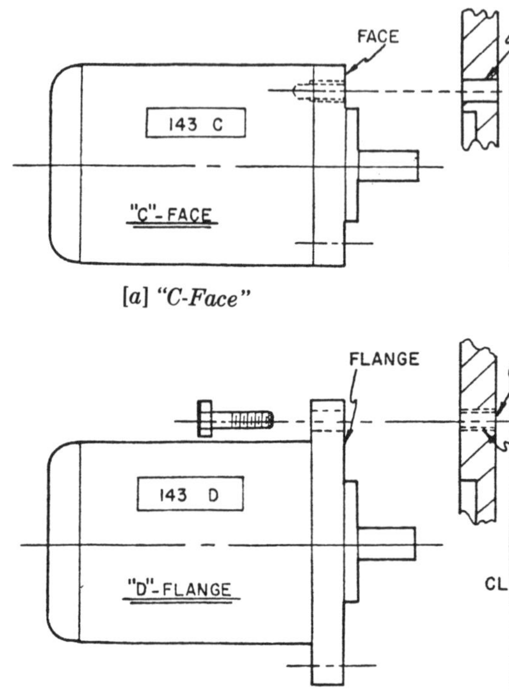

[a] "C-Face"

[b] "D-Flange"

Fig. 2-7. End Mounting for Motors

Regardless of the type of base or mount it is critical to firmly tighten the motor to the machine. Common torque values for machine bolt fasteners should always be observed. Slots in motor bases are to have flat washers over them, and lock washers should always be used with bolts. Motors and the machines they power tend to vibrate and proper securing of motors to machines will help keep parts from becoming loose, allowing belts to slip and alignments to alter.

****

CONSTRUCTING EXPERIMENTAL MOTORS

Know what makes the electric motor operate. Construct your own DC or AC electric motor. Follow the plans found in the Appendix of this manual or the Instructor Manual for this exciting project.

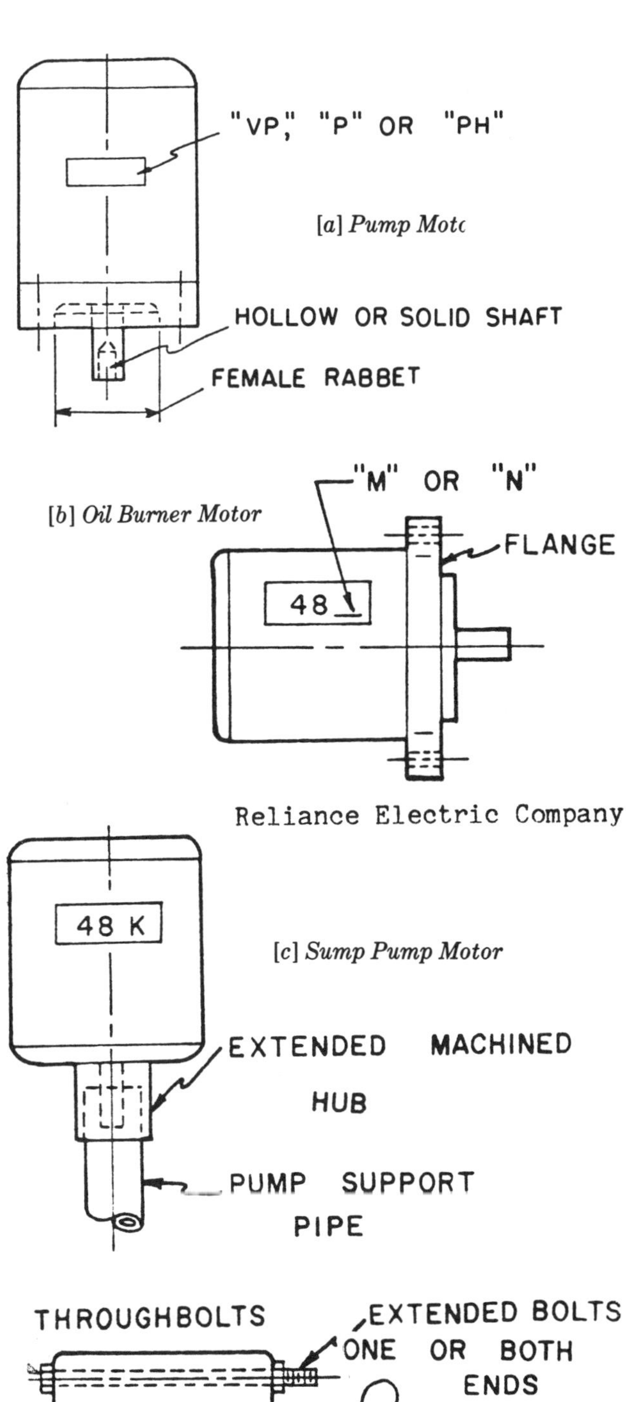

Fig. 2-8. Definite Purpose Motors

END BELLS AND ENCLOSURES

End bell configuration and the type of enclosure are extremely important factors of motor selection due to the location or environment. Basically, motor enclosures can be classified into two general types: (1) open enclosure, and (2) totally enclosed. Motors with open enclosures, Figure 2-9, allow the easy flow of air to be passed over windings and bearings which must be cooled because they get hot under normal use. Motors with total enclosures allow little or no air to pass directly over the windings and cooling takes place by air passing over the exterior parts of the motor housing and the radiation of heat to cooler surrounding air.

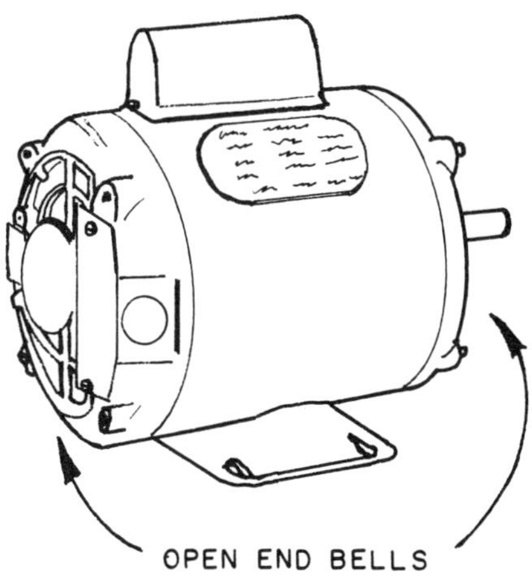

Fig. 2-9. An Open Enclosure Motor

The less expensive open enclosure motor cannot be used under the following environmental conditions: (1) dust, especially explosive dust such as feed; (2) submersion in liquids and other high levels of moisture; (3) exposure to corrosive materials, such as fertilizers, cement and acids; and (4) sparkproof situations as with certain chemicals and petroleum vapors. Totally enclosed motors must be used when these conditions exist.

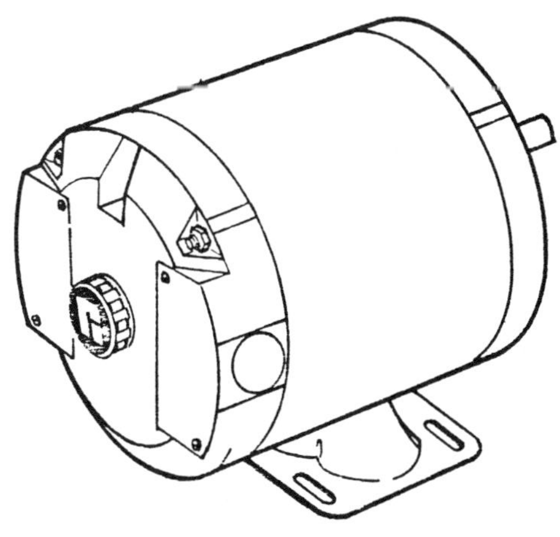

Fig. 2-10. A Totally Enclosed Motor

A typical totally enclosed motor is illustrated in Figure 2-10. Because of its tight enclosure it will withstand these adverse environmental conditions. No air enters the motor from the outside for cooling. With this condition, to cool the totally enclosed motor manufacturers must use higher quality insulation on motor windings and leads to guard against higher operating temperatures. Therefore, totally enclosed motors will cost more per horsepower power unit.

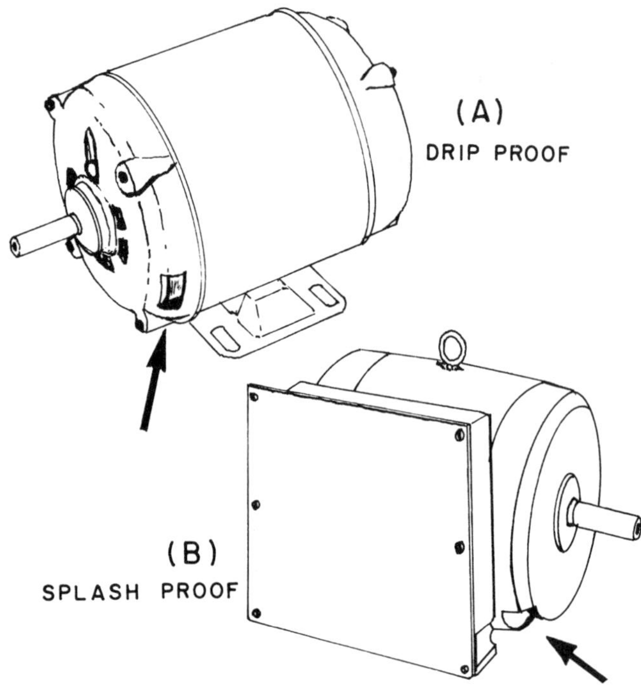

Fig. 2-11. End Bell Configurations of Typical Open Enclosure Motors

There are two types of open enclosure motors. They are the drip proof, Figure 2-11 (a), and the splash proof, Figure 2-11 (b). The main difference between these two are the end bells. The end bell of the drip-proof motor allows a free transfer of air into and through the motor. Slots are found close to the bearing housing to provide easy cooling of the bearings. This style of enclosure is designed for indoor use only, in clean air surroundings and where there is no danger of splashing liquids. Objects and liquids which fall upon these motors should be at an angle no greater than 0-15 degrees downward from the vertical.

The splash-proof motor has fewer openings for cooling than the drip proof. Therefore, it is designed for non-toxic (basically water only) splashes. It may be used outdoors, but common sense protection should be used to protect it from the weather elements. For example, if used as a water pump power unit, a small box-like building with a roof should cover the pump and motor.

End bell and enclosures are a primary consideration in motor selection. Motors which are of an open enclosure style, that are misused in dusty conditions, for example, will have their useful expected life greatly reduced. If a totally-enclosed motor is needed for the existing environmental conditions, it should be obtained even if the cost is more. The more expensive motor is frequently the least expensive in the long run.

BEARINGS, SHAFTS AND LUBRICATION

A motor shaft is a round rod protruding through the center of the motor. The shaft will protrude out one end or both, (as in the case of a double-shaft motor) pass through the center of the rotor, and be fixed in place by bearings fitted into the motor's end bell parts. Bearings and shafts are critical for efficient motor operation motors. The slightest breakdown of bearings and shafts will cause excessive overheating of parts, resulting in reduced useful work, wasted electricity, overloading of the motor, excessive heat build-up in windings and last a "burned-out" motor. Simple and common sense maintenance, minor repair and proper operation in driving machinery will help prevent these problems.

To understand how to care for bearings and shafts, first note how they function. The two basic types of bearings are sleeve and anti-friction or ball bearings. The sleeve type bearing is illustrated in Figure 2-12. The sleeve is a pressed-in bushing, matched with an inside diameter (I.D.) approximately one to two thousandths of an inch larger than the shaft's outside diameter (O.D.). A clearance of approximately one thousandths of an inch is excellent for holding an "oil film" between the two metallic parts and for allowing for proper lubrication.

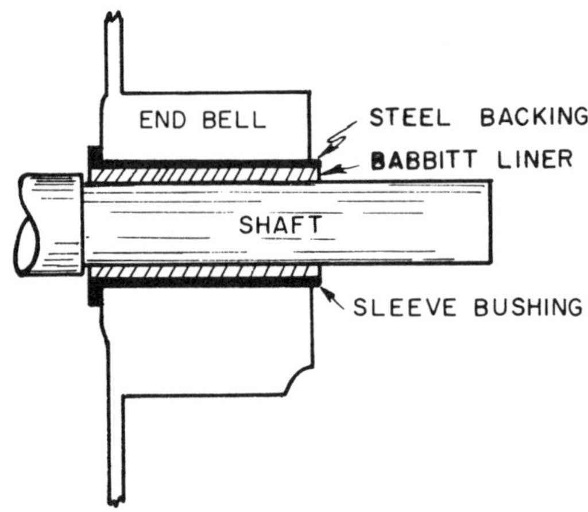

Fig. 2-12. Sleeve Type Bearing

The motor's shaft and bearings align the relative position of the rotor to the stator poles. This clearance is extremely small and critical, usually three to five thousandths of an inch. Therefore, if a shaft and/or a sleeve bearing wears a few thousandths, there will be a drag of the rotor on the stator poles, causing serious problems to motor parts resulting in a "burned-out" motor. It would appear that sleeve bearings, because of one metallic piece physically sliding over the surface of another metallic piece separated only by an oil-film, would be inferior to the anti-friction or ball bearing. This is not necessarily true. After all, millions of fractional horsepower motors have sleeve bearings and last for years when properly used and serviced on recommended intervals. There are several advantages to motors with sleeve bearings: (1) they are less expensive to purchase; and (2) when bearings need replacing, there is less expense than with anti-friction bearings. In comparison, when ball bearings go bad, the bearing encasement spins within the recess of the end bell and the entire end bell must be replaced along with the bearing resulting in a costly repair bill.

Lubrication is essential for both sleeve and anti-friction or ball bearing style of bearings. Figure 2-13 shows a yarn material used to store oil lubricants. The oil moves by capillary action through the window (hole in side of bushing) and keeps an oil film between the two metallic parts. The sleeve or bushing is made with a steel backing

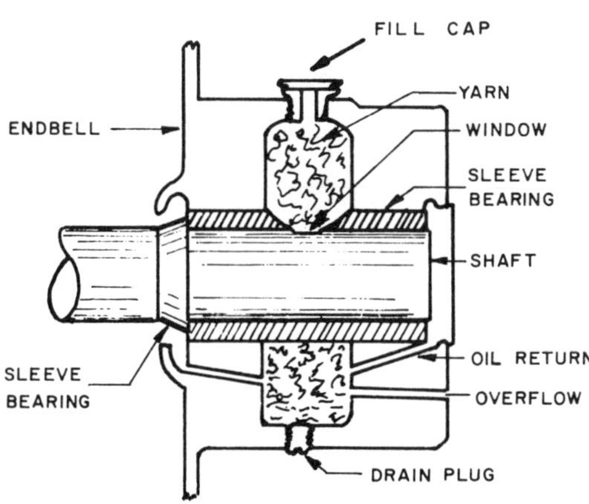

Fig. 2-13. Detail of Sleeve Type Bearing

and with a babbitt liner. Babbitt is porous in nature and with groove slottings at an angle from the window hole, oil readily flows to the surface of parts. The top one-half of a sleeve bearing is illustrated in Figure 2-14. Oil can follow the grooves from the window opening to lubricate the bearing parts. Proper motor mounting must be done to permit oil flow into the window to lubricate the shaft.

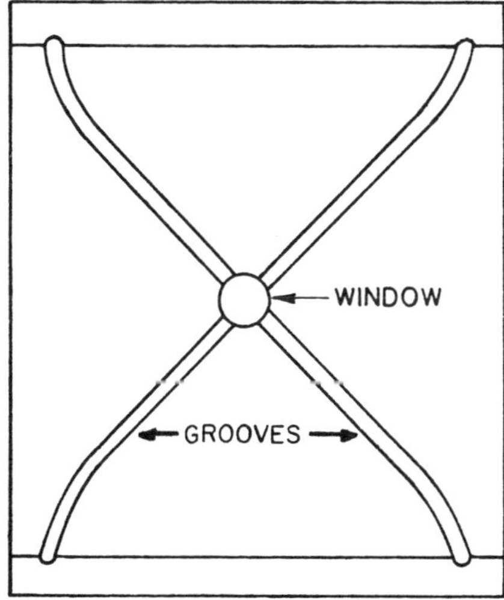

Fig. 2-14. Sleeve Bearing Grooves and Window

Anti-friction or ball bearings can be divided into two types, Figure 2-15. The sealed type (a) must be removed from the motor or machine to be lubricated but the lubricated type (b) can be lubricated through the filler plug with a zerk and grease gun. When the motor requires periodic lubrication, special care must be taken to mount the oil filler caps upward permitting gravity flow of oil. Special oil retention designs have been made for sleeve bearing motors permitting them to be mounted in all positions. Sealed bearing motors can be mounted in any position. Anti-friction bearing, regardless of lubrication style, rotate very easily, and would normally have less drag or friction than sleeve type bearings.

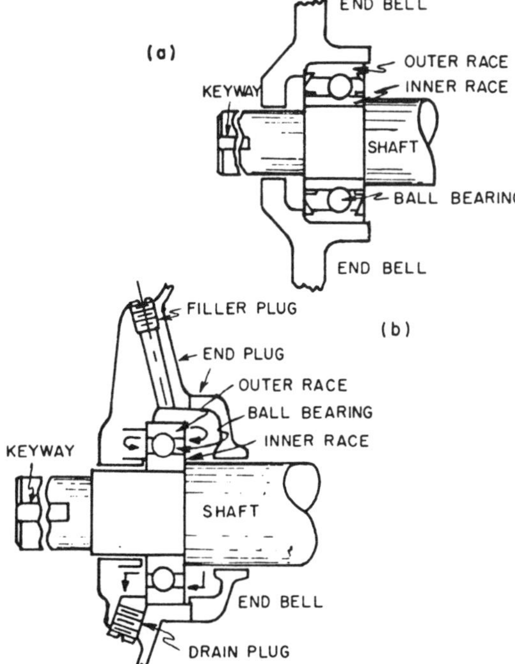

Fig. 2-15. Anti-Friction Ball Bearings

Anti-friction bearings are made of an inner race (which has its inside fixed to the motor shaft) and an outer race (which has its outside held firmly in place by the end bell enclosure). Numerous round balls, as in the case of the ball bearing assembly are rolled in a groove on the outside of the inner race and the inside of the outer race. Therefore, the inner race moves with the motor shaft and the outer race "stands still". The balls would be moving about one-half the speed of the inner race. Rolling a round pencil back and forth between the hands will demonstrate how these anti-friction bearings provide free movement between the two surfaces. The cross section of an anti-friction bearing is sketched in Figure 2-16.

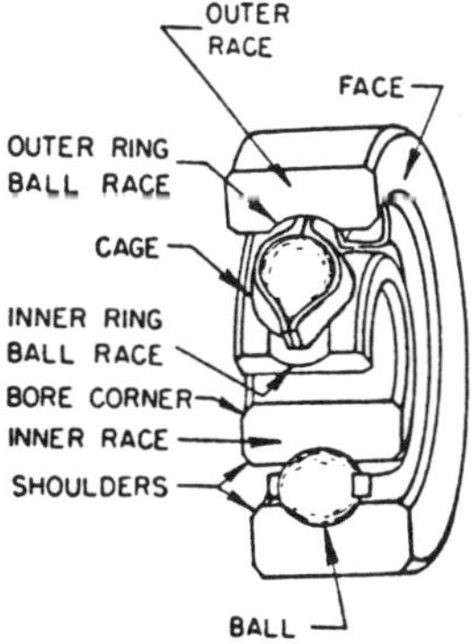

Fig. 2-16. Parts of an Anti-Friction Bearing

Although ball bearings are the most popular type of anti-friction bearings, other bearing styles, such as needle, roller or taper rollers are used on some electric motors.

Gearheads utilize sleeve and ball bearing but also tapered roller, needle thrust and drawn-cup full complement needle bearings. The needle bearings have smaller rollers than the 'roller' bearings. The needle bearings must be used with a hardened steel shaft because the shaft is the inner race for the bearing. Operating speeds must be lower for needle bearing than roller or ball. Needle bearing when used on motor shafts create more noise than the ball bearings. Bearing selection is based on the application.

Heavier alignment loads, as caused by belts and chains pulling sideways, can be placed on anti-friction bearings. When these bearings wear out special tools are needed for removal from the frame. A bearing separator and puller set and a several ton press will be needed to pressure the inner race over the shaft. When bearings go "bad", sometimes the seizure of the races to the balls or rollers will cause the outer race to "turn" in the end of the bell housing, ruining it also. This phenomenon is called "spun bearings". If the bearing spins on the inner race to the motor's shaft, the shaft may be damaged and a turning lathe required for re-dressing the shaft before the bearing is pressed on. Sleeve bearings are best for axial loads, or loads which equal forces distributed in the full circle of turning. When side forces, such as a V-belt, exceed 20 pounds excessive wear may be placed on sleeve bearings. Frequently the user of a sealed bearing motor does not inspect the motor often enough. However, the motor still needs the same attention in terms of keeping it clean (inside and out), providing proper overload protection, correct load alignment, and other maintenance checks which good precautionary practices dictate in caring for any motor.

Flat and keyed shafts must be kept in good condition. Motors commonly have a keyway in the PTO shaft(s), as shown in Figure 2-17.

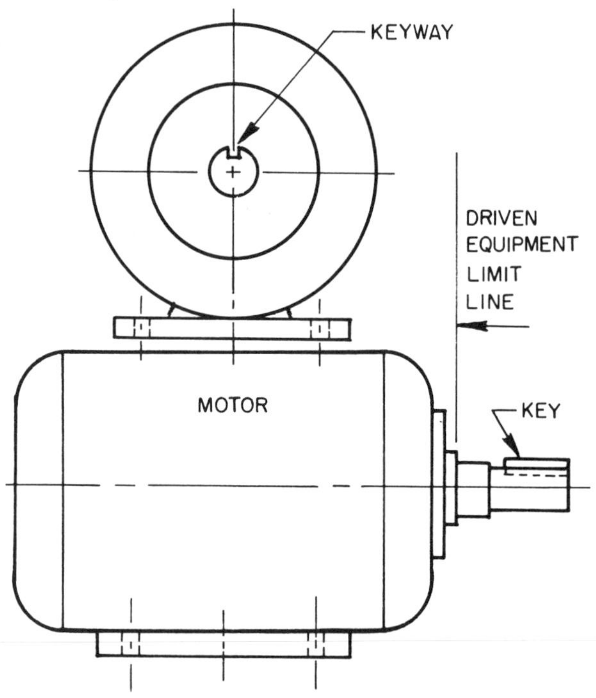

Fig. 2-17. Typical Keyway Grooved in the PTO Shaft

Common sizes of key stock such as 3/16 x 3/16", 1/4 x 1/4", 5/16 x 5/16" etc. are designed to attach a driving fixture such as a pulley or gear to the PTO shaft. See appendix for further explanation of keyway designs. The key stock in place would have one-half in the shaft and one-half in the driving fixture, as illustrated in Figure 2-18.

A tapped and threaded hole is usually above the keyway on the driving fixture to help hold the key stock and driving fixture to the shaft.

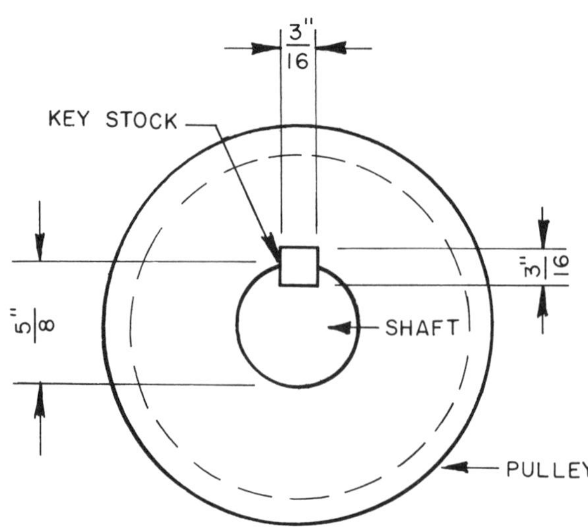

Fig. 2-18. Three-Sixteenth Inch Key Stock in Place on a Five-Eighth Inch Shaft

Shafts exposed to the elements and the abuse of being used can become damaged. When motors are disassembled for internal cleaning or service, the driving fixture such as pulleys or couplers must be removed. Mechanical cleaning of shafts with a mill file and emery cloth will often help in the removal of driving fixtures. Set screws on the driving fixtures should be fully removed, penetrating oil saturated in the set screw hole and an appropriate puller used to remove the fixture. Most PTO shafts will have a centered depression on the outer end as a convenient place to center pivot a wheel or gear puller set. Shaft material on most motors is mild steel subject to bending. Keep pullers in a balanced position when removing pulleys. See Figure 2-19. The pull or applied force should always be as close to the center of the driving fixture as possible. Pneumatic impact tools are excellent gear puller drivers because the impacting helps break rust loose and hold the gear puller set in place.

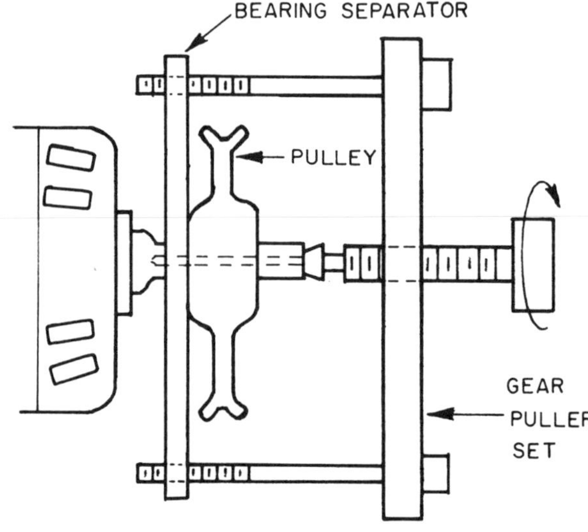

Fig. 2-19. Removing a Typical Pulley Using a Bearing Separator and Gear Puller Set

After removal of driving fixture, thorough cleaning of the shaft and filing of the keyway groove may be necessary for restoring to original shape.

COOLING SYSTEMS

Cooling of an electric motor is necessary. The motor, as with all other energy converting machines, gets hot as it converts electrical energy into a rotating force. The electric motor is 60 to 70 percent efficient in converting energy and losses are largely expelled in the form of wasted heat. Heat must be carried off or severe damage will occur to the internal motor, especially the windings. A basic physics principle states that heat flows from high concentrations to low concentrations. Therefore when a motor is hotter than its surroundings, heat will naturally try to flow away. The heat flowing away will readily be absorbed in the cooler surrounding air, as in Figure 2-20. However, the natural flow of the heat generated from the motor usually is not sufficient to keep the motor cool enough for continuous operation. Therefore, support cooling systems are added to satisfactorily accomplish the task of cooling. Cooling of most motors is done by one of two methods or a combination. They are: (1) forced air convection currents, and (2) radiation.

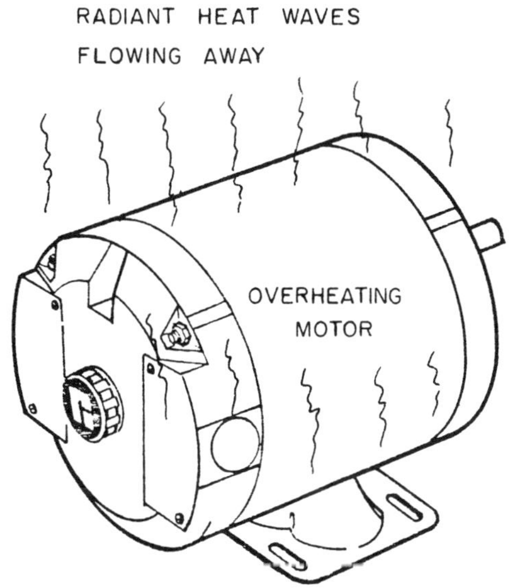

Fig. 2-20. Heat of Motor Operating Temperatures Will Flow Naturally to Cooler Surroundings.

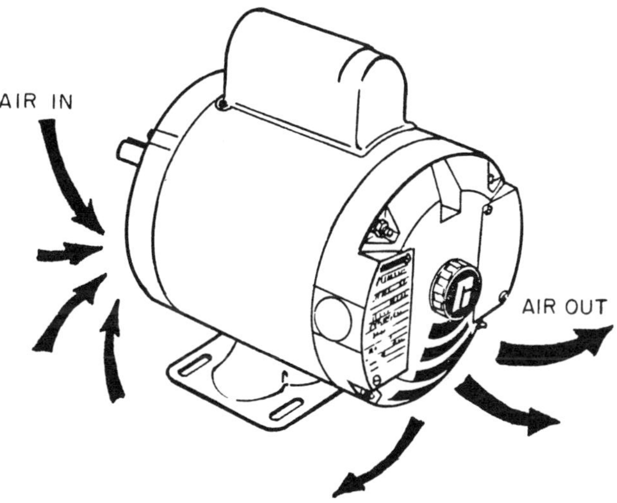

Fig. 2-21. A Drip-proof Motor with Fans Moving Forced Air Convection Currents for Cooling Purposes

In the case of drip-proof or splash-proof motors, discussed under End Bells and Enclosures, the heat build-up is cooled by forced air convection currents, see Figure 2-21. Fans are placed on one or both ends of the motor's rotor to move air in one direction. As cooler air from the outside surroundings is forced through the motor, it picks up heat being generated principally from the winding and convects the heated air to the outside atmosphere. The motor's nameplate will indicate the design of the motor's ability to cool. An explanation of temperature rise in Unit III, NAMEPLATE INFORMATION, will help in understanding the testing and rating schemes for indicating the cooling ability of motors. Note Figure 2-22 for theoretical temperature values of a motor when cooling. Basically, a motor's ability to cool is: (1) the function of passing air over the windings to withstand heat. Windings are insulated coils of looped conductors. If they lose their thin coating of varnish insulation, they will short out and be ruined.

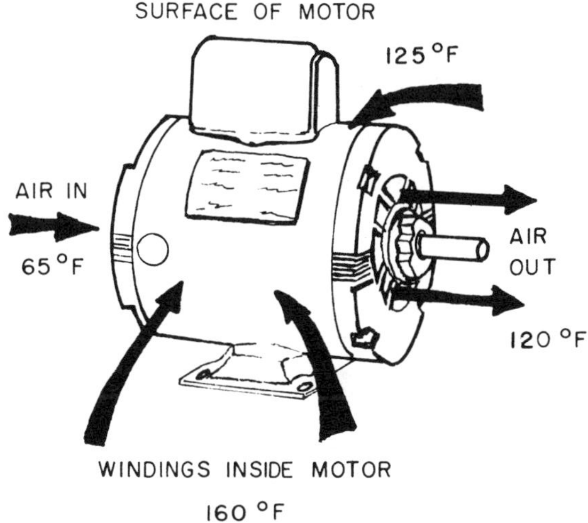

Fig. 2-22. Theoretical Temperature Values of a Motor when Cooling

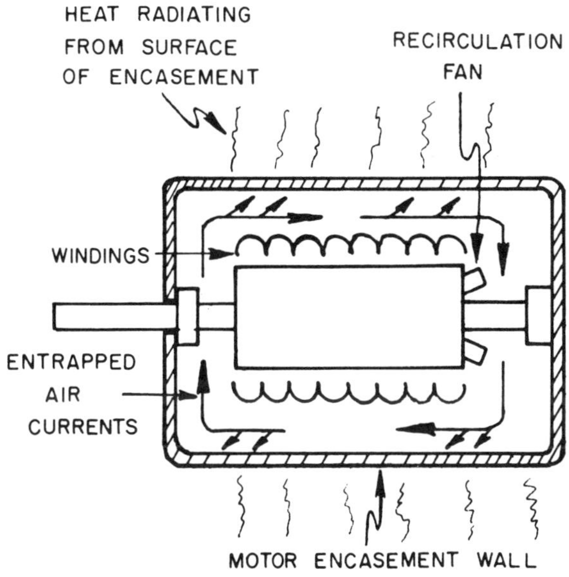

Fig. 2-23. The Internal Cooling of a Typical Totally Enclosed Motor by Radiation to the Atmosphere

Radiation is the other major theory of cooling a motor. The totally enclosed motor cannot have air currents passing directly through it and, therefore, requires special attention for cooling. Radiation is the primary method of cooling used on the totally enclosed motor. A re-circulation of the entrapped air inside the motor helps conduct the heat from the windings to the inside of the encasement wall so it may transfer through the wall and finally radiate away to the cooler surrounding air. Radiation cooling is illustrated in Figure 2-23.

Cooling accomplished by the radiation technique is not an adequate cooling system for normal air surrounded motor conditions. It is, however, used for submersible water pump motors and is very adequate because water temperatures below earth are about 55 degrees F. and constantly moving cool water over the outer encasement conducts the heat away from the motor. Moving water is a much better absorber of heat than is air. Manufacturers add additional cooling to the outside encasement of the motor. In Figure 2-24 the motor is the same as Figure 2-23, except a surface forced air system has been added. The fan pulls outside air in, directing it over the outer surface of the encasement to more effectively carry off heat radiated through the encasement wall.

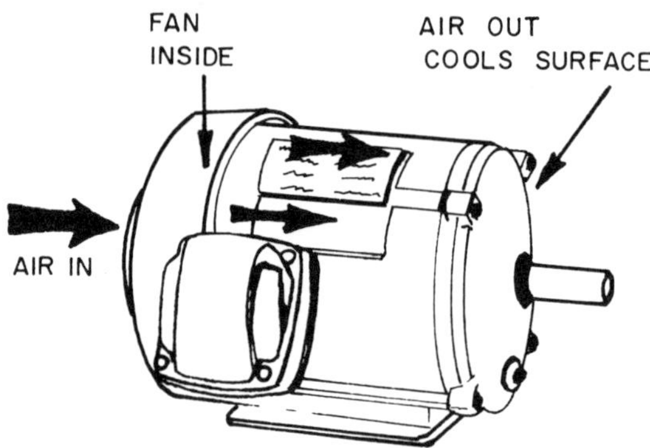

Fig. 2-24. A Forced Air System Blowing Over the Surface of the Outer Motor Encasement.

One more step can be taken to obtain additional cooling characteristics, that is the addition of radiant fins as shown in Figure 2-25. The fins allow for a much greater effective area to be exposed to outside air and the air blast of the external surface cooling system just described. The re-circulated entrapped air from inside the motor has more metal in which to transfer the heat from the windings. This additional metal is called a heat sink, and acts as a holding reservoir for heat until it can be radiated to the surfaces of the fins and literally blasted away by the external cooling fan.

Adequate cooling of the motor is very necessary. If not done, serious problems develop such as: (1) a breakdown of the insulation of the winding conductors; (2) overloading of the motor due to partial shorts in the windings; and (3) a higher amperage draw resulting in a greater voltage drop. A voltage drop results in lower rpm, more inductive slip; and, therefore, less cooling by the fan air blast. The slower the motor runs, the hotter the motor gets and more insulated winding conductors break-down. In other words, this vicious cycle starts over again and continues to repeat itself until the motor windings "burn-out". The bearings during this sequence leading to "burn-out" overheat, often expand to a binding condition, and cause even heavier loads and heat build-up to be placed on the motor. As motors get hot the oil tends to boil away, leaving dry bearings. All these problems are easy to avoid by following simple maintenance suggestions: (1) provide for exacting overload current protection; (2) keep motors clean inside and outside (3) lubricate motor bearings; and (4) pay particular attention to alignment of driving systems and mounting positions. More details will be covered in Unit XII, ELECTRIC MOTOR MAINTENANCE.

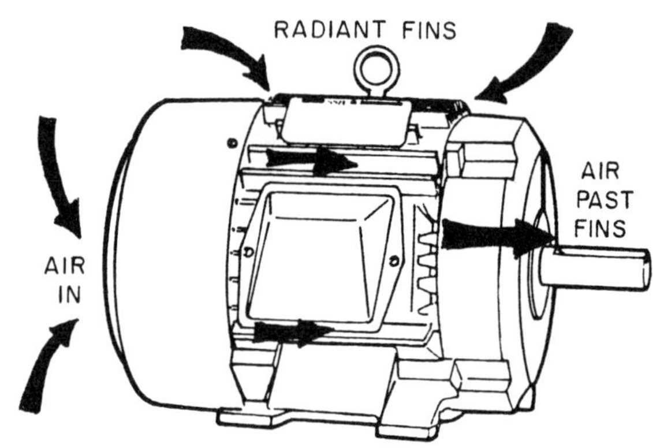

Fig. 2-25. A Totally Enclosed Motor with Radiant Fins to Increase its Cooling Effectiveness

MOUNTING POSITIONS

The mounting position of a motor is usually dictated by the machine it drives. Four common positions are shown in Figure 2-26.

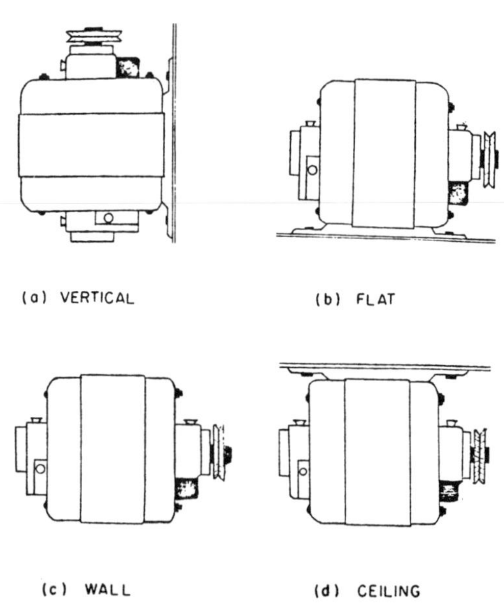

Fig. 2-26. Motor Mounting Positions

The flat position is the most popular mounting position. The ceiling and wall positions are similar to the flat, because the rotor, the shaft through the rotor and PTO shaft are in a horizontal plane. Motors with sleeve bushings as shown previously in Figures 2-12 and 2-13, are quite sensitive to mounting position. If positioned wrong, these motor bearings, may prematurely lose their lubrication or simply not lubricate properly. Generally, motors with sleeve type bearings should be mounted with oil fillers upward. Some newly engineered motors have oil retention designs which allow any position mounting of sleeve bearing motors. One of the big advantages of the sealed lubricated ball bearing motor is that it can be mounted in any position. It is especially useful for mounting in the vertical positions with the PTO shaft up or down. The weight of the rotor is considerable for most motors, yet the ball bearings will withstand the end thrusts from the weight of the rotor providing downward pressures. Along with the mounting position, the space around the motor is critical. All electric motors are air cooled, therefore, enough physical room for air currents from forced circulation for cooling radiated heat produced by the motor or produced by the machine the motor is driving should be provided.

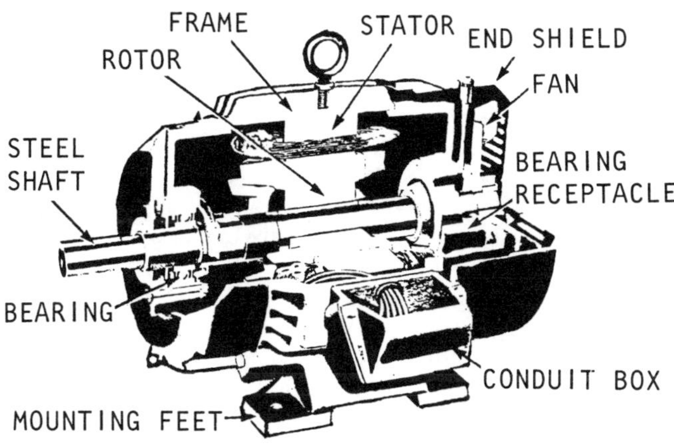

Fig. 2-27. Physical Features for Parts Identification

External features of a motor have been identified, and some internal parts have been exposed in Figure 2-27. The nameplate on a motor lists information about electrical aspects of the motor, code data for the physical features and special information which identifies the motor to the manufacturer. These data will be explained in Unit III, NAMEPLATE INFORMATION.

****

CONSTRUCTING EXPERIMENTAL MOTORS

Increase your electric motor expertise by constructing experimental electric motors. DC motors could be series, shunt or compound wound. There are two plans for constructing an AC synchronous motor. Refer to the Appendix of this manual and the Instructor Manual for construction and operation details.

DEFINITION OF TERMS

ANTI-FRICTION BEARINGS: Reduces the force that slows down motion when metal surfaces touch. In motors, anti-friction bearings are usually the ball or roller type.

AXIAL LOAD: Having equal forces distributed in a full circle of turning. There are no significant on-side forces.

BABBITT: A soft, silvery, anti-friction alloy composed of tin with small amounts of copper and is porous in nature. It is used for a liner on many sleeve bearings.

BALL BEARINGS: A part of a machine in which the moving parts revolve or slide on rolling metal balls so there is very little friction.

BASE: Holds or acts as a support for the electric motor body.

BEARING: A part of a machine on which another part turns or slides so that there is little friction. Two types of bearings are used in motors: anti-friction and sleeve.

CONVECTION: To pass heat to the air, around the motor by using air currents.

DRIP-PROOF ENCLOSURE: An open enclosure style of end bell designed to not allow splashing liquid or objects to enter the slots when falling at an angle of not greater than 15 degrees from the vertical. Generally, drip-proof enclosure motors are used for indoor purposes.

DRIVEN: Receives directed motion from the driver. The driven is the pulley, gear or sprocket attached to the machine being powered.

DRIVER: To set or direct motion. The driver is the pulley, gear or sprocket attached directly to the motor's shaft and provides the power.

ENCLOSURE: Act of confining or surrounding. The style of end bells dictate the type of enclosures; either open enclosure or totally enclosed.

END BELLS: The ends of the motor which may or may not have air slots in them, depending upon type of enclosure. The end bells carry the bearing and provide for alignment of the rotor to stator.

END SHIELDS: See end bells.

FRACTIONAL HORSEPOWER: Motors that are generally one horsepower or less but may be up to three horsepower. All frame sizes with two digits, such as NEMA 42 to 66, are fractional horsepower motors.

FRAME: Houses the working parts of an electric motor such as windings, stator and rotor.

HEAT SINK: Additional metal which acts as a holding reservoir for heat.

HORSEPOWER (hp): Unit of power in the U.S. customary system, equal to 745.7 (746) electrical watts; mechanically 33,000 foot-pounds per minute or 550 foot-pounds per second.

INSULATION: Material to minimize the passage of heat or electricity into or out of the electric motor.

INTEGRAL HORSEPOWER: Larger motors which are one-horsepower and larger. They are designated by three digits NEMA frame numbers from 143 to 505.

KEYWAY: Slot in a shaft, pulley, gear or hub for receiving square key stock or a Woodruff key.

MOUNT: The part of the motor that attaches the motor to a piece of equipment. Common types are rigid, resilient or flange mountings.

NATIONAL ELECTRICAL MANUFACTURER'S ASSOCIATION (NEMA): An organization which standardizes numerous sizes, configurations and quality of materials for the manufacture of electrical equipment including motors.

OIL FILM: A thin layer of oil used to lubricate between moving parts.

OPEN ENCLOSURE: An enclosure with slots in the end bells to allow air to be passed through the motor and cool the motor parts. There are two types of open enclosure: drip proof and splash proof.

POWER TAKE OFF: The shaft which protrudes out of one or both ends of a motor, engine or gear case which is used to drive or power machinery.

PREFIX: Usuallly a letter added before a set of basic numbers.

RADIATION: Waves such as heat emitted from a source such as a motor.

RESILIENT MOUNT: Mounting which has a technique for reducing vibration.

RIGID MOUNT: Mounting which is inflexible.

ROTOR: Rotating part of the induction electric motor.

SHAFT: A long, generally cylindrical bar, which rotates and transmits power. (See power take off)

SLEEVE BEARING: Thin metal alloy bushing used as a bearing to reduce friction on the shaft of an electric motor.

SLIP: The difference between the speed of the rotating stator magnetic field and the rotor is known as slip and expressed as a percentage of a synchronous speed. Slip generally increases with an increase in load.

SPLASH-PROOF ENCLOSURE: An open enclosure style of end bell designed to not allow splashing liquids or objects to enter slot when falling at an angle no greater than 100 degrees from the vertical. This enclosure is used for both indoor and outdoor applications, but with outdoor usage protection is needed against the weather elements.

STATOR: The part of an induction motor's magnetic structure which does not rotate. the stator is made up of laminations with a large hole in the center in which the rotor can turn; there are slots in the stator in which the winding for the coils are inserted.

STATOR FRAME: The part of the motor to which stator poles are mounted.

STATOR POLES: Iron or other metallic plates with insulated conductors wrapped around them. May be man-made electro-magnets or permanent magnets.

SUFFIX: Usually a letter added after a set of basic numbers. For example, the "E" could be considered a suffix to 147E.

TOTALLY ENCLOSED: Style of end bell enclosure that has no cooling slots to move air through the motor. Therefore, the motors windings are cooled by radiation and convection currents.

TEAO: Totally Enclosed, Air Over
TEFC: Totally Enclosed, Fan-Cooled
TENV: Totally Enclosed, Non-Ventilated
TEPV: Totally Enclosed, Pipe-Ventilated

WINDINGS: Insulated conductors wound or looped to make a coil.

ZERK: Adapter on the grease inlet of a motor which fits a grease gun.

NOTES

CLASSROOM EXERCISE II-A

External Features of an Electric Motor

1. When a motor is used as an __________ power unit, it is necessary to have the motor secured to drive the piece of equipment.

2. Two common styles of motor mountings and bases are ________________ and ________________.

3. When smaller motors are the rigid base, they are usually cast as part of the motor stator frame.
   True or False

4. The most popular style of motor mounting for fractional hp motors is the ________________ mount.

5. The main advantage of the resilient mounting is: ____

6. Explain the "D-flange" motor mounting and give one specific application of use. ________________
   ____

7. Motor enclosures can be classified into two general types:

   a. ____

   b. ____

8. List four (4) environmental conditions that would require a totally enclosed motor enclosure:

   a. ____

   b. ____

   c. ____

   d. ____

9. Explain the difference between drip-proof and splash-proof open motor enclosure.

   ____

   ____

   ____

10. Match the following motor environmental conditions with required exclosure types or styles: (1) totally enclosed (2) drip-proof and (3) splash proof.
    __________ A. Submersed pump motor.
    __________ B. Fan motor where moisture could drip straight down from ceiling.
    __________ C. Feed mill grinder.
    __________ D. Furnace fan motor.
    __________ E. Presence of petroleum vapors.
    __________ F. Liquids could splash on motor.

11. The two most common types of bearings are ________________________ and ________________.

12. The least expensive bearing for use in less restrictive operating conditions is the ____________ type.

13. The motor's shaft and bearings align the relative position of the ____________ to the __________.

14. Proper bearing lubrication is the establishment of an ______________________ of approximately one-thousandths of an inch thick.

15. Common anti-friction bearings include sleeve bearings. True or False.

16. Anti-friction or ball bearings are divided into two types according to lubrication system:

    a. ____

    b. ____

17. __________________ bearings are recommended for axial loads or loads which have equal forces distributed in the full circle of turning while a __________ type bearing would be recommended for a motor having excessive side forces.

18. Sealed bearing motors require little or no attention to service or maintenance. True or False.

19. Explain the purpose of a shaft keyway:____

    ____

    ____

20. List recommended procedures for removing a pulley from a motor shaft when motor is being dis-assembled for internal cleaning.

    a. ____

    b. ____

    c. ____

    d. ____

21. The basic physics principle that applies to motor cooling states: ______________________

    ____.

22. The natural flow of heat generated from a motor usually is not sufficient to keep the motor cool enough for continuous operation. True or False

23. Cooling of most motors is done by ______________________ and ____________________

24. Open enclosure type motors are cooled by the ______________________________ method.

25. Basically, a motor's ability to cool is: a.______________________________________

    b. ____

26. Overheated motors often short out and are ruined because the thin coating of __________ has been burned away.

27. ____________________________ is the primary method of cooling used on the totally enclosed motor.

28. Cooling characteristics of totally enclosed motors can be improved by the addition of radiant fins referred to as ____.

29. List three (3) problems that may develop if adequate cooling of a motor is not provided:

    a. ____

    b. ____

    c. ____

30. List the four (4) common mounting positions for electrical motors and one specific example of use of an electric motor for each position.

	Position	Use
a.	________________	________________
b.	________________	________________
c.	________________	________________
d.	________________	________________

31. A motor mounted in the vertical position should have ball bearings. True or False

32. The ________________ on an electric motor lists information about electrical aspects, code data for physical features and special information identifying the motor to the manufacturer.

33. You have probably been involved with an electric motor or two which has failed. Complete the following table about these motors.

MOTOR	WAS BEING USED ON A	CAUSE OF FAILURE
1	________________	________________ ________________
2	________________	________________ ________________
3	________________	________________ ________________
4	________________	________________ ________________

LABORATORY EXERCISE II-A

External Features of an Electric Motor

Locate three (3) electric motor applications in your home, farm, business location or school shop and identify the external features discussed in Unit II.

EXTERNAL FEATURE OR CHARACTERISTICS	MOTOR		
	1	2	3
1. Motor Use			
2. Mounting or Base			
3. Enclosure (End Bell)			
4. Bearing			
5. Shaft			
6. Lubrication			
7. Cooling System			
8. Mounting Position			

NAMEPLATE INFORMATION

The electric motor has a nameplate which provides information for the person selecting, servicing and repairing electric motors. This information can be divided into two categories the electrical features and the physical features. The electrical features are standard for most motors regardless of the manufacturer while many of the physical features are standardized by the National Electrical Manufacturers Association (NEMA). The nameplate also provides identification numbers and letters supplied by the manufacturer for identification purposes, note nameplates Figure 3-1.

ELECTRICAL FEATURES

HORSEPOWER

The size of the motor is designated by horsepower (hp). Fractional horsepower motors have NEMA frame numbers 42 to 66. The size is usually less than one horsepower but may include sizes up to three hp. Integral horsepower motors have NEMA frame numbers 143 to 505. The sizes are generally larger than two horsepower but may include sizes down to one hp. The size of motor will be listed as a

MARATHON ® ELECTRIC
WAUSAU, WISCONSIN 54401
MODEL KE254TTDR7026EG – R26W | FRAME 254T
TYPE TDR | DES | PH 3 | CODE G | INS CL B | ENC DP
DUTY CONT | MAX AMB °C 40 | SHAFT END BRG 309 | OPP END BRG 207
SUITABLE FOR USE ON 208V
AT 42 MAX AMPS
VOLT 230/460 | AMP 40/20
RPM 1750 | HZ 60 | NEMA NOM. EFF | NOM PF | MAX CAP KVAR
HP 15 | SF 1.15 | CORR AMP
LOW VOLTAGE LINE | HIGH VOLTAGE LINE
3 LEADS | 6 LEADS | 9 LEADS | 6 LEADS | 9 LEADS

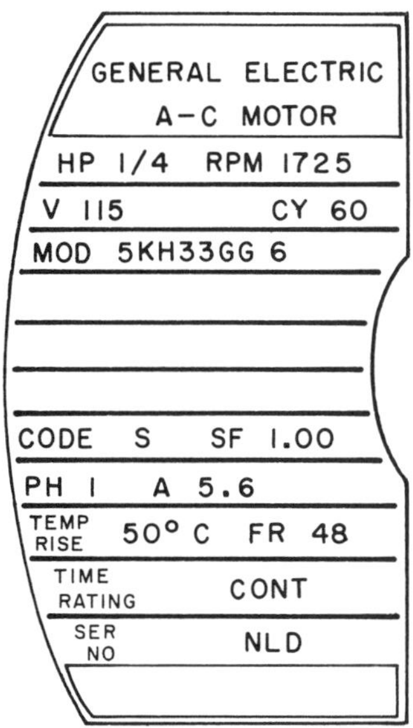

RELIANCE | R.P.M. D.C. MOTOR
FRAME | HP | TYPE TR | IDENTIFICATION NO.
VOLTS | AMPS. | R.P.M. | DUTY | AMBIENT
FIELD VOLTS | FIELD MAX.AMPS. | FIELD AMPS. | WINDING
INSULATION | POWER FACTOR
MADE IN U.S.A. 613-5-AF
RELIANCE ELECTRIC COMPANY / CLEVELAND, OHIO 44117

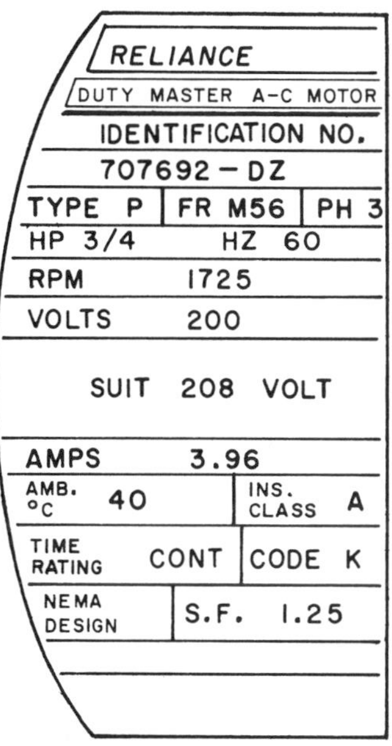

RELIANCE | DUTY MASTER A.C. MOTOR
FRAME 213T | TYPE P | INS.CLASS B | IDENTIFICATION NO. P21J8311B
HP 7.5 | R.P.M. 1760 | VOLTS 230/460 | AMPS. 21.0 / 10.5 | HZ 60 | S.F. 1.0
DESIGN B | CODE H | PHASE 3 | LOW VOLTS | HIGH VOLTS
DRIVE END BEARING 35BC02XPP3M
OPP DRIVE END BEARING 30BC02XPP3M
AMB. 40° C | DUTY CONT.
EFF. INDEX K
MADE IN U.S.A. 613XS
RELIANCE ELECTRIC COMPANY / CLEVELAND, OHIO

Fig. 3-1. Electric Motor Nameplates

fraction or whole numbers and fraction. One horsepower is equivalent to 746 watts (W). Wattage calculation for direct current (DC) motor is different than the alternating current (AC) motor, note formulas (a) and (b).

Electrical Horsepower Formulas

(a) DC: $W = \dfrac{V \times I}{746 \text{ watt/hp}}$

(b) AC: $W = \dfrac{V \times I \times \text{Power factor}}{746 \text{ watt/hp}}$

The mechanical definition of hp is the movement of 33,000 lb-ft per minute or 550 lb-ft per second, note formulas (a) , (b) and (c).

Horsepower Formulas

(a) 1 Hp = 33,000 lb-ft / minute

(b) 1 Hp = 550 lb-ft / second

(c) $1 \text{ Hp} = \dfrac{\text{Pull or Push in Lbs} \times \text{Length of Lever Arm in Ft} \times 2\text{ "Pi"} \times \text{RPM}}{33{,}000}$

or

$= \dfrac{\text{Lb} \times \text{Arm-Ft} \times 2 \times 3.14 \times \text{rpm}}{33{,}000}$

Formula (a) and (b) represents hp expressed as linear movement and (c) as rotational motion. Because an electric motor has a rotating shaft, the ability to develop torque is listed on motor performance curves. Torque involves push or pull in ounces or pounds and the radius at which the force is applied and is measured in inches or feet. With torque data available, the horsepower can be calculated by the formulas (a) and (b).

Horsepower and Torque Formulas

(a) $\text{Hp} = \dfrac{\text{Torque (lb-ft)} \times \text{rpm}}{5252}$

(b) $\text{Torque(lb-ft)} = \dfrac{\text{Hp} \times 5252}{\text{rpm}}$

Formula (a) is derived from formula (c) above and is also the accepted dynamometer formula. With torque unknown use formula (b). Torque has been indicated as rotation of the electric motor shaft but torque exists whether there is rotation or not. Torque and hp will be discussed in more detail in UNIT IX, ELECTRIC MOTOR SELECTION and PERFORMANCE TESTING.

TYPE OF CURRENT

Alternating current is symbolized by the sine wave illustrated in Figure 3-2. There is a maximum positive (+) value, a maximum negative (-) and a zero point when the AC fluctuates. Electric motors operating on DC will have current flowing in only one direction.

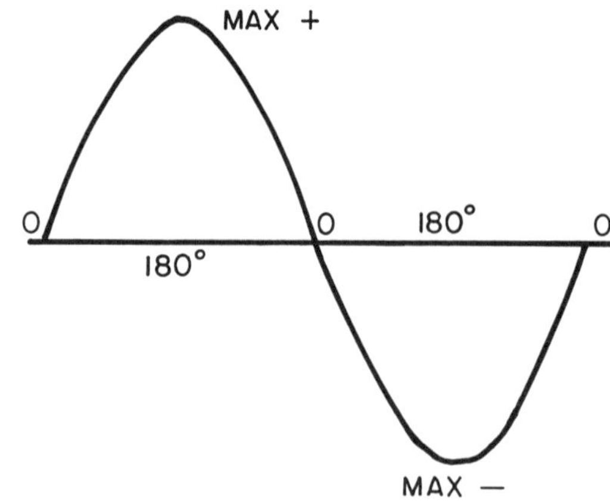

Fig. 3-2. AC Sine Wave

The AC electric motor, is the most common in the home, on farms, in processing plants and in businesses because AC is the commonly generated electrical service. Motors being driven by DC, Figure 3-3, must have a special power source such as a battery or DC rectified from AC.

Fig. 3-3. AC and DC Motors

PHASE

The sine wave in Figure 3-2 represents single-phase power because of its 180 degrees of positive value electrical travel and 180 degrees of negative value electrical travel. Electric motors can be designed for single- or three-phase power. Three-phase power is illustrated in Figure 3-4.

Three-phase power has a maximum positive value every 120 degrees or three in 360 electrical degrees. If comparing "phases to pistons" in a two-stroke cycle engine the changing from single- to three-phase adds two more power impulses per 360 degrees. Additional information about phases will be supplied in UNIT IV, PRODUCTION of ELECTRICAL ENERGY.

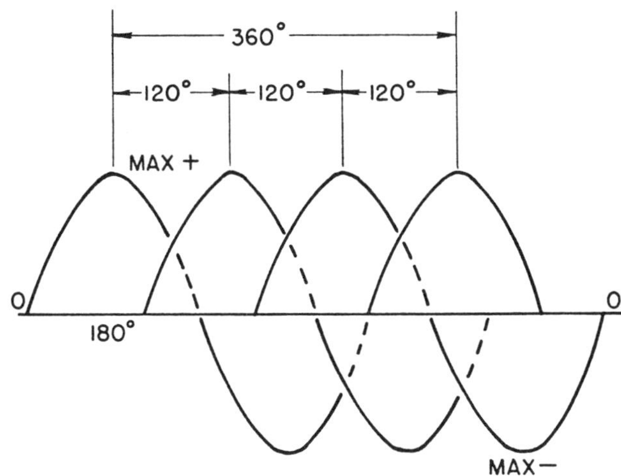

Fig. 3-4. Three-Phase Power

CYCLE OR HERTZ

If the electric motor has been manufactured for sale in the North American continent, it will be designed for 60 cycles or Hertz (Hz) per second. Referring to Figure 3-2, the AC sine wave has a maximum positive and negative per cycle. If the generated electrical current is multiplied by 60 positive values and 60 negative values, there is 60 Hz current which is often spoken of as frequency. Countries other than the United States and Canada often generate 50 Hz current. Hertz values within + or - 5 percent of the rated Hz will not damage the motor. This value will be used later to calculate electric motor revolutions per minute (rpm).

REVOLUTIONS PER MINUTE

The rotational speed of the electric motor is recorded rpm or slip speed and is calculated at motor full load. The speed of the motor is dependent upon; (1) the electrical Hz, (2) the number of poles or pairs of poles in the stator windings, and (3) the number of seconds per minute. Two rotational speeds are frequently discussed for induction run motors; synchronous and full load. Synchronous speeds are theoretical and can be calculated. The rotational full-load speeds are dependent upon a variety of factors caused primarily by design and electrical features. To determine the operational rpm use either: (1) a revolution counter and stop watch; (2) speed-o-meter tachometer; (3) an electronic tachometer; and (4) a vibration tachometer as illustrated in Figure 3-5.

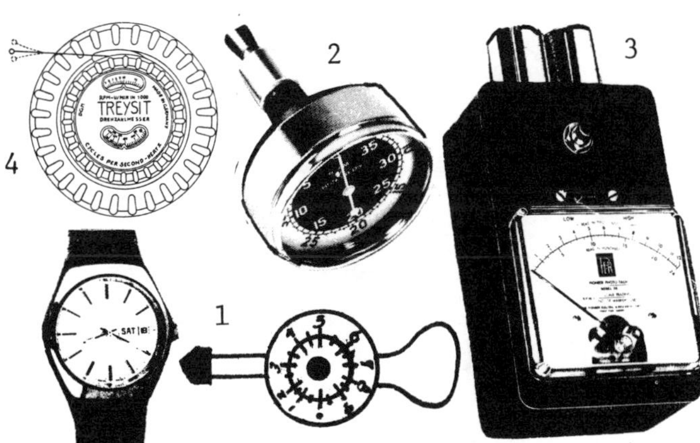

Fig. 3-5. Devices for Determining RPM

Fig. 3-6. Stator Poles in Running Windings

The arrows, Figure 3-6, point to a running winding on the motor stator. Opposite this pole is the other pole for this pair. Ninety degrees around the motor frame is another pair of running windings in the stator for this specific motor. Another stator pole is identified by the arrows in Figure 3-7. These are starting windings which usually have fewer turns of a smaller conductor. This motor has four starting windings or two pair and are centered where the other stator winding coils meet. These windings are essential for starting the rotor and will be discussed further in UNIT IV, PRODUCTION OF ELECTRICAL ENERGY. To calculate the synchronous rpm, use the following formula:

$$\text{Synchronous Speed} = \frac{60 \text{ Hz} \times 60 \text{ sec/min}}{\text{prs of stator poles}}$$

$$= \frac{3600}{1}$$

$$= 3600 \text{ rpm}$$

The stator in Figure 3-6, has two pairs of poles and with two as the divisor, the synchronous speed becomes 1800.

Fig. 3-7. Stator Poles in Starting Windings

The rpm of an induction run motor can be determined if the number of poles can be counted and the Hz is known. Now that synchronous rpm can be calculated it's easy to identify that motor nameplates do not list 3600, 1800 and 1200 but rather 3450, 1725, and 1140 rpm or numbers near these values. The reduction in speed is attributed to slip which can vary from 2 to 13 percent. Slip is caused by electrical conditions within the stator and rotor plus friction. If all the electrical energy going into the motor produced work the electric motor would be 100 percent efficient. Friction and heat losses prevent 100 percent efficiency. There will be friction between the bearings and rotor shaft plus resistance from the fan blades as cooling air is moved in and through the motor. Heat losses involve eddy-currents and hysteresis and an explanation of these electrical conditions is beyond the scope of this manual. The grade of steel in making the motor rotor and stator and the use of laminated metal also influence the heat losses, note Figure 3-8.

$$\text{Slip \%} = \frac{\text{Syn. rpm - Act. rpm}}{\text{Synchronous rpm}} \times 100$$

$$= \frac{1800 - 1725}{1800} \times 100$$

$$= 4.2$$

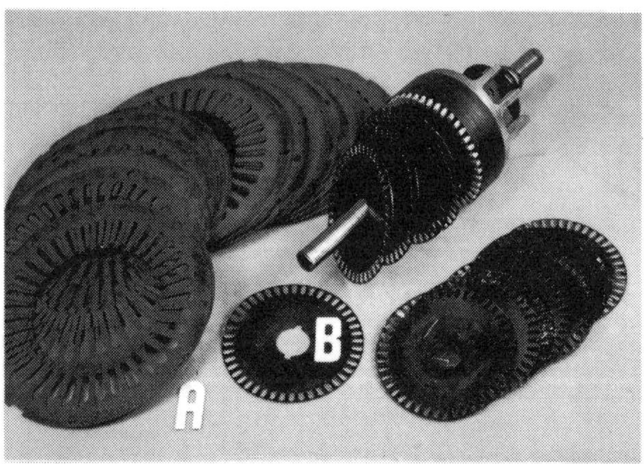

Fig. 3-8. Stator [a] and Rotor [b] Laminations

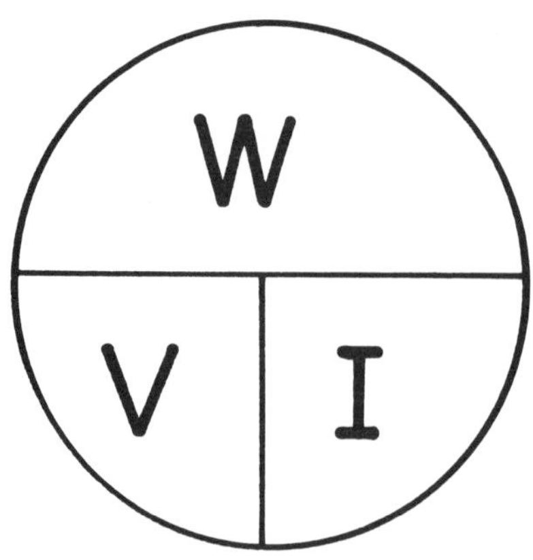

Fig. 3-9. Power Formula Measured in Watts

THERMAL PROTECTION

Electric motors placed under load in excess of full load will draw extra amperes and overheat the windings in the motor. If continued for an extensive time, the motor will be damaged. Protection can be designed into the motor in the form of a manual or automatic reset thermal device which has normally closed (N/C) contacts.

[a] Automatic Reset Thermal Overload

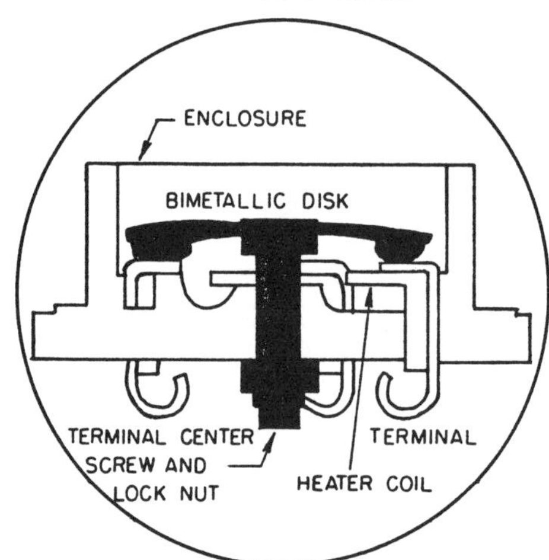

[b] Manual Reset Thermal Overload

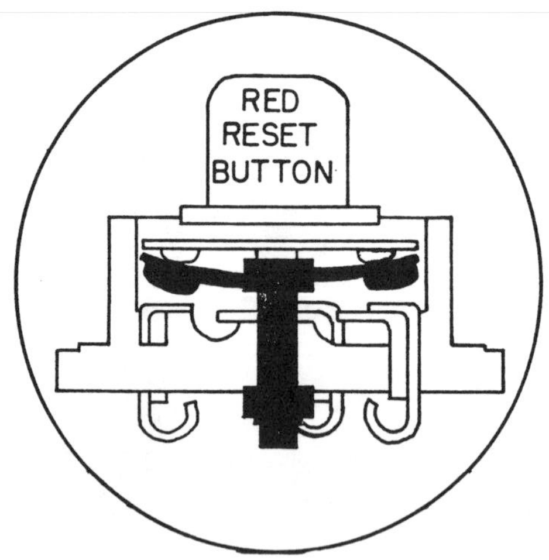

Fig. 3-10. Thermal Overload Devices

VOLTS

The transformer voltages (V) will be 120 or 240 under most conditions but could be 208 volts. If the power source is from a delta transformer it will supply 120 or 240 volts. The wye "Y" or star transformer will supply 120 and 208 volts.

AMPERES

The amperes (A or I) will be listed for the motor's rated current at full load. Some electric motors can be operated on either 120 or 240 volts and are called dual-voltage motors. If the ampere rating on 120 volts is 7.0 A, the motor will require only 3.5 A on 240 volts. In both electrical connections, the power which is measured in watts (W) is the product of I x V and equals 840 W. The ampere value for the motor is inversely proportional to the applied voltage.

When there is an overload on the motor, the heater coil under the contact points get hotter and the movable bimetallic disk unit will expand snapping from convex to concave and break the contact surface shutting off the motor. When the bimetallic unit cools, the surfaces will make contact again and the motor will start automatically, note Figure 3-10 (a). The manual reset type breaks contact by the same method but will not permit the motor to run again until reset by an operator. The reset button is normally red, Figure 3-10 (b). A motor that continues to overheat should be checked for a problem or the problem could be the driven machine. If the problem is not the driven machine, a lack of ventilation, lubrication or other unexplained friction, it could be assumed that the motor is being expected to handle a load greater than its rated size. The ammeter is the measuring instrument to detect overloading conditions. The thermal overload device can be the shape of a disk, bimetallic strip or other configuration but regardless of style will perform the same function. The thermal overload unit is positioned in the motor circuit so that current going to the windings must pass through it and when overloaded, it disconnects the line power service.

[*a*] *Manual Reset on End Bell*

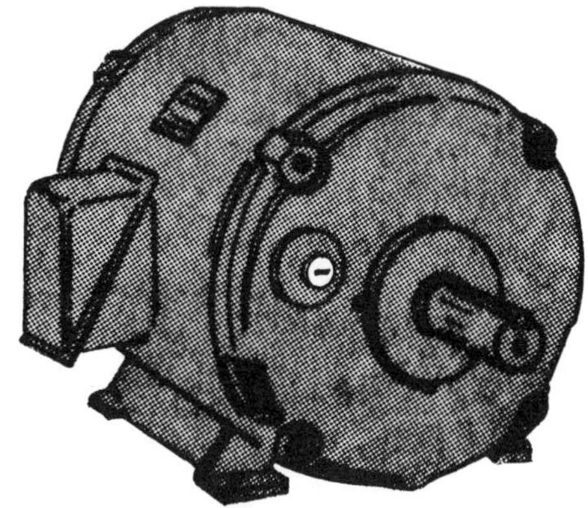

[*b*] *Manual Reset on Side Mounted Box*

[c] *Automatic Reset Inside End Bell* [*not visible from outside*]

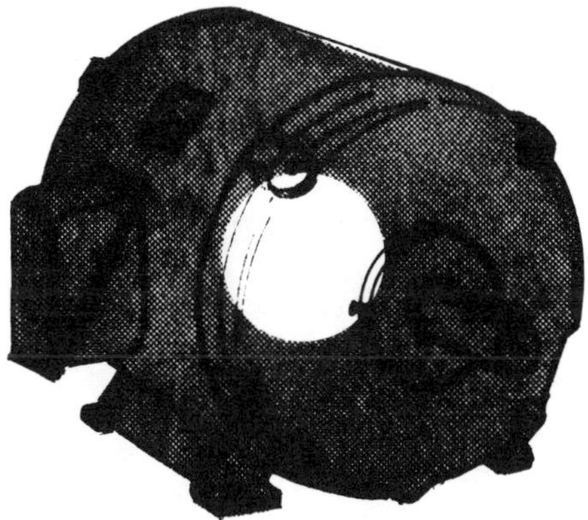

Fig. 3-11. Manual and Automatic Reset Thermal Overload Buttons

The following terms are found on motor nameplates which have thermal overload protection; thermoguard, thermal protection and overload protection. The electric motor can tolerate temporary overloads, beyond the rated amperage, without causing damage to the motor. The thermal overload device will be designed to carry at least a 15 percent overload but not more than 25 percent overload before opening the circuit and disconnecting the motor. Observe the outside of a motor to locate the thermal overload button. Note in Figure 3-11, the Manual and Automatic Reset Thermal Overload Buttons. All thermal protectors used on motors are the time delay action type.

Sometimes there is a letter following this designation which is a company assigned symbol indicating the type of overload relay system. For example, Table 3-15, the relay systems are designated by Q, P, U, S, X, and Z by the Marathon Electric Manufacturing Corporation. The position of the thermal overload device in the motor circuit is illustrated in Figure 3-12. The overload heating coil is placed in series with the running windings of a single voltage motor, Figure 3-12 (a). Trace the circuits to verify the statement. Dual voltage motors are those which can be made to operate on either voltage, 115 or 230 for example, by changing leads. The instructions are usually on the inside of the coverplate or on the nameplate. The motor in Figure 3-12 (b) is a dual voltage motor that has been connected to 115 volts. Trace these circuits. Both running windings and the starting winding are in parallel and the heater coil of the overload device is in series with R4 and R3 running winding, therefore, the heating coil detects only the current flow through one of the two running windings. When an overload is detected, the normally closed (N/C) contacts of the protector break line service to the windings.

The motor, Figure 3-12 (c) is connected for 230 V service. Trace the circuits. The running windings, R1 and R2 are in series with R3 and R4. The starting winding is parallel with running winding R1 and R2. Also the R3 and R4 running winding is in series with the starting winding. The heating coil of the overload device is in series with all of the circuits and all the current flows through the device when connected to 230 volts. When connected to 115 volts only one-half of the total running current flows through the heater. The heater unit carries the same current in dual-voltage motors whether connected to 115 or 230 volts. A dual-voltage motor could be rated for 8 A at 115 V and 4 A at 230 V. The calculations for current flow through the overload coil at 115 V would be 1/2 x 8 I x 115 V = 460 W and for 230 V, it would be 4 x 230 = 460 W. If not wired in this fashion, the motor manufacturer and/or consumer would need interchangable thermal overload devices for motor operation at different voltages.

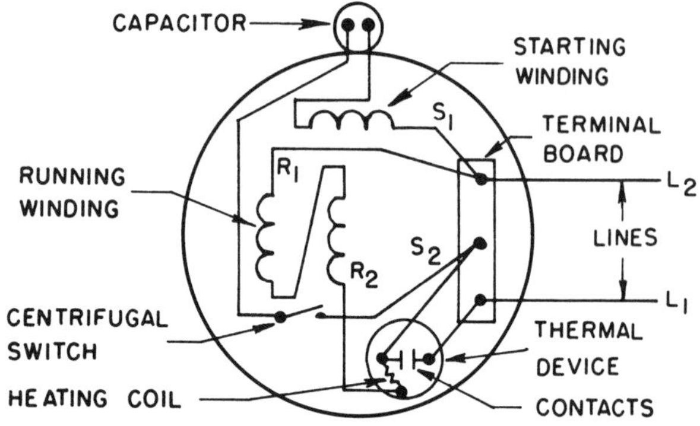

[a] *Single-Voltage Motor, 230 Volt Connection*

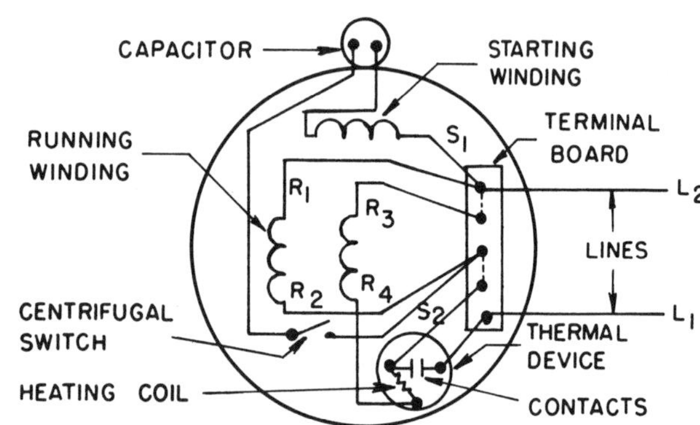

[b] *Dual-Voltage Motor, 115 Volt Connection*

CONNECTION

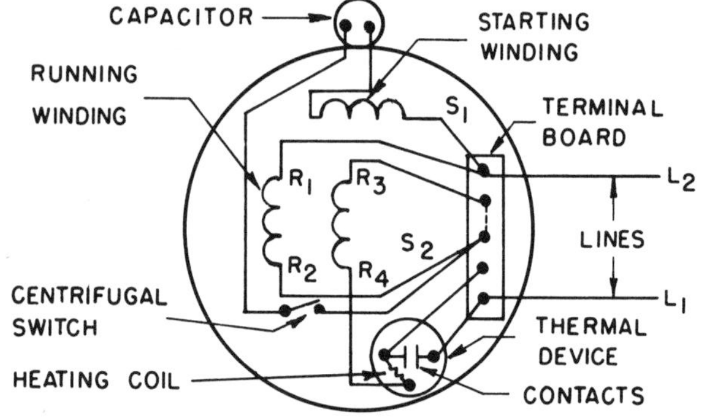

[c] *Dual-Voltage Motor, 230 Volt Connection*

Fig. 3-12. Thermal Overload Devices in Electric Motor Circuits

CODE

The code symbol will be a letter which designates the current required to start the electric motor. The technique to determine the current is to measure the volts and amperes when the motor rotor is not permitted to turn (locked rotor). The value is reported in KVA per hp and is calculated by dividing the product of line voltage times the amperage value by 1000. This code value represents the first power surge required to start the motor and is to be considered when determining over-current protection for electric motors. Code letters and KVA values per hp are listed in Table 3-1. To determine the starting current for a 1 hp single-phase electric motor which has an "F" rating of 5.0 KVA per hp use the following formula.

$$\text{Amperes} = \frac{\text{KVA/hp x hp x 1000}}{\text{V x 1.0 (1-phase value)}}$$

$$= \frac{5.0 \times 1 \times 1000}{240 \times 1.0}$$

$$= 20.83$$

Table 3-1. Code Letters for Locked-Rotor KVA

CODE LETTERS FOR LOCKED-ROTOR KVA

CODE LETTER	KVA PER HORSEPOWER
A	0 - 3.14
B	3.15 - 3.54
C	3.55 - 3.99
D	4.00 - 4.49
E	4.50 - 4.99
F	5.00 - 5.59
G	5.60 - 6.29
H	6.30 - 7.09
J	7.10 - 7.99
K	8.00 - 8.99
L	9.00 - 9.99
M	10.00 - 11.19
N	11.20 - 12.49
P	12.50 - 13.99
R	14.00 - 15.99
S	16.00 - 17.99
T	18.00 - 19.99
U	20.00 - 22.39
V	22.40 - and up

Except polyphase wound-rotor motors.

If the electric motor selected were three-phase, use this formula:

$$\text{Amperes} = \frac{\text{KVA/hp x hp x 1000}}{\text{V x 1.73 (3-phase value)}}$$

$$= \frac{5.0 \times 1 \times 1000}{240 \times 1.73}$$

$$= 12.04$$

Since the phase values are 120 degrees out of phase the line current is 1.73 times the phase current.

SERVICE FACTOR

The service factor (SF) provides a numerical number listing the amount of overload that an electric motor can withstand at its rated Hz and voltage. If a 1.0 value is listed on the nameplate, the motor can be safely operated only at its rated hp. Typical SF for AC motors are; 1.15, 1.20, 1.25, 1.35, and 1.4. Assume that the watts output needed for a load is 520 and a 0.5 hp motor is available. The 0.5 hp x 746 watt/hp = 373 watts which indicates the motor is inadequate for the load demand. The required 520 W divided 373 W = 1.39 or an additional 40 percent output is needed. The capability of a 0.5 hp motor with a 1.4 SF is the product of 0.5 hp x 1.4 SF or 0.70. Thus, the decision to buy a 0.5 hp motor with a 1.4 SF or a 0.75 hp motor wih a 1.0 SF is an economic decision based on motor purchase price. There are other factors in motor selection but the wattage factor must be satisfied.

If the nameplate has SFA, it stands for service factor ampere. This rating is the current drawn by the motor when operated at service factor load. For example, a motor with a 5.0 A full load value with a SF of 1.2 would have a SFA of 6.0 A.

TIME

The time rating (duty) of a motor can be continuous or have a specific time listed on the nameplate as 5, 15, 30, or 60 minutes. If a motor is labeled 15 it can be operated at its rated full load for only 15 minutes without overheating. Some nameplates have both a service factor and time rating but this is not common for fractional hp motors. The continuous time rated motor would be the best. Whether an electric motor has a SF of one or greater and a definite time rating or a continuous rating depends upon a number of design features and the insulation class.

EFFICIENCY INDEX

Efficiency index, a more recent electrical factor is identified by a letter code. The index letters and efficiency values are enumerated in Table 3-2.

Table 3-2. NEMA Efficiency Index Letter Code

NEMA EFFICIENCY INDEX LETTER CODE TABLE		
EFFICIENCY INDEX	MINIMUM *1 EFFICIENCY %	NOMINAL *2 EFFICIENCY
A	95.0	
B	94.1	95.0
C	93.0	94.1
D	91.7	93.0
E	90.2	91.7
F	88.5	90.2
G	86.5	88.5
H	84.0	86.5
K	81.5	84.0
L	78.5	81.5
M	75.5	78.5
N	72.0	75.5
P	68.0	72.0
R	64.0	68.0
S	59.5	64.0
T	55.0	59.5
U	50.5	55.0
V	46.0	50.5
W	less than 46.0	

*1. Minimum efficiency (guaranteed) is the value that the manufacturer guarantees all motors of that rating will meet or exceed. One motor manufacturer defines "minimum" as the value that 95% of their motors will meet.

*2. Nominal efficiency is the average value of a large number of motors of the same make and model. Individual motors can vary widely from this average.

Supplied by Westinghouse Electric Corporation

It has been estimated that 58 percent of United States' electric energy goes to drive more than 750 million motors. These motors range in size from very small to 200 hp and the larger motors account for the majority of the energy consumed. The following data on energy efficient motors have been supplied by the Westinghouse Electric Corporation.

The efficiency of a motor is the ratio of its output to its input. It is a measure of the effectiveness the motor has for converting electrical power into mechanical output. The formula for calculating efficiency is:

$$\text{Efficiency \%} = \frac{746\text{ W x hp output}}{\text{watts input}} \times 100$$

Efficiency will always be less than 100 because of losses which are converted to heat. The losses are: (1) core; (2) stator I squared R; (3) friction and windage; and (4) stray load losses. These losses have always been present but now the motor manufacturer feels pressure to design and produce an energy efficient motor. To accomplish this the following changes can be made: (1) special grade of low loss lamination steel; (2) reduction in lamination thickness; (3) increases in the stator and rotor core lengths; (4) increase in the amount of copper used in the stator winding; (5) low-resistance rotor design; (6) smallest practical air gap; and (7) better design of the ventilation fan.

The modification in design, engineering and materials necessary to produce higher-than-standard efficient motors means a higher than standard production cost. Energy efficient motors demand higher prices and over a period of time the lower operation cost and savings in energy will justify the expenditure. The years for payback can be determined by using the the formula listed in NEMA Standards Publication MG10-1977. NEMA has established minimum efficiency percentages and nominal efficiency with efficiency index letters that can be expected from a given motor design. The full-load efficiency of a motor, when operating at rated voltage and frequency, shall not be less than the minimum efficiency value identified for the motor. Purchase of an energy efficient motor would not be practical when motors are used only intermittently.

The remaining facts and discussion of the nameplate are related to physical features of an electric motor.

PHYSICAL FEATURES

INSULATION CLASS

Electrical conductors in a motor must be insulated from each other and the coils of conductors must also be insulated from each other to prevent short-circuiting and leakage. The stator has coils of wire which are insulated from each other and also from the frame of the motor, Figure 3-13 (a).

Wound rotors for electric motors have coils of wire in slots which have leads coming out to the commutator as illustrated in Figure 3-13 (b). Insulation can be fiberglass materials, mylar film, adhesive glass cloth tapes, acrylic resin, polymerizing polyester varnish plus others and their quality determines the value of the motor's insulation. The insulation classes are A, B, F and H. When a motor

Fig. 3-13.Insulation in the Stator and Rotor

operates in a temperature exceeding the insulation limit, excess heat begins to break down the insulating material. This breakdown can be rapid or gradual. Excess heat carbonizes the insulation causing it to char, become brittle and flake off. An accepted rule states that for every 10 degrees Celsius (C) increase in continuous operating temperature, over the insulation limit, the motor life is cut in half. Refer to Figure 3-14, for the relationship of operating temperature to motor life for insulation classes A, B, F and H. To insure an understanding on the use of Figure 3-14, project a vertical line from the bottom scale at the 200 degree C point and determine the anticipated hours of motor life if it has A, B, F, or H insulation. The answer should be 100, 360, 1800 and 18,000 hours, respectively. Likewise, if a motor life of 10,000 hours is desired (vertical scale) with Class A insulation, the temperature must not exceed 130 degrees C, with Class B, 157 degrees C, with Class F, 177 degrees C and with Class H, 207 degrees C. As motors operate, there will be a temperature rise and the insulation quality will determine the motor life.

TEMPERATURE RISE

The allowable temperature rise in the motor windings is determined by the quality of insulation material. The early standards stated that motors which could withstand 105 degree C would be designated as Class A. Three temperature limits generally determine the maximum operating values; (1) ambient; (2) rise by resistance; and (3) hot-spot allowance. Refer to Table 3-3. for the C values for each class of insulation.

*Table 3-3. Temperature Limits for Alternating Current Motors**

TEMPERATURE LIMITS	TEMPERATURE IN ° C			
	CLASS A	CLASS B	CLASS F	CLASS H
Ambient Temperature	40	40	40	40
Rise by Resistance	50	80	105	125
Service-Factor Margin	10			
Hot-Spot Allowance	5	10	10	15
Total Temperature	105	130	155	180

*Motor with a service factor of 1.0 and drip proof exposure.

The term 40 degree C ambient means the temperature of the space around the motor and the value is the same regardless of the class of insulation. Temperature rise by resistance is determined by measuring the resistance in the conductor at room temperature and again after running the motor, this could be a 40 degree C rise. The increase in resistance is an exact measure of the rise in temperature. Hot-spot allowance is allotted because the temperature is not the same in all parts of a coiled conductor.

In looking at one nameplate, it might say temperature rise, 50 degree C, with a continuous time rating. This would indicate the motor can withstand an additonal 50 degree temperature rise above the surrounding 40 degree C ambient temperature. This must be a motor with Class A insulation, Table 3-3, which would indicate there is still a 15 degree C safety factor. Another nameplate reveals an ambient degree C of 40 with a continuous time rating. This was from an older motor and the 40 degree C measured by a thermometer on the windings corresponds in value to a 50 degree C increase based on rise in resistance. This motor also had Class A insulation. If a 40 degree C ambient rated motor is to be operated in an area with a temperature greater than 40 degree C a few decisions must be made; (1) buy a motor with a better class insulation; (2) expect the motor to have a shorter life; or (3) decrease the motor load. Likewise, if the same 40 degree C motor were to be operated in a cooler area, it would be possible to; (1) buy a motor with lower quality insulation; (2) expect a longer motor life; and (3) increase the motor load. When the nameplate lists a temperature value, it must be added to the ambient value. If, however, the temperature values of 105, 130, 155 and 180 were used on the nameplate the authors feel the consumer would be more confident of the exact temperature value. Likewise, if the insulation class letters; A, B, F and H were used, the consumer would have to correlate the letters to C values. If motors with temperature rise and ambient figures are being compared use the service factor as a basis for comparison. Temperature rise is probably the most vague point on the electric motor nameplate. Because of these points of confusion, manufacturers are more often replacing the temperature rise rating with ambient temperature rating, insulation class and service factor.

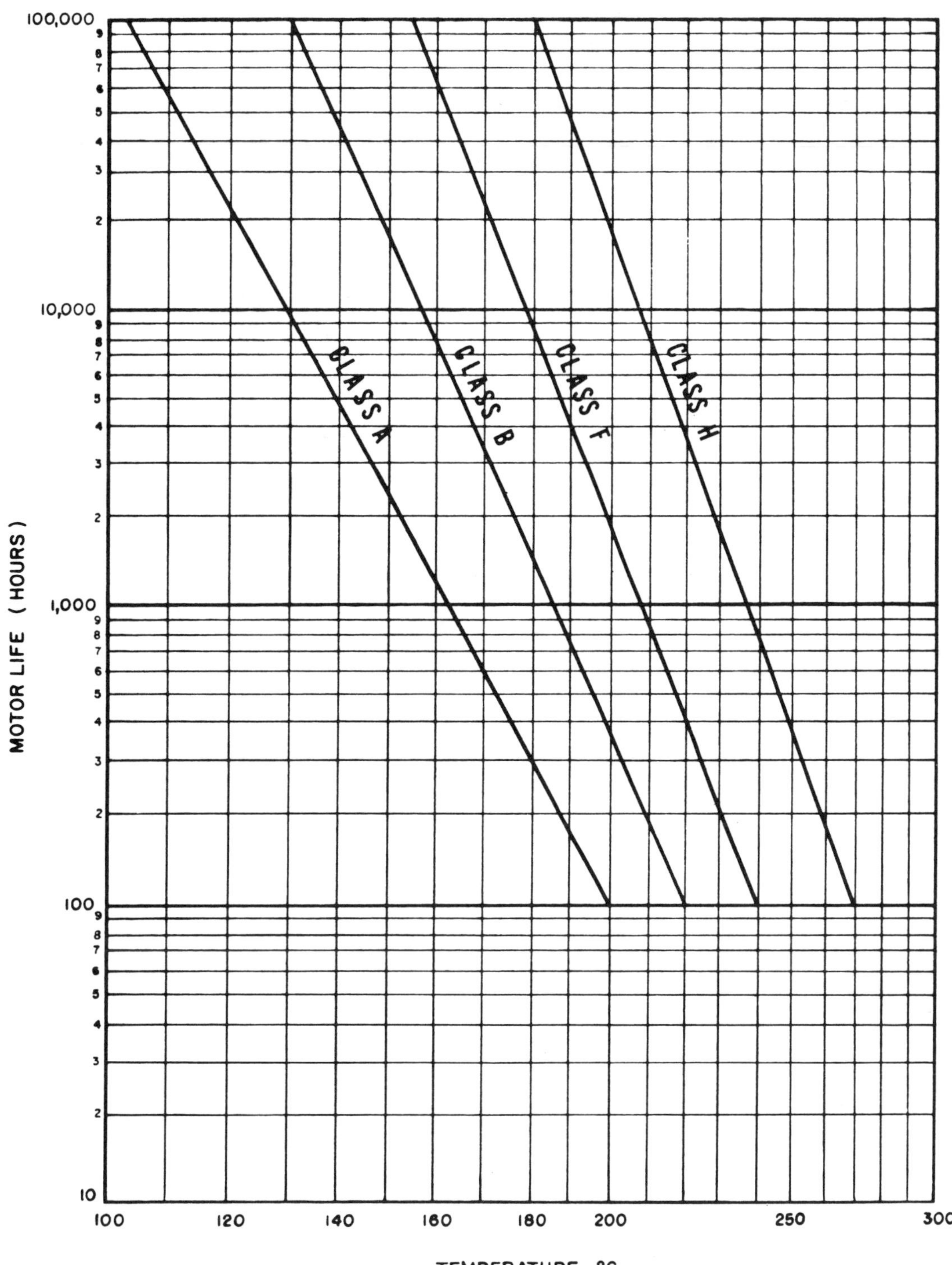

Fig. 3-14. Relationship of Operating Temperature to Motor Life by Classes of Insulation

NEMA DESIGN

NEMA has design standards about locked rotor (starting current), starting torque, breakdown torque and full-load slip. Letters A, B, C, D and F are used t describe three-phase squirrel cage integral hp motors. The following tables from the NEMA standards are found in the Appendix: (1) Three-Phase Squirrel Cage AC Motor Design Standards for Locked-Rotor Torque and Breakdown Torque, MG 1-12.35; and (2) Three-Phase Squirrel Cage AC Motor Design Standards for Locked Rotor Current, MG 1.12.34.

The design letters representing fractional hp single-phase motors are O and N, and L and M represent the integral hp motors. These standards provide the locked-rotor currents for 115 and 230 volt. The following tables are also in the Appendix; (1) Single-Phase Fractional Horsepower Locked -Rotor Current, MG 1-12.32; and (2) Single-Phase Integral Horsepower Locked-Rotor Current, MG 1-12.33; and the locked-rotor torque values for single-phase, general purpose motors, MG 1-12.31.

Table 3-4. Three-Phase Squirrel Cage Integral Horsepower Motors

DESIGN	STARTING CURRENT	STARTING TORQUE	BREAK-DOWN TORQUE	FULL LOAD SLIP
A	Normal	Normal	High	Low
B	Low	Normal	High	Low
C	Low	High	Normal	Low
D	Low	Very High	----------	High
F	Very Low	Low	Very Low	Above Normal

FRAME

Frame "size" and "number" are the same. Frame numbers are either two or three numbers. Two-digit numbers generally indicate fractional hp motors and three-digit numbers are for integral hp motors. As the physical size of the motor gets larger so does the frame number. There may be motors of the same hp in different frame number classifications, note Table 3-5.

Table 3-5. Frame Numbers for Some Common Fractional and Integral Horsepower Motors.

	FRAME NUMBER*	HORSEPOWER SIZE
FRACTIONAL	42, 48	1/4, 1/3, 1/2, 3/4
	56, 56J, 56H, 56C, 56HZ	1/4, 1/3, 1/2, 3/4, 1, 1-1/2, 2
	66	1, 1-1/2
INTEGRAL	143T, 145T	1, 1-1/2, 2, 3, 5, 7-1/2, 10
	254U, 256U	15, 20
	284TS, 286TC	25, 30

Consult NEMA Tables and motor mfg. order catalogs to obtain information on other frame numbers: 182, 184, 213, 215, 284, 286, 324, 326, 364, 365, 404, 405, 444, 445, 449, 504 and 505.

Table 3-6. Fractional Horsepower Motor Frame Suffix Letters

LETTER	IDENTIFICATION DESCRIPTION
C	Face mounting.
G	Gasoline pump motor.
H	Indicates a frame having a larger "F" dimension.
J	Jet pump motor.
K	Cellar drainer and sump pump motors.
L	Motors for home laundry equipment.
M	Oil burner motors.
N	Oil burner motors.
Y	Special mounting dimensions. (see manufacturer)
Z	All mounting dimensions are standard except the shaft extension.

Table 3-7. Integral Horsepower Motor Frame Suffix Letters.

LETTER	IDENTIFICATION DESCRIPTION
A-1	D.C. Motor
A-2	Some detail dimensions for D.C. motor may differ from those of an A.C. motor having the same frame designation numbers.
C	Face mounting on drive end.
D	Flange mounting on drive end.
P	Vertical hollow and solid shaft motors.
PH	Vertical hollow and soild shaft motors.
S	Standard short shaft for direct connection.
T	New standardized shaft, Set in 1964.
U	Different detail dimensions, Set in 1953.
V	Vertical mounting.
Y	Special mounting dimensions.
Z	All mounting dimensions standard except shaft extension.

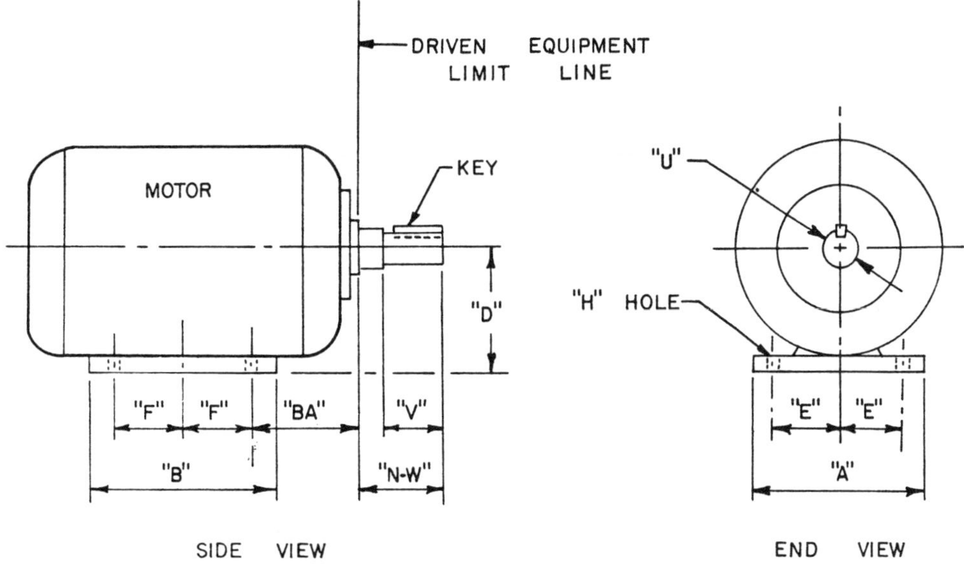

PARTIAL NEMA TABLE

FRAME NUMBER	KEY WIDTH	KEY THICKNESS	KEY LENGTH	"A"	"B"	"D"	"E"	"F"	"BA"	"H"	"N-W"	"U"	"V"
56	3/16	3/16	1-3/8	....	....	3-1/2	2-7/16	1-1/2	2-3/4	11/32 SLOT	1-7/8	5/8	....
56 H	3/16	3/16	1-3/8	....	....	3-1/2	2-7/16	2-1/2	2-3/4	11/32 SLOT	1-7/8	5/8	....
143	3/16	3/16	1-3/8	7	6	3-1/2	2-3/4	2	2-1/4	11/32	2	3/4	1-3/4
145	3/16	3/16	1-3/8	7	6	3-1/2	2-3/4	2-1/2	2-1/4	11/32	2	3/4	1-3/4
143 T	3/16	3/16	1-3/8	7	6	3-1/2	2-3/4	2	2-1/4	11/32	2-1/4	7/8	2
505	3/4	3/4	7-1/4	25	23	12-1/2	10	9	8-1/2	15/16	8-5/8	2-7/8	8-3/8

Fig. 3-15. Motor Mounting Dimensions *Reliance Electric Company*

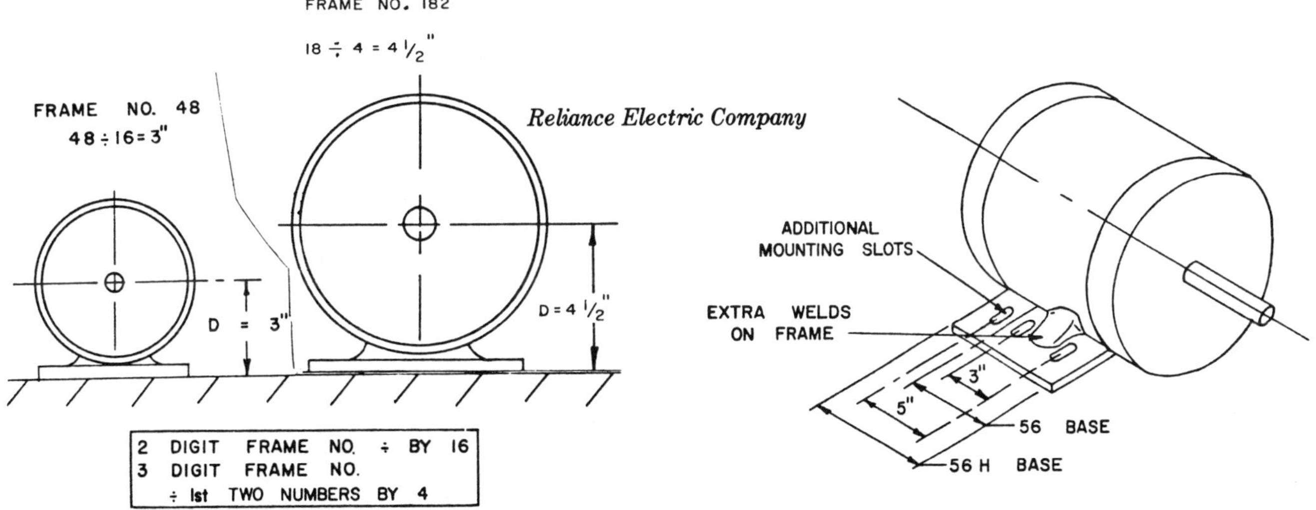

Reliance Electric Company

Figure 3-16. "D" Dimension Determination

Fig. 3-17. "H" Base Special Standard

Frame dimensions have been established for foot-mounted AC motors. The standardized dimensions are; length and width of the base, length and width of the mounting holes, diameter of the output shaft, length of the output shaft and the distance from the base bottom to the center of the output shaft. These dimensions are diagrammed in Figure 3-15 illustrating motor a side and end view plus six frame examples. There are approximately 40 frames for foot -mounted AC motors, note Appendix for more details.

The "D" dimension is from the bottom of the frame to the center of the output shaft. This distance can be calculated easily by remembering the following rules; in the two digit frame, divide the number by 16, and in the three-digit numbers, divide the first two digits by 4. Study the sketches in Figure 3-16.

The frame number can also have letters. The letters after the numbers are "suffix" letters and those before the numbers are called "prefix" letters. Two suffix letters, H and T were noted in Figure 3-15. These letters are explained in Table 3-6 for fractional hp motors and in Table 3-7 for integral hp motors.

Development of motors for special application with non-standard mounts made the use of suffix letters necessary. One example is the special "H" base illustrated in Figure 3-17.

Motors are usually foot mounted but in some applications, there will be an end mount. Foot mounting is the least expensive and the mount will have legs (called feet) or a base plate. Small motors have a rolled steel cylinder body and a base plate welded to it. Larger motors will have a cast frame and the feet are part of the body casting. The dimensions for these mounting holes are listed in Figure 3-15 and in a table in the Appendix. Foot mounted frames are classified as rigid but on some appllications flexibility is desired when a motor is subjected to shock loads. This flexibility was achieved by the development of the "U" shaped cradle base plate with resilient mounting rings, note Figure 3-18. With all the standards and conformity by manufacturers, there are still exceptions to the rule. An accessory which has been developed for motors with belt drives is the sliding base, note Figure 3-19.

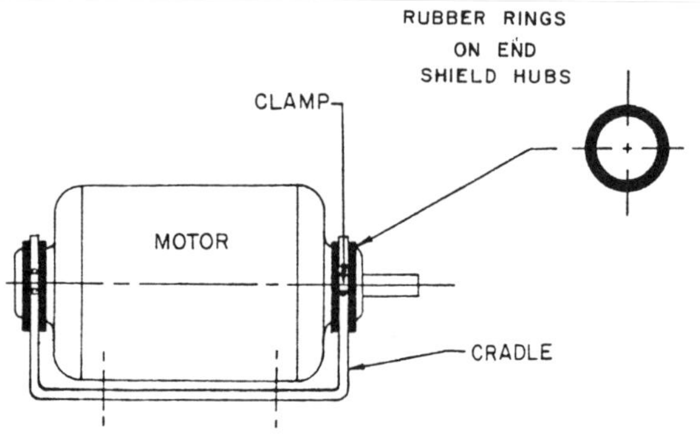

Fig. 3-18. Resilient Motor Mount

Reliance Electric Company

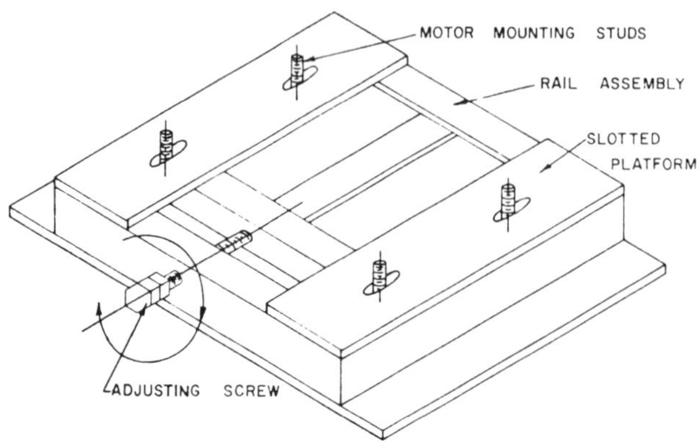

Fig. 3-19. Sliding Base For Motors

ENCLOSURES AND END BELL VENTILATION

The electric motor stator, rotor and windings must have satisfactory protection against gases, vapors, dust, liquid, weather, temperature changes, and falling objects, note Table 3-8. NEMA has standards for different protective enclosures and methods of cooling. Categories used to identify the enclosures are "open" and "closed" as listed in Table 3-9.

An open drip-proof motor, Figure 3-20, has ventilation openings of 15 degrees or more from the vertical. This is the most common type motor for applications away from liquids and dust. Open splash-proof motors are protected from dripping and splashing liquids for 100 degrees or more from the vertical. These motors frequently have baffles in the ventilation openings. The open guarded motor has a screen over the ventilation openings to prevent entry of foreign objects. Open pipe ventilated motors, usually 100 hp, have a duct bringing clean fresh air into the motor from a distant point. The open externally ventilated motor has a fan mounted on the case and the air is taken from the immediate area. Weather protected motors can be exposed to rain, snow and dust. Type I motors have some protection from the elements but Type II are designed to operate without any other protection. The different types of enclosed motors are listed in Table 3-9 and an example in Figure 3-21. These motors are used in extremely dusty and corrosive conditions. There will not be any outside ventilating air for cooling the internal motor parts with the totally-enclosed non-ventilated motor enclosure (TENV) type motors. If the motor has a fan inside, it moves the air and the heat is conducted through the case and radiated to the air. The totally-enclosed fan-cooled enclosure (TEFC) motor, see Figure 3-22, has heavy fins on the outside which provide a greater surface area for faster elimination of heat. These motors will probably have a better class of insulation. Totally-enclosed pipe ventilated (TEPV) motors have cooling air brought in and carried away. This motor would be 100 hp or larger as would the totally enclosed and is designed to withstand internal explosions. If there were an explosion, it would be confined inside the motor and not cause further explosion outside in an explosive atmosphere where the motor is operating . Explosion-proof motors are used where there are petroleum products, liquid or gases. Dust-ignition-proof motors are used where dust might accumulate on the motor. Motors operating in these hazardous conditions must meet all the standards of the Underwriters' Laboratory (UL).

Table 3-8. Hazards to Electric Motors

CONDITIONS	CONTAMINANTS

High Humidity
    Excessive moisture or steam in ventilating air
Outdoor installation
    Direct sunlight
    Wind-driven rain, sleet, or snow
Abnormal temperatures
    Ventilating air too hot
    Ventilating air too cold
    Mounting surface too hot
Limited ventilation
	Airborne gases or vapors
	    Water vapor
	    Corrosive gases
	    Chemical or solvent vapors
	
	Airborne dust
	    Abrasive
	    Conducting
	    Magnetic
	    Corrosive

HAZARDS	

Explosive or inflammable gases
 	
Explosive or inflammable dust
	Liquid droplets
	    Water
	    Corrosive liquid
	    Solvents
	    Lubricants

Table 3-9. Protective Enclosures For Electric Motors *Reliance Electric Company*

OPEN	ENCLOSED
Drip-proof	Totally-enclosed (T.E.N.V.) Totally-enclosed fan-cooled (T.E.F.C.)
Splash-proof	
Guarded	Totally-enclosed fan-cooled guarded Totally-enclosed pipe-ventillated
Open pipe-ventilated	Totally-enclosed water-to-air cooled Totally-enclosed air-to-air cooled
Open externally-ventilated	
Weather protected (type I, type II)	Explosion-proof Dust-ignition-proof

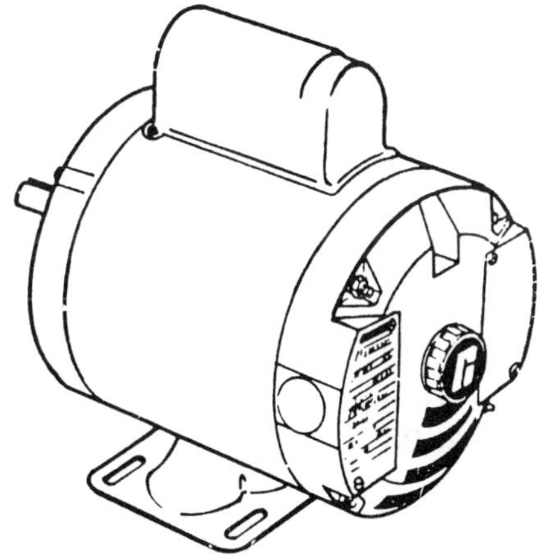

Fig. 3-20. Open Drip-Proof Motor Enclosure

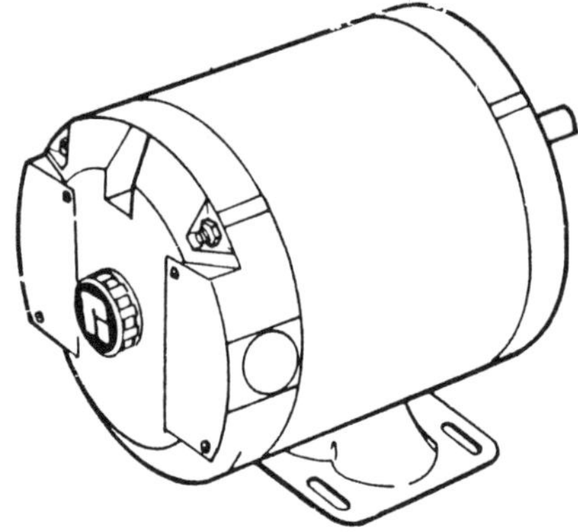

Fig. 3-21. Totally-Enclosed Non-Ventilated Motor Enclosure [TENV]

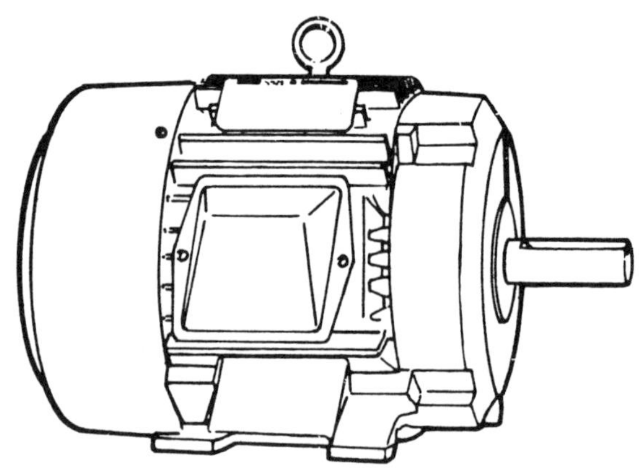

Fig. 3-22. Totally-Enclosed Fan-Cooled Enclosure [TEFC]

SHAFT DIAMETER

The output shaft for the motor is standardized by NEMA for motor frame size, note Table 3-10, for a partial listing and the Appendix for greater detail.

Table 3-10. Shaft Diameters For Foot-Mounted Motors

MOTOR FRAME SIZE	SHAFT SIZE, IN.
56	5/8
143, 145	3/4
182, 184	7/8
213, 215	1-1/8
254U, 284TS	1-3/8
284, 324S	1-5/8
326U, 364TS	1-7/8
324T, 364U	2-1/8
364T, 365T	2-3/8

MANUFACTURERS DESIGNATIONS

TYPE

Type designates the product name, for example: split phase-start, induction run; capacitor-start, induction run; permanent-split capacitor; synchronous; three-phase and many others. The letters which identify the types of single- and three-phase induction run AC motors for a number of companies are found in Table 3-11. The term, types, is fairly consistent with electric motor manufacturers.

BEARINGS

Some nameplates will list the replacement bearings. There will be a listing for the drive end and then the opposite drive-end bearing. One company may have numbers like 208 and 206 but another may have longer numbers like 35BCO2XPP3M·and 30BCO2PP3M. These will be found only on larger electric motors.

The remaining data on the nameplate can be contributed to the manufacturer and is established by them and not NEMA. The terms used by them will be; type, identification number, serial number, model number, style and form. Each company will have their own terminology. Standardization is more difficult but an attempt will be made to help identify the types of motors by these data.

IDENTIFICATION NUMBER

Identification number on a standard motor produced by Reliance Electric Company will be the model number if prefixed by M/N. If the motor is not standard, it will have a serial number. Another company will use serial number and style to identify their motors and still another will use part number, serial and type.

Each motor manufacturer has its own numbering system. As product lines are expanded and more motors are manufactured the code system will need to be altered. Each company will have identification symbols for their motor electrical type. The symbols in Table 3-12 were being used by the Marathon Electric Manufacturing Corporation in the 80's.

Table 3-11. Single- and Three-Phase Induction Run AC Motors

TYPE OF MOTOR NAMEPLATE SYMBOL*	COMPANY								
	GOULD	BODINE	DAYTON	MARATHON	DOERR	RELIANCE	GE	BALDOR	WESTING-HOUSE
Split Phase-Start	SP	SI	4K, 5K, 6K	S	SN-TENV S-OPEN SF-TEFC	SP	KH	S	FH, FHT
Capacitor-Start	CS	CS	4K, 5K, 6K	C	KN-TENV K-OPEN KF-TEFC	CS	KC	L	FJ, FZ
Capacitor-Start Capacitor-Run	CP	CCI or DI	4K, 5K, 6K	B	IN-TENV T-OPEN TF-TEFC	CH	KCR		FT
Permanent-Split Capacitor	CX or CM	CI	3M	A	C-OPEN CF-TEFC	CL	KCP		FLL
<u>SYNCHRONOUS</u> Capacitor Split	None None	CY SY	None None	None None	None None		SC SH		
Three-Phase Squirrel Cage	SC	PP	2N, 3N	T	PN-TENV P-OPEN PF-TEFC	P	K	M	FS, LIFE LINE TEE
High Efficiency					PN-TEN P-OPEN PF-TEFC				FT, FHT

*These symbols are found on motor nameplates. Blanks indicate the company does not manufacture such a motor or the authors were unable to secure the conclusive data.

ENCLOSURES

Each motor has an enclosure type and the letters used by Marathon Electric Manufacturing Corporation are listed in Table 3-13.

Other essential information for the identification of Marathon Electric Manufacturing Corporation fractional hp model numbers is in Table 3-14.

Using the information from the previous Tables 3-12, 3-13, and 3-14, the model numbers and letters explain features of the motor in Figure 3-23.

Identification of Marathon Electric Corporation integral hp model numbers is more complicated. The following data, Table 3-15, Identification of Marathon Electric Integral Horsepower Model Numbers, helps identify Marathon Electric Corporation motors. With the basic data identified, interpret the identification of an integral hp motor in Figure 3-24.

*Table 3-12. Motor Electrical Type Symbols**

SYMBOL	ELECTRICAL TYPE
A	Perm.-Split Cap.
B	Cap.-Start, Cap.-Run
C	Cap.-Start, Ind.-Run
K	3-Phase Wound Rotor
M	3-Phase Syn. Motor
R	Rep.-Start, Ind.-Run
S	Split Phase
T	3-Phase Sq. Cage Ind.-Run
U	Universal
X	Repulsion Induction

**Marathon Electric Manufacturing Corporation*

*Table 3-13. Types of Motor Enclosures**

LETTER	TYPE OF ENCLOSURE
D	Drip proof
E	Explosion Proof -- TENV
F	TEFC (Totally Enclosed Fan-Cooled)
G	Explosion Proof, TEFC
O	Open
S	Splash Proof
T	TENV (Totally Enclosed Nonventilated)
V	Blower Ventilated
W	Weather Protected

**Marathon Electric Manufacturing Corporation*

*Table 3-14. Identification of Marathon Electric Fractional Horsepower Model Numbers**

POSITION	DESIGNATED DATA
1.	DATE CODE -- Year
2.	OVERLOAD DESIGNATION P -- Single Phase Manual Q -- Single Phase Automatic R -- Special, Not UL Approved S -- Special, Not UL Approved T -- 150 Frame Manual U -- Single or 3-Phase Manual or Automatic, Not UL Approved V -- No Thermal Protection but with UL Recognition for Motor Construction W -- Single Phase, Permanent-Split Capacitor, UL Approved
3.	DATE CODE -- Month
4.	NEMA FRAME NUMBER
5.	ELECTRICAL TYPE -- See Table 3-12
6.	SPEED -- In Hundred of Full-Load RPM
7.	TYPE ENCLOSURE -- See Table 3-13
8.	ENGINEERING DESIGN -- These numbers are issued in numerical sequence as designs are created for that particular frame type and enclosure. These can be 2, 3, or 4 digits.
9.	MINOR MODIFICATION -- Electrical of Mechanical
10.	SERIAL NO. OF DESIGN -- Issued in alphabetical sequence as designs are created.
11.	FACTORY -- W--Wausau E -- Earlville

**Marathon Electric Manufacturing Corporation*

Symbol	P	Q	A	48	S	17	D	2086	A	B	W
Identification Number Position	1	2	3	4	5	6	7	8	9	10	11

	EXPLANATION
P	Year Mfg.
Q	Single-Phase, Auto Reset
A	Month Mfg.
48	NEMA, Frame Number
S	Split Phase
17	1700 RPM (In Hundredths of Full Load)
D	Drip Proof
2086	Engineering Design
A	Electrical of Mechanical Modification
B	Serial Number of Design
W	Factory, Wausau

Fig. 3-23 Fractional Horsepower Motor Description

*Table 3-15. Identification of Marathon Electric Integral Horsepower Model Numbers**

<table>
<tr><th>POSITION</th><th colspan="4">DESIGNED DATA</th></tr>
<tr><td>1</td><td colspan="4">DATE CODE -- Year</td></tr>
<tr><td>2</td><td colspan="4">RELAY DESIGNATION
Q -- Automatic Overload Protector
P -- Manual Overload Protector
U -- Single or 3-Phase Manual or Automatic. UL Approved
S -- Thermally Protected, Manual Reset, Non UL Recognized Protector and Motor Construction
X -- Thermally Protected, Manual Reset, Non UL Recognized Protector and Motor Construction
Z -- Thermally Protected, Manual Reset, Non UL Recognized Protector but UL Recognized Motor Construction</td></tr>
<tr><td>3</td><td colspan="4">DATE CODE -- Month</td></tr>
<tr><td>4</td><td colspan="4">NEMA FRAME NUMBER --</td></tr>
<tr><td>5</td><td colspan="4">ELECTRICAL TYPE -- See Table 3-12</td></tr>
<tr><td>6</td><td colspan="4">TYPE ENCLOSURE -- See Table 3-13</td></tr>
<tr><td>7</td><td colspan="4">CONSTRUCTION -- R -- Rolled Steel; S -- Cast Iron; L -- Aluminum</td></tr>
<tr><td>8a</td><td colspan="4">ENGINEERING DESIGN AND SPEED -- These numbers are issued in numerical sequence as designs are created for that particular frame, type and enclosure. With four digits, first digit is the design and the last three digits designate pole and speed.</td></tr>
<tr><td>8b</td><td>Numbers</td><td>Pole and Speed</td><td>Numbers</td><td>Pole and Speed</td></tr>
<tr><td></td><td>001 -- 025
301 -- 325
601 -- 625</td><td>2-pole -- 3600 RPM</td><td>076 -- 100
376 -- 400
676 -- 700</td><td>6-pole -- 1200 RPM</td></tr>
<tr><td></td><td>026 -- 075
326 -- 375
626 -- 675</td><td>4-pole -- 1800 RPM</td><td>101 -- 125
401 -- 425
701 -- 725</td><td>8-pole -- 900 RPM</td></tr>
<tr><td>9</td><td colspan="4">ELECTRICAL MODIFICATION -- Use in Sequence, A, B, C, etc.</td></tr>
<tr><td>10</td><td colspan="4">MECHANICAL MODIFICATION -- Use in Sequence, A, B, C, etc.</td></tr>
<tr><td>11</td><td colspan="4">FACTORY -- W -- Wausau; E -- Earlville</td></tr>
</table>

**Marathon Electric Manufacturing Corporation*

SYMBOL	L	X	D	143T	C	D	R	7	033	A	A	W
IDENTIFICA-TION NUMBER POSITION	1	2	3	4	5	6	7	8		9	10	11

EXPLANATION	
L	Year
X	Thermal Protection, Manual Reset, Non UL Protector
D	Month
143T	NEMA Frame Number
C	Capacitor-Start-Induction Run
D	Drip Proof
R	Rolled Steel Construction
7	Engineering Design
033	Four-Pole, 1800 RPM
A	Modification, Electrical
A	Modification, Mechanical
W	Factory (Wausau)

Fig. 3-24. Identification of an Integral Horsepower Electric Motor

SYMBOL	T	G	S
IDENTIFICATION NUMBER POSITION	1	2	3

EXPLANATION

T -- Three Phase

G -- Explosion Proof -- TEFC

S -- Cast Iron Frame

Fig. 3-25. Marathon Electric Manufacturing Corporation Motor Type Numbers

MODEL NUMBERS

Many nameplates will also list a type number, in addition to, the model number. The Marathon Electric Manufacturing Corporation has a five-digit type number. The first position information is in Table 3-12, and the second position information is in Table 3-13. The third position designates the basic construction with "A" representing aluminum, "R" indicates rolled steel frame and "S" a cast iron frame. The fourth position represents electrical details and the fifth position has details on the shaft length and position (vertical or horizontal) and type of base. A typical type number is explained in Figure 3-25.

It is beyond the scope of this manual to list the identification systems for all companies. It is important to realize that systems do exist and if details are needed to contact the specific manufacturer.

Knowledge of the identification letters for motors is advantageous when studying and replacing them. The nameplate contains the electrical and physical features of electric motors. Some features are NEMA Standards, others UL while others are used only by the manufacturer. The electric motor must be matched to the electrical service which will be discussed in UNIT IV, PRODUCTION OF ELECTRICAL ENERGY.

****

CONSTRUCTING EXPERIMENTAL MOTORS

Want to construct some simple Experimental Electric Motors? If so, go to the Appendix and review the list of materials, the detailed plans and follow the procedure for constructing and operating the motors. Plans and instructions are available for:

1. Series, shunt or compound wound DC motor with two-pole armature.

2. Series or shunt wound DC motor with three-pole armature.

3. Similar to No. 1 and 2 with a different configuration.

4. AC synchronous motor with vertical rotor.

5. AC synchronous motor with horizontal rotor.

Plans for motors 1 and 3 are in the Appendix of this manual and the plans for motors 2, 4 and 5 are in the Instructor Manual for Electric Motors - Principles, Controls, Service and Maintenance.

DEFINITION OF TERMS

ALTERNATING CURRENT (AC): A current which reverses its direction of flow.

AMBIENT TEMPERATURE: The temperature of the surrounding cooling medium, such as gas or liquid, which comes into contact with the heated parts of the motor. The cooling medium is usually the air surrounding the motor. The standard NEMA rating for ambient temperature is not to exceed 40 degrees celsius.

BIMETALLIC: Sensing device made of two metals fastened together with different coefficients of expansion. One metal will have a different expansion rate than the other causing the metal bar to bend.

BREAKDOWN TORQUE: Minimum torque developed with rated voltage without an abrupt drop in speed.

CELSIUS: Degrees C = (Degree F - 32) x 5/9.

CONDUCTOR: Materials allowing current to move through it easily. Copper and aluminum are the common metals used for electrical conductors. Copper is used in electric motors.

CYCLE: A complete flow of current in each direction as in AC.

DIRECT CURRENT (DC): A current that does not change its directions of flow or alternate.

EDDY CURRENT: Localized currents induced in an iron core by alternating magnetic flux. These currents translate into losses (heat) and their minimization is an important factor in lamination design.

FAHRENHEIT: (Degree C x 9/5) plus 32.

HERTZ: Metric term representing cycles per second or frequency.

HORSEPOWER (hp): A unit of power equal to 746 theoretical watts in electricity and 33,000 foot pound per minute in mechanical energy.

HYSTERESIS LOSS: The resistance offered by materials to becoming magnetized (magnetic orientation of molecular structure) results in energy being expended and corresponding loss. Hysteresis loss in a magnetic circuit is the energy expended to magnetize and demagnetize the core.

INSULATION:Material which will not permit the flow of current.

LOCKED ROTOR: Holding the rotor so it will not turn when the motor is energized.

POWER FACTOR: A measurement of the time phase difference between voltage and current in an AC circuit.

REVOLUTIONS PER MINUTE (rpm): The number of times per minute the shaft of the motor (machine) rotates.

SERVICE FACTOR (SF): A numerical value listing the amount of overload a motor can withstand at rated load.

SERVICE FACTOR AMPERE (SFA): Current drawn by the motor when operating at the service factor load.

SHORT-CIRCUIT: A connection between two lines or energized parts of such low resistance that excessive current flows, as in the case when a hot and neutral conductor makes contact.

SINGLE-PHASE: The alternation of an electrical current from a potential of zero, to a positive charge, back to zero, to a negative charge, and back to zero again during one cycle.

SYNCHRONOUS SPEED: The speed of the rotating magnetic field set up by the stator winding of an induction-run motor. In a synchronous motor the rotor locks into step with the rotating magnetic field, and the motor is said to run at synchronous speed.

TEMPERATURE RISE: Some electrical energy losses in motors are converted to heat causing some of the motor parts to heat up when the motor is running. The heated parts are at a higher temperature than the air surrounding them thereby causing a rise above room (ambient) temperature.

THERMAL PROTECTOR: An overheating protective device responsive to motor temperature which protects the motor against dangerous overheating due to overload or failure to start. This protection is available as either manual or automatic reset.

THREE-PHASE: The alternations of three, single-phase electrical currents overlapping each other by one-third during a cycle.

TORQUE: Torque is a turning effort in ounce/feet or pound/feet (force x distance).

WATT: A unit of electrical power or the rate of using electrical energy. A watt is equal to one ampere at one volt. Watt = V x I x power factor for AC loads. AC circuits with inductive loads (motors and transformer welders) will have a power factor (pf) value of less than 1.0. AC resistance loads have a pf of 1.0, therefore the formula is usually written as W = V x I.

NOTES

CLASSROOM EXERCISE III-A

Nameplate Information

Describe the features of an electric motor based on the identified information on this mulitpurpose nameplate.

1. Mfg. Trademark: Mighty Motor Mfg. Co.					
2. Identification or Mfg. Reference No. or	Model No.	3. Serial No.	4. Type		
W Q A 48 S 17 D 2086			S		
5. Drive End Bearing		6. Opp. Drive End Bearing			
Sleeve		Sleeve			

MANUFACTURER'S INFORMATION

1. ______________________________
2. ______________________________
3. ______________________________
4. ______________________________
5. ______________________________
6. ______________________________

<table>
<tr><td colspan="8">1. Mfg. Trademark: Mighty Motor Mfg. Co.</td></tr>
<tr><td colspan="4"></td><td></td><td></td><td>7. HP</td><td>8. RPM</td></tr>
<tr><td colspan="4">W Q A 48 S 17 D 2086</td><td></td><td></td><td>1/3</td><td>1725</td></tr>
<tr><td>9. Volts</td><td>10. Ampere</td><td>11. Phase</td><td>12. Hz</td><td>13. SF</td><td>14. SFA</td><td>15. Code</td><td>16. NEMA Nom. Eff.</td></tr>
<tr><td>115</td><td>6.1</td><td>1</td><td>60</td><td>1.0</td><td></td><td>R</td><td>U</td></tr>
<tr><td colspan="6"></td><td colspan="2">17. Thermal Protection</td></tr>
<tr><td colspan="6"></td><td colspan="2">Auto</td></tr>
<tr><td colspan="3"></td><td colspan="4"></td><td></td></tr>
<tr><td colspan="3"></td><td colspan="4"></td><td></td></tr>
<tr><td colspan="3">18. Low Voltage Connection</td><td colspan="4">19. High Voltage Connection</td><td></td></tr>
<tr><td colspan="3"></td><td colspan="4"></td><td></td></tr>
</table>

ELECTRICAL FEATURES

7. ______________________________
8. ______________________________
9. ______________________________

10. ______________________________
11. ______________________________
12. ______________________________

13. ______________________________
14. ______________________________
15. ______________________________

16. ______________________________
17. ______________________________
18. ______________________________
19. ______________________________

1. Mfg. Trademark: Mighty Motor Mfg. Co.

W Q A 48 S 17 D 2086

20. Ins. Class	21. NEMA Design	22. Duty or Time	23. Temp. Rise	24. AMB° C	
B	0	Cont.		40	

25. End Bell

Open Drip Proof

26. Rev. Inst.

27. Frame

PHYSICAL FEATURES

20. ______________________________

21. ______________________________

22. ______________________________

23. ______________________________

24. ______________________________

25. ______________________________

26. ______________________________

27. ______________________________

CLASSROOM EXERCISE III-B

Electric Motor Applications

1. Convert the motor rating of 192 ounce-inches to pound-inches.

2. Convert the motor rating of 192 ounce-inches to ounce-feet.

3. Convert the motor rating of 12 pound-inches and 16 ounces-feet to pound-feet.

4. What is the starting torque in ounce-inches produced by an electric motor which has a 4-inch diameter pulley and a scale reading of 96 ounces?

5. What is the horsepower size of an electric motor operating as 1750 rpm that has 192 ounce-inches of torque?

6. What torque in pound-feet can be expected from a 3/4 hp motor operating at 1725 rpm?

7. What is the synchronous rpm for an electric motor with three pair of poles operating on 60 hertz?

8. What is the maximum horsepower rating of a one-third horsepower motor with a 1.25 service factor?

9. A motor had a synchronous rating of 1200 rpm but when tested, it operates at only 1116 rpm. What slip percent does this equal?

10. How many watts of electrical power would a motor use in 10 hours if rated for 120 volts and 5.0 ampere and had a 1.0 power factor?

11. Determine the starting current for a 2 hp motor operating on single-phase 240 volts, with a code rating of "R."

12. Electric motors have insulation classes with the temperature rating in Celsius. Convert these values to Fahrenheit.

CLASS	C	F
A	105	____
B	130	____
F	155	____
H	180	____

13. What frame number would be needed for an electric motor when the machine to be driven has a mounting plate with four holes with a spacing of 5" for side and 5-1/2" for end spacing.

14. What clearance would be expected from the center of the drive shaft and bottom of the base if the electric motor has a 326S frame number?

15. What suffix letter would be needed for the motor frame if buying a 1/3 hp motor for a basement sump pump?____________

CLASSROOM EXERCISE III-C

MOTOR MODEL NUMBER ANALYSIS

Below are two electric motor model numbers. Refer to Tables 3-12, 3-13, 3-14, 3-15 and Figures 3-23, 3-24 3-25, and the manual to identify the meaning of each letter, number, or series of numbers. Circle the answer, fractional or integral horsepower, for each model number.

a.

POSITION	K	V	M	56	C	17	D	2089	A	B	W

1. ______
2. ______
3. ______
4. ______
5. ______
6. ______
7. ______
8. ______
9. ______
10. ______
11. ______

Is this a fractional or integral horsepower motor? Circle your choice.

b.

POSITION	J	Q	A	215T	B	D	R	7	076	A	A	W

1. ______
2. ______
3. ______
4. ______
5. ______
6. ______
7. ______
8. ______
9. ______
10. ______
11. ______
12. ______

Is this a fractional or integral horsepower motor? Circle your choice.

LABORATORY EXERCICE III-A

Complete the following table on the electric motors supplied for this laboratory activity. A motor may not have all the necessary information.

Features	Motor 1	Motor 2	Motor 3
Horsepower			
Voltage			
Ampere			
Phase			
Hertz			
SF / SFA			
RPM			
Duty or Time			
Amb. Degree C			
Temperature Rise			
Code			
Thermal Protection			
Ins. Class			
NEMA Nom. Eff.			
NEMA Design			
Drive End Bearing			
Opp. D.E. Bearing			
End Bells			
Frame			
Type			
Mfg. Reference No.			
Identification No.			
Model No.			
Serial No.			
Manufacturer			

LABORATORY EXERCISE III-B

Identification of Motor Parts

SPLIT-PHASE MOTOR

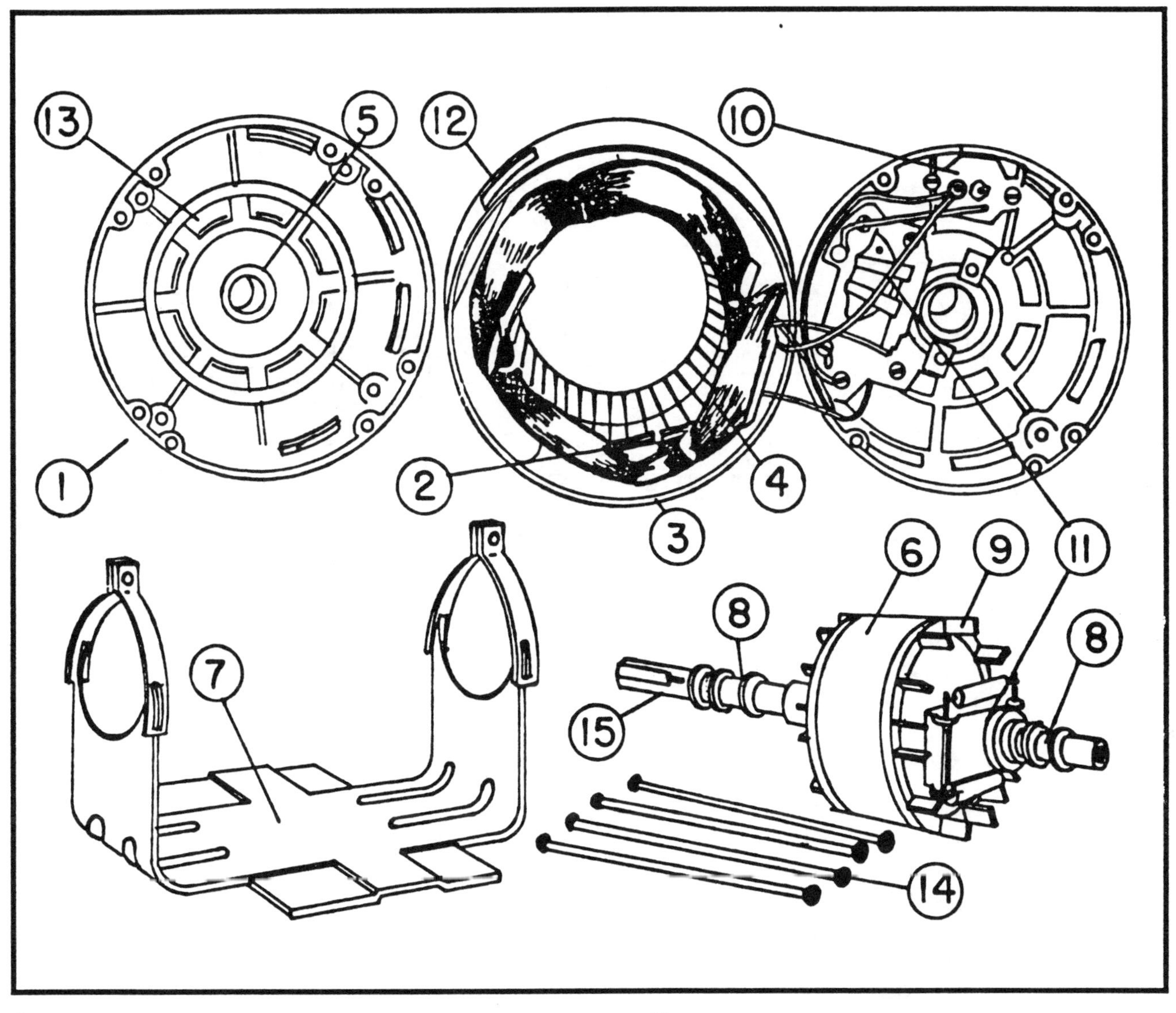

1. ______________________

2. ______________________

3. ______________________

4. ______________________

5. ______________________

6. ______________________

7. ______________________

8. ______________________

9. ______________________

10. ______________________

11. ______________________

12. ______________________

13. ______________________

14. ______________________

15. ______________________

LABORATORY EXERCISE III-C

Identification of Electric Drill Parts

UNIVERSAL MOTOR

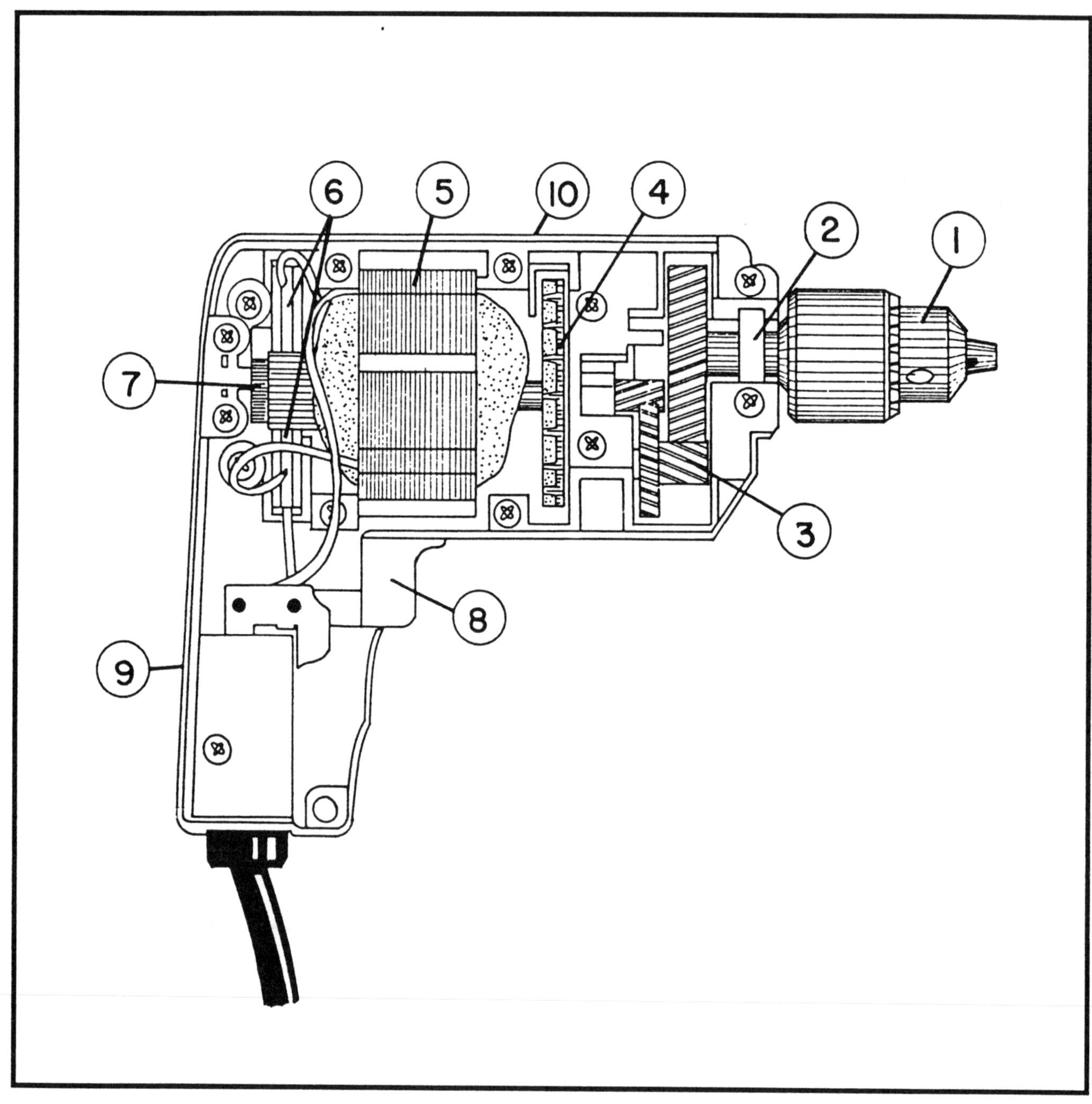

1. ______________________
2. ______________________
3. ______________________
4. ______________________
5. ______________________
6. ______________________
7. ______________________
8. ______________________
9. ______________________
10. ______________________

PRODUCTION OF ELECTRICAL ENERGY

The nameplate provided information about the electrical features required to operate an electric motor. Electrical energy generated must match the electrical requirements of the motor. Electrical motors can be designed to operate on either DC or AC. With DC, the voltage usually remains the same and current flows in the same direction. With AC, the voltage varies and current reverses its direction producing a maximum "positive" and "negative" value each revolution of the generator, refer to Figure 3-2. Generators can be turned by power sources such as: (1) internal combustion engines; (2) steam dynamos; (3) gravity flow water; and (4) geothermal water. Various types of fuels are also used to change water to steam, such as liquid fossil fuel (petroleum), solid fossil fuels (coal) and nuclear fuels.

DIRECT CURRENT

A machine driven by an outside rotational force converting mechanical energy into electrical energy is called a generator. The production of DC will be explained by the simplified single loop generator, note Figure 4-1. The DC generator has a commutator which is used to reverse the connections to the rotating loop conductors at the instant the current in the loop conductors is reversed. The resulting output current flow is always in one direction because of the split commutator design. When there is need for a variable speed motor, the use of DC is advantageous.

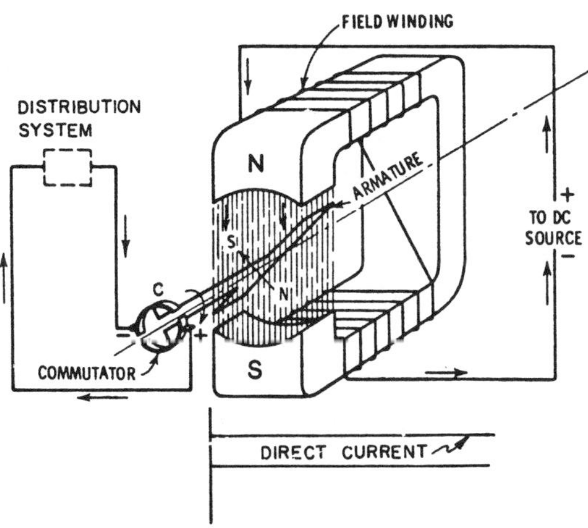

Fig. 4-1. Production of Direct Current

ALTERNATING CURRENT, SINGLE-PHASE

The main advantage of AC has been its generation at lower voltages and then the use of step-up transformers to increase the voltage for transmission. The transformer can again be used to step-down the voltage for consumer use. When the voltage is stepped-up, the current or amperage is decreased and small conductors can be used. At the consumer's point the voltage is reduced and amperage increased and the conductor size adjusted accordingly. A simplified simple loop generator will be used to illustrate the production of AC electricity, Figure 4-2. A magnetic field must first be established between the north and south poles in the generator. This is accomplished in the field winding coil with the direct current which is said to "excite" the field. The single loop rotates on axis "x" and has leads connected to slip rings "S". Brushes "B" slide on the slip rings while electrical current to the consumer flows through the distribution system. When the loop rotates, conductors "D" and "E" have a voltage induced in them causing the current to flow to the consumer. The loop is not cutting the magnetic lines of flux between the N-pole and S-pole when in line with the lines of flux or when perpendicular to the lines of flux. The loop must move through the magnetic lines of flux "L", to have a voltage produced. The direction of the voltage depends on the direction of the field and direction of the loop movement. Strength of the generated voltage depends upon the speed of the loop rotation and strength of the magnetic field between the N- and S-poles. Let's study one complete rotation of the generator loop. At the starting point, see Figure 4-2, the loop is not moving, and the dotted side is

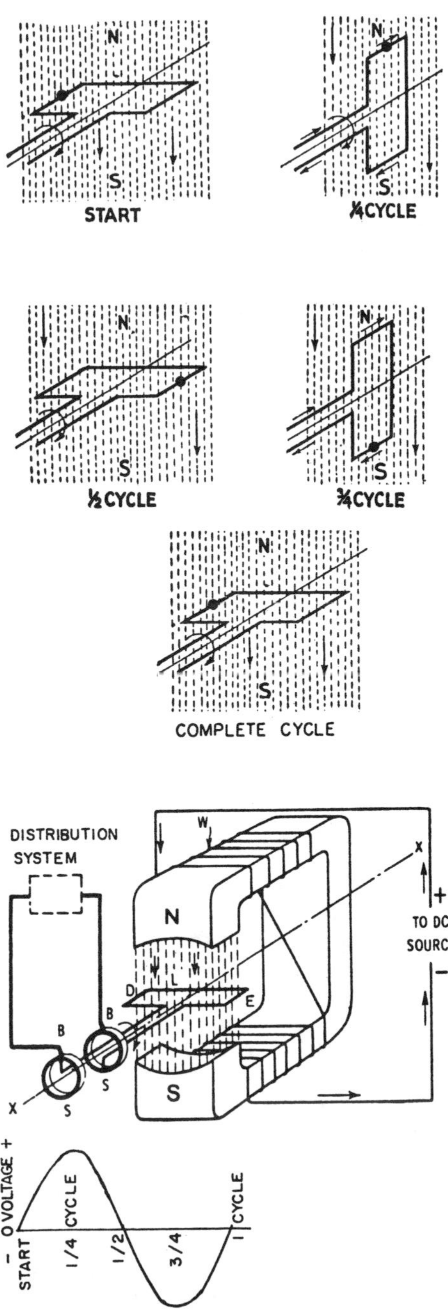

Fig. 4-2. Production of Single-Phase Alternating Current.

parallel in the field and the voltage is zero. The voltage rises to the maximum positive value at 1/4 cycle as the dotted side is moving perpendicular through the field and to the right. The voltage decreases to zero at the 1/2 cycle, note the dotted side is moving downward and parallel to the field. As the loop continues to rotate the dotted side moves to the left, and the voltage reverses. The 3/4 cycle produces the maximum negative value and the motion of the dotted side is perpendicular to the field. The last 1/4 distance of rotation completes the 360 degrees of travel. The dotted side moves perpendicular through the field and returns to the zero voltage value. Study the sine wave positions below the loop illustrations. This is single- phase AC generation and if the loop rotates 60 times per second, there is 60 Hz or 60 cycles per second current being produced. In the design of a standard generator there are two constants, 60 Hz and 60 seconds/minute and there are two variables, pairs of poles and armature rpm. Calculate the armature rpm if the generator has 30 pairs of poles and produces 60 Hz current.

$$\text{RPM} = \frac{60 \text{ Hz x } 60 \text{ Sec/Min}}{30 \text{ Pairs of Poles}}$$

RPM = 120

If a slower generator armature speed were desired, for example 90 rpm, calculate the number of pairs of poles needed to produce 60 Hz.

$$\text{PAIRS OF POLES} = \frac{60 \text{ Hz x } 60 \text{ Sec/Min}}{\text{RPM}}$$

PAIRS OF POLES = 40

What would be the Hz output of a generator with 20 pairs of poles that is operating at 180 rpm?

$$\text{Hz} = \frac{\text{Pairs of Poles x RPM}}{60 \text{ Sec/Min}}$$

$$\text{Hz} = \frac{20 \text{ x } 180}{60}$$

Hz = 60

The features influencing generator design are: (1) poles; (2) rpm; (3) sec/min; and (4) Hz. These terms will appear again when the rotation of electric motors is explained.

ALTERNATING CURRENT, THREE PHASE

Electrical current could be produced by putting two, three or even more single-phase generators together. Three-phase current is easily produced, transmitted, adapted to production applications and is especially advantageous for motors. Three simplified loop generators can be combined on the same shaft to produce three-phase current, note Figure 4-3. The three loops rotate in the same magnetic field, have slip rings and the consumers are connected in circuits A, B, and C. The current producing unit is now called an alternator and each phase peak is 120 degrees from the prior one, note Figure 4-3. With the three-phase

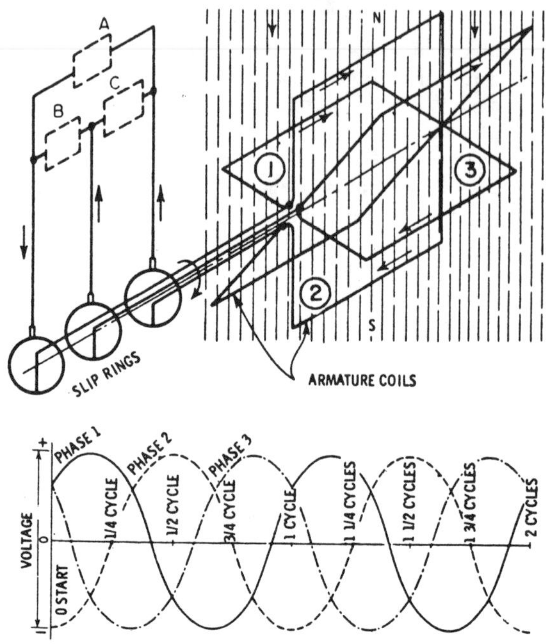

Fig. 4-3. Production of Three-Phase Alternating Current.

alternator and standard wiring connections three single-phase circuits are produced: AB, AC, and BC. The electrical service on the load side can be single- or three-phase. The three-phase service is distributed on transmission lines from the generating plant unless a phase converter is used on single-phase generated current at the load site.

VOLTAGE SUPPLIED

The voltage at the service entrance is dependent upon the electrical service and the transformer type. A transformer has an iron core with two coils of insulated copper wire immersed in oil. The oil acts as an insulator to cool the windings, note Figure 4-4. The power source goes to the primary windings and the secondary windings to the load. Transformers are rated in kilovolt-ampere, KVA.

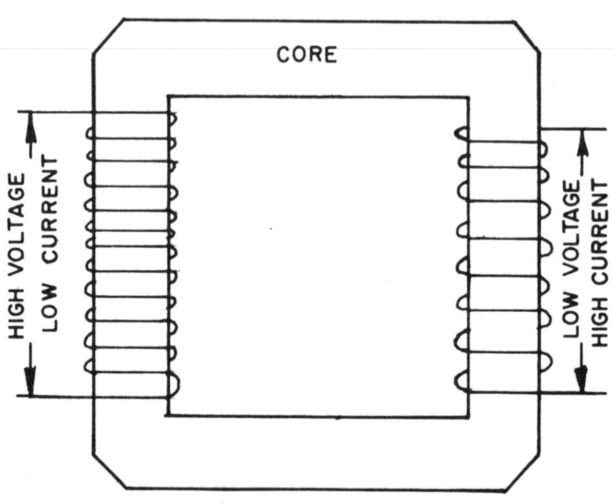

Fig. 4-4. Basic Transformer

With single-phase service there will be one transformer. In some areas, three-phase power will be coming from three different transformer cases making the identification of three-phase power easy. A recent design is to place the transformer winding for all three-phases in one case. Primary (p) and secondary (s) voltages are directly proportional to the turns (T) in the primary and secondary coils. Primary and secondary current is inversely proportional to the number of primary and secondary coils. The direct proportion of volts to turns can be satisfied with the equation: $Vp/Vs = Tp/Ts$. The power pole voltage is frequently 7200 or 2400 volts. If the primary coil has 1200 turns and 7200 volts, to determine the number of turns in the secondary coil if 240 volt service is desired the following calculation is needed: $7200/240 = 1200/Ts$. There would be 40 turns in the secondary coil. For 120 volt service, the secondary coil would be center tapped with 20 turns on each side. Current will flow in the secondary when a load is connected to the secondary due to the induced voltage in the secondary. The inverse proportion of current to turns can be calculated with the equation: $Is/Ip = Tp/Ts$. If an electric load requires twenty amperes in the secondary, the amperes in the primary coil can be solved by: $20/Ip = 1200/40$. The amperes in the primary would be 0.67 A. It is assumed that the primary and secondary ampere turns are equal as determined by: $Ip\ Tp = Is\ Ts$. The Ip value (0.666) x Tp value (1200) = 800 and the Is value (20) x Ts value (40) also equals 800.

Likewise the product of volts and amperes for motors at a low (l) voltage is equal to volts and amperes at the higher (h) voltage as determined by: $Vl\ Il = Vh\ Ih$. An example of this would be the volt and ampere nameplate data of 230/460 V and 21.0/10.5 ampere found on a motor. The Vl value (230) x Al (21.0) equals 4830 W and the Vh value (460) x Ah (10.5) also equals 4830 W. The most common electrical service is 120/240 V, 60 Hz and single-phase. There is also the possibility of 120/208 V. Motors of three-quarter hp and smaller will often be on 120 V service, and larger hp motors will commonly be on 240 V. If possible, motors larger than 2 hp should be provided with three-phase service.

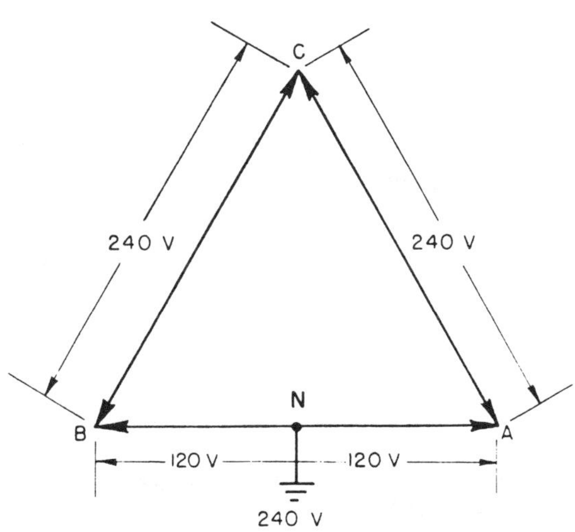

Fig. 4-5. Grounded Center Tap Delta System

DELTA TRANSFORMER

The three-phase Delta transformer has 120/240 V connections with the 120 V connection from a delta point to the grounded center tap, Figure 4-5.

The three-phase delta power source can have three transformer cases as illustrated in Figure 4-6. The center tap, N, is an earthen ground and becomes the neutral conductor for the regular wiring system. Connect A, B and C for 240 V three-phase service. Single-phase, 240 V, service can be obtained by connecting A and B, B and C, or A and C. Single-phase 120 V service is obtained by connecting A and N or B and N. Phase C is called the wild phase and is never used with the neutral because it would result in 180 volts. Although the delta transformer is the most commonly used, there is another style transformer that provides 120/208 service.

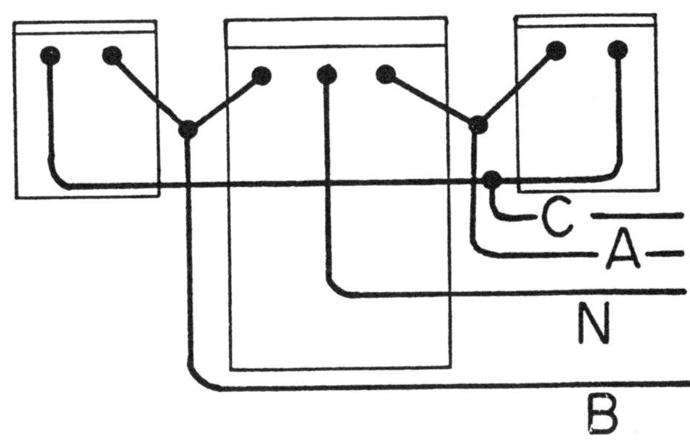

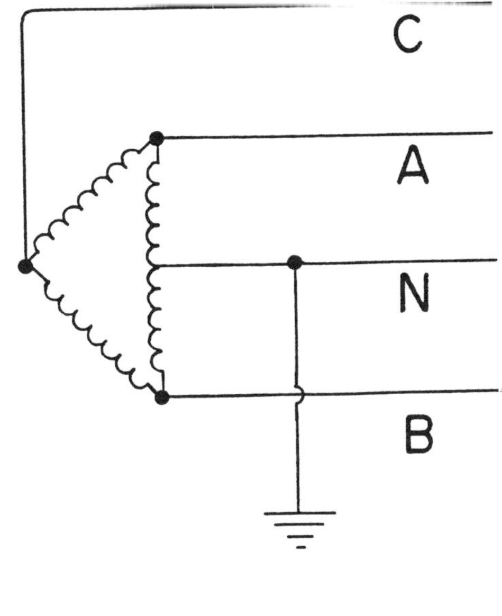

Fig. 4-6. Three-Phase Delta Transformer Power Source

WYE "Y" OR STAR TRANSFORMER

The three-phase Wye transformer has 120/208 V connections with the 120 V connection from the center tap point, study Figure 4-7. The "Y" connection usually results in a four conductor service, one wire being common to all transformers and grounded to form a neutral. The voltage obtained by connecting A to N, B to N and C to N is 120 volts. The voltage between A and B, B and C or A and C is not 240 V but the square root of 3 x 120 volts or 208 volts. Refer to Figure 4-8 for this transformer configuration. The chief advantage of this system is that the 120 volt load can be balanced on all three transformer configurations. The main disadvantage is that single-phase, high voltage is only 208 volts which can cause problems for operating larger electric motors and transformer type AC arc welders if not wired correctly.

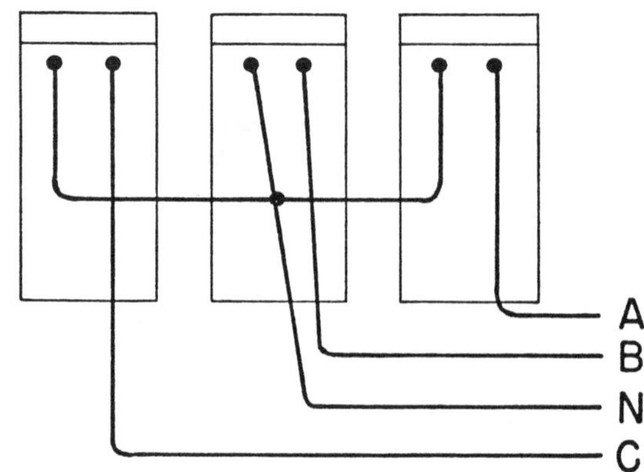

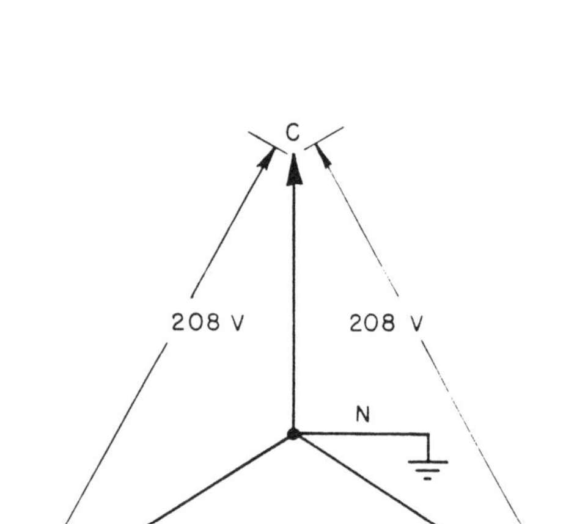
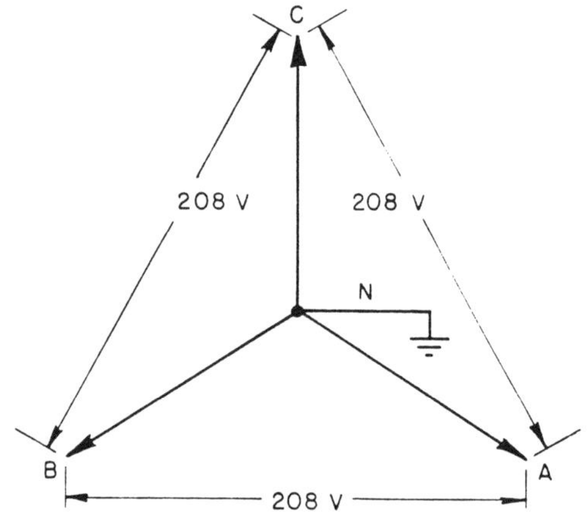

Fig. 4-7. Grounded Wye or Star System

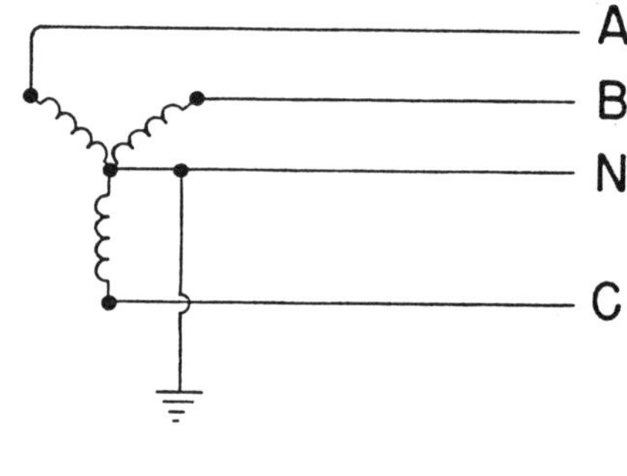

Fig. 4-8. Three-Phase Wye Power Source

This service is widely used in school and office buildings where the major load is for lighting and where the need for three-phase power is of lesser importance. If three-phase service is needed and only single-phase is available, phase converters are available. Three types of phase converters available are the capacitor type, auto-transformer capacitor converters and the rotary three-phase converter.

This knowledge of the electrical service, phase and voltage is important because the motor must match the service provided. Satisfactory motor performance is generally obtained over a range of plus or minus 10 percent from rated voltage and a plus or minus five percent from rated frequency. This accounts for the fact that motors are rated for 115/120 volts. Variations from the rated voltage and frequency will change the rated motor torque. Torque developed by a motor is approximately proportional to the square of the voltage and inversely proportional to the square of the frequency. Proper motor voltage is important for hard starting loads because at 80 percent of its rated voltage, a motor develops only 64 percent of the torque that is developed at the rated voltage.

A generator or alternator converts mechanical energy into electrical energy. These terms and components; armature, commutator, field windings, frame with poles, magnetic field, magnetic poles, magnetic lines of flux were used to describe the generation of electrical energy. These same or similar terms will be used to describe the components and explain the operation of electric motors in Unit V, MOTOR CLASSIFICATION AND OPERATION.

DEFINITION OF TERMS

ALTERNATOR: Rotating electrical machine that produces AC and is often called an AC generator.

DELTA ( ): Configuration for wiring of a three-phase transformer.

GENERATOR: Rotating electrical machine that provides a source of electrical energy. It converts mechanical energy into electrical energy.

GENERATED CURRENT: Current produced by a generator.

GEOTHERMAL: Energy from steam resulting from ground water contacting heated rocks below the earth's surface.

KILOVOLT-AMPERE (Kva); Rating for transformers.

TRANSFORMER: A device for increasing or decreasing the voltage and current as electricity passes through it by induction.

WINDINGS, PRIMARY: Windings in a transformer connected to the power source.

WINDINGS, SECONDARY: Windings in a transformer connected to the service for the consumer.

WYE (Y): Configuration for wiring a three-phase transformer. Looks like a letter "Y". It is sometimes called a start transformer.

NOTES

CLASSROOM EXERCISE IV-A

Production of Electrical Energy

1. The two types of generated electrical energy are ____________________ and ____________________.

2. Direct current is produced with a ____________________ and alternating current with an ____________________.

3. Sketch the scope pattern for AC and DC.

AC DC

4. One revolution of a single loop generator represents __________ electrical degrees and the voltage produced has one maximum __________ value and one maximum __________ value. This represents __________ phase current. Adding two more loops to the generator will result in __________ phase current.

5. The voltage output of a generator depends upon: ____________________ and ____________________.

6. Complete the following table used in manufacturing generators:

Generator	Seconds/Min	Hz	Pairs of Poles	RPM
A	60	60	____	30
B	60	___	60	50
C	60	50	____	100
D	60	60	240	____

CLASSROOM EXERCISES IV-B

Transformer Types, Volts, and Phases

1. Transformers are rated in ______________________________.

2. Identify the following transformer parts.

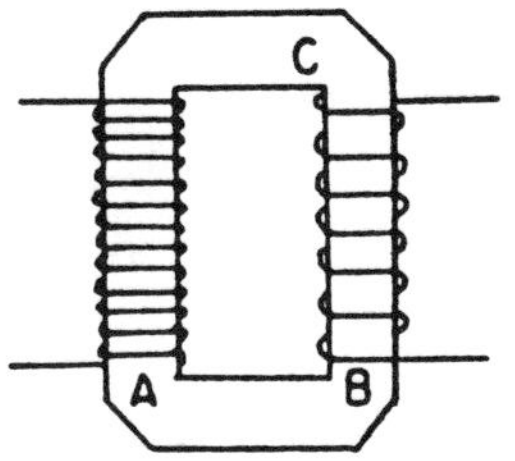

A. ____________________ coil

B. ____________________ coil

C. ____________________

D. Coil with high volts and low current. Left? ____ Right? ____

E. Coil with low volts and high current. Left? ____ Right? ____

3. Identify the step-up and step-down transformer.

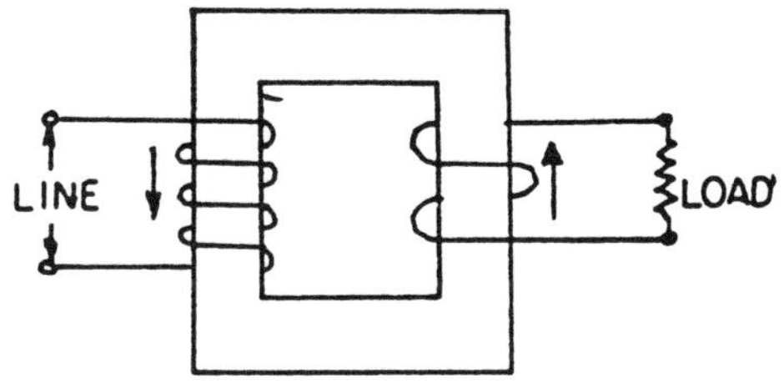

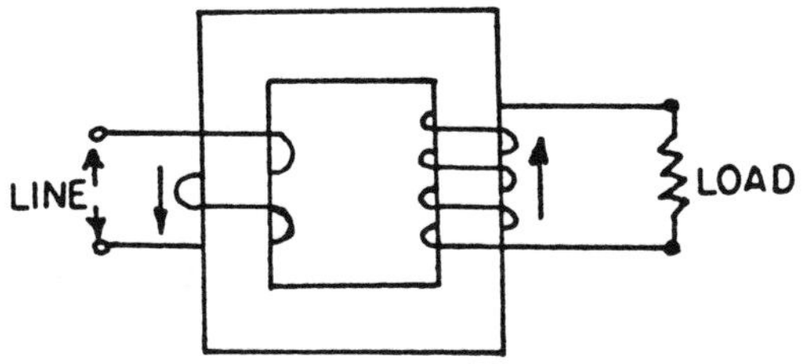

A. ____________________ B. ____________________

4. The primary to secondary coil turn ratio for 3A is ____________________ and for 3B is ____________________.

5. Combining the transformer turns, volts, and ampere relationship results in the following formula:

$$\frac{V_p}{V_s} = \frac{I_s}{I_p} = \frac{T_p}{T_s}$$

The electrical distribution line had 2400 volts, the transformer primary turns were 260, the service voltage was 120, and the electric motor load was 12 ampere. Complete the following table:

________	________	________
________	________	________

The primary to secondary turn ratio is ______________________________.

6. Recalculate the table of information when the service voltage is 240 with the other values remaining the same.

The primary to secondary turn ratio is ______________________________.

7. An electric motor on 120 volts has a 20 ampere rating. If the amperes are inversely proportional to the volts, what would be the ampere rating on 240 volt service?

$$\frac{Vl}{Vh} = \frac{Ih}{Il} =$$

8. Complete the table of volts and phases for the following delta transformer connection combinations:

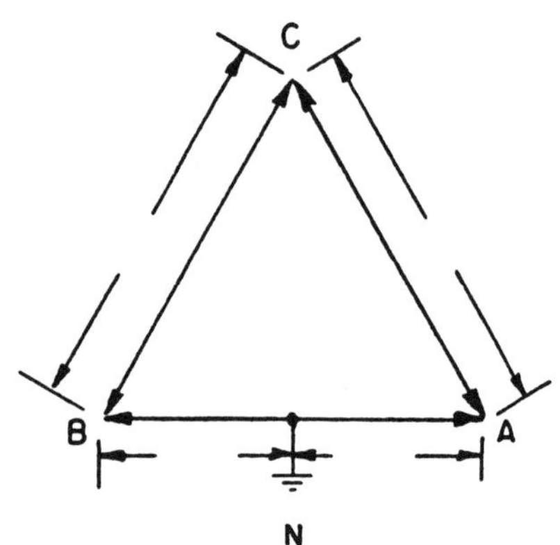

Combination	Volts	Phase
A-B	_____	_____
B-C	_____	_____
A-C	_____	_____
B-N	_____	_____
C-N	_____	_____
A-B, B-C, C-A	_____	_____

9. Complete the table of volts and phases for the following "Y" transformer connection combinations:

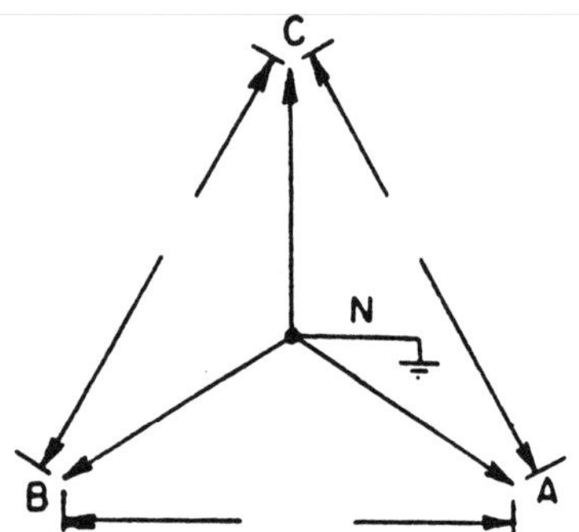

Combination	Volts	Phase
A-B	_____	_____
B-C	_____	_____
A-C	_____	_____
A-N	_____	_____
B-N	_____	_____
C-N	_____	_____
A-B, B-C, C-A	_____	_____

MOTOR CLASSIFICATION AND OPERATION

A machine that converts electrical energy into rotary mechanical energy is called a motor. Electric motors can be classified according to type of electrical service, either AC or DC. With AC service there can be either single- or three-phase motors. There are universal motors which can operate on either AC or DC.

MOTOR CLASSIFICATION

In the flow chart, Figure 5-1, the relationship of AC motors can be traced. The name or terminology used for motors is related to the electrical service, physical features and the rotation starting techniques. Single-phase motors are the most common and can be synchronous or induction run. The most frequently used electric motor is the AC, single-phase induction run type. The name, induction, refers to the electrical current being induced from the stator to the rotor. AC can be induced and DC can be produced by rectifying AC. Both AC and DC can be generated as explained in UNIT IV.

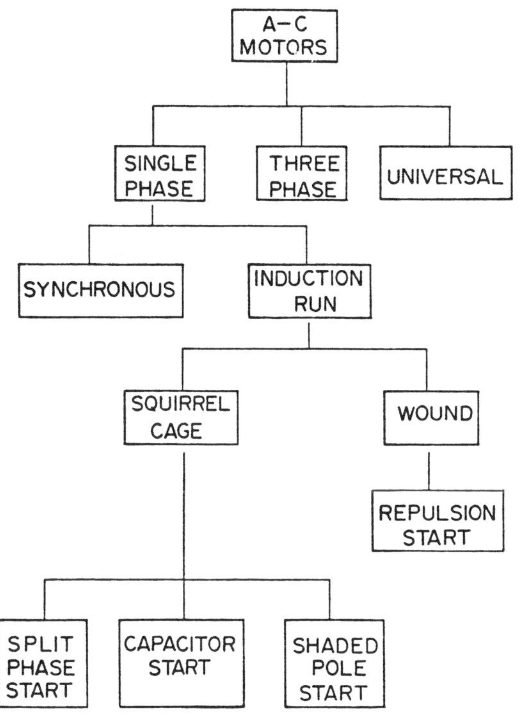

Fig. 5-1. Types of AC Motors

MOTOR PARTS AND CIRCUITS

An induction motor is one in which the magnetic field in a rotor is induced by currents flowing in the stator. Induction run motors have a rotor which is either squirrel cage or the wound style. The squirrel cage rotor motor is more economical to produce but the wound rotor, repulsion start-induction run motor, has the ability to start a greater load. The squirrel cage rotor, Figure 5-2 (a) is not really a cage because of the laminated metal in the center. The white aluminum metal rods that run lengthwise function as electrical conductors. These rods are supported by end rings. The rotor poles are formed by the steel laminations. This configuration, without the steel laminations, does form a cage and is the source of the name. The wound rotor, Figure 5-2 (b), has many strands of copper conductor wound around the shaft coming out to the commutator segments. The stator of the electric motor has windings which are called poles. There are running and starting windings, Figures 3-6 and 3-7. The stator gets its name because it remains stationary. The electrical current that causes the rotor to turn is induced from the stator winding. One set of windings helps start the rotor and the other set keeps the rotor turning after starting. Squirrel cage rotor type motors have starting systems called: (1) split-phase start, (2) capacitor-start and (3) shaded pole-start. There are three types of capacitor motors: (1) capacitor-start, induction run, (2) capacitor-start, capacitor run (two value capacitors) and (3) permanent-split capacitor. These motors run by current being energized into the running windings. The running windings can be coils of copper conductors, or coils of copper conductors with a capacitor (s). The completed names have now been derived: (1) split phase-start, induction run (2) capacitor-start, induction run, (3) shaded pole-start, induction run, (4) capacitor-start, capacitor run and (5) permanent-split capacitor.

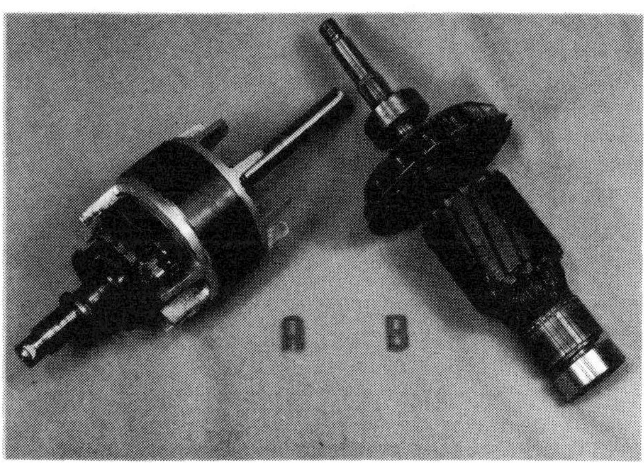

Fig. 5-2. Squirrel Cage and Wound Rotor

Additional motor types are the shaded pole and synchronous. Three synchronous types are hysteresis, reluctance and permanent magnet. Repulsion-start is a wound rotor motor, Figure 5-1. There are two other similar motors, repulsion induction and repulsion, with a wound rotor. Wound rotor motors are more expensive to produce and have been replaced in many applications with the two-value capacitor. The Baldor Electric Company was still producing this motor in 1982. There are two types of three-phase motors; induction run and synchronous. Induction run motors can have either squirrel cage or wound rotors.

MAGNETISM AND ELECTRICITY

Magnetism has been mentioned in the development of electricty and the generation of DC, and AC, single- and three-phase current. An understanding of magnetism is also essential when explaining how an electric motor operates. A permanent magnet has a north pole and a south pole and between these poles lines of flux represent the magnetic field. Permanent magnets may be bar or horseshoe shaped, note Figure 5-3. The magnetic field with

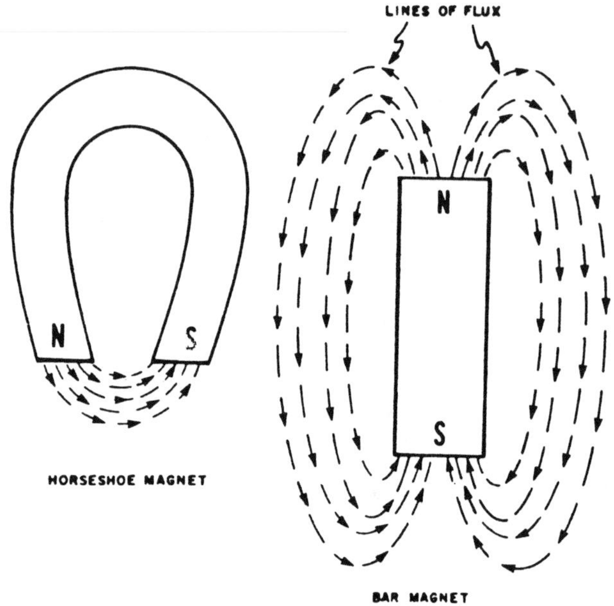

Fig. 5-3. Horseshoe and Bar Magnets

the greatest number of lines of flux is the strongest. The arrows indicate that the lines of flux leave the magnet at the north pole, travel through space and re-enter the south pole. The presence of this magnetic field can be demonstrated by placing a magnet under a glass and sprinkling iron filings on top of the glass, Figure 5-4 (a). The iron filings will form lines corresponding to the lines of flux. Placing a directional compass near the end of a permanent magnet will prove the presence of the magnetic field and identify the poles, note Figure 5-4 (b). The blued end of the compass is the north pole and the plain end is the south pole. It is interesting to note that the earth's geographic N-pole is the magnetic S-pole and the compass N-pole is attracted that direction. The geographic S-pole is the earth's magnetic N-pole. Place two bar magnets under a glass as in Figure 5-5 and again sprinkle iron filings on the top.

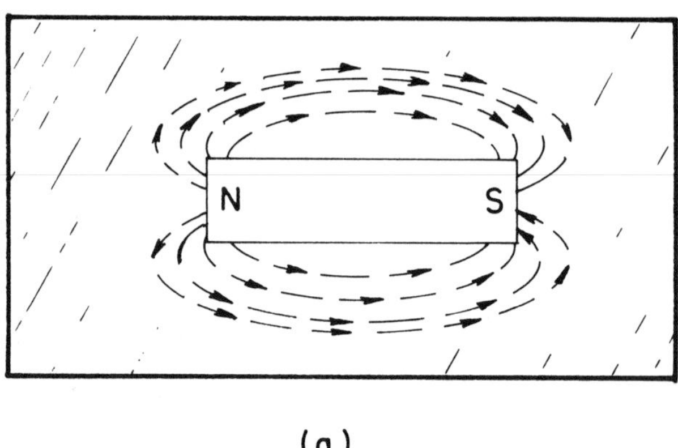

(a)

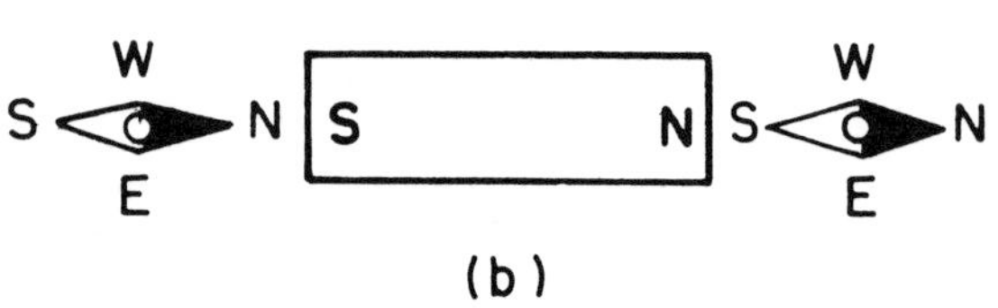

(b)

Fig. 5-4. Demonstrating Lines of Flux in a Magnetic Field.

In Figure 5-5, (a) the N- and S-poles attract each other (unlike poles attract) and this can be observed by looking at the lines of flux on the iron filings. The repelling action of the two N-poles (like poles repel) can be observed in Figure 5-5 (b) by the action of the iron filings.

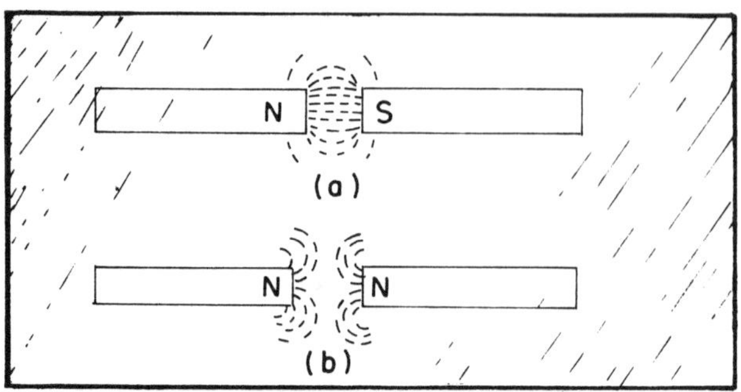

Fig. 5-5. Magnetic Poles

A magnetic field is present around a conductor when current flows through the conductor as sketched in Figure 5-6. Stop the current flow through the conductor and the magnetic field disappears. There are two rules about the flow of current and electrons. The right hand rule states that current flows from positive to negative and this will be the concept discussed and displayed in this manual. The left hand rule states that the electrons flow from negative to positive.

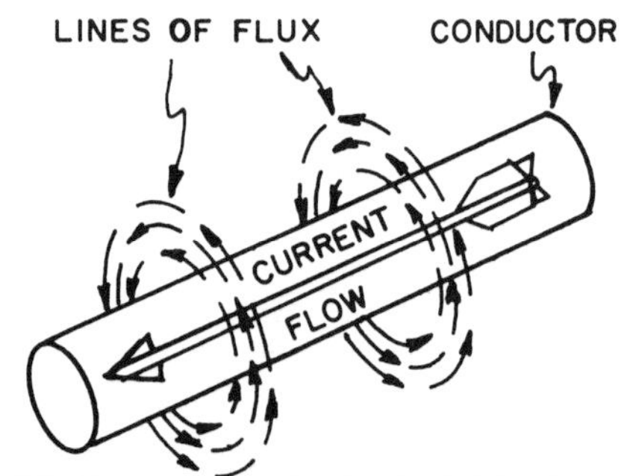

Fig. 5-6. Magnetic Flux Around a Straight Conductor

Shape a conductor into a coil, see Figure 5-7. The lines of flux which were concentric around the conductor form a magnetic field that has a N- and S-pole. Shaping the conductor into coils and increasing the number of coils will increase the strength of the magnetic field. To increase the strength of the magnetic field even more, place a metal laminated core (armature) inside the coiled conductors. This increase is possible because the metal offers less resistance to the lines of flux than does air.

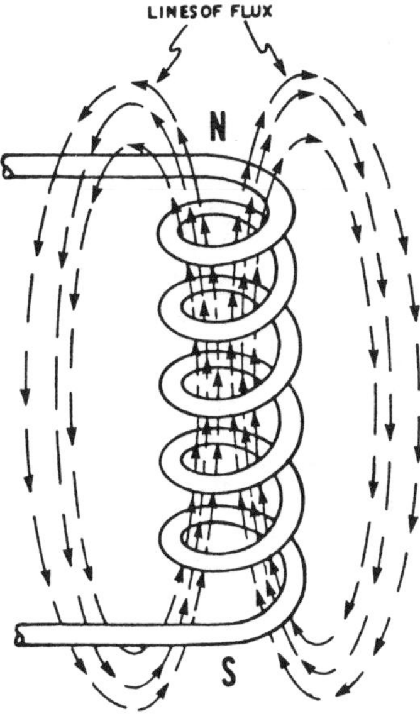

Fig. 5-7. Magnetic Flux Around a Coiled Conductor

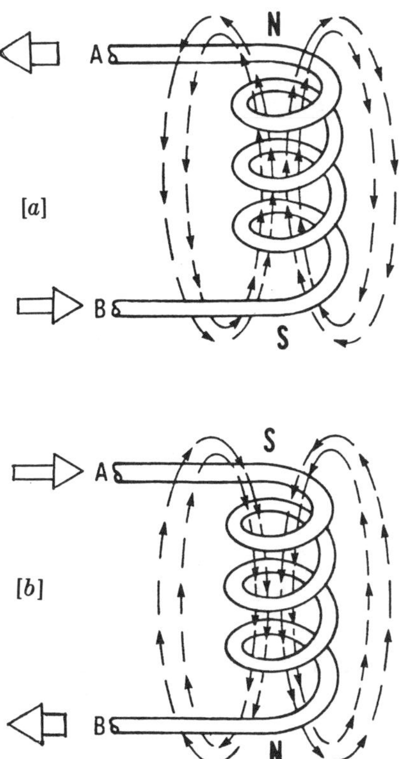

Fig. 5-8. Electro-magnetic Coils

Remember, when a current passes through a conductor a magnetic field is produced around the conductor. It is possible to have a stationary magnetic field and pass a conductor through the field, or a stationary conductor and move a magnetic field past it, and in either situation, there will be a current flow through the conductor. This current flow is called induced current, because it did not result from a direct connection to a voltage source. Thus, current can be induced or generated as illustrated in Figures 4-1, 4-2, and 4-3.

The permanent magnets have their lines of flux but the coiled conductors do not have lines of flux until connected to a source of electrical current. Based on this principle, the coils in Figure 5-8 are called electro-magnetic coils. If current is flowing from B to A, the top of the coil is the N-pole (a). Reverse the current flow, A to B and the S-pole is now at the top of the coil (b).

The reversal of current flow can be accomplished with a DC battery by changing the coil lead ends from one battery terminal to the other. The reversal is done automatically at the AC generating power plant because the frequency changes 60 times a second, and the terminology for this is 60 cycle or 60 Hz. This change of current flow, Figure 3-2, is 120 times per second. Sixty times per second the current flows in a positive "+" direction and sixty times per second in the negative "-" direction. The amount of current flow is dependent on load. It would be difficult for a large number of amperes to be flowing in a positive direction one moment and in a negative direction the next moment. Lower amperage values will be used in Figure 5-9, AC Time-Current Graph. The current in amperes may peak at 15 amperes but then decrease to zero. It is important that this wave be identified by some mumerical value, therefore, the sine wave can be identified by a certain percentage of the peak value. The need for this value is in the calculation of a load in watts. Assume there is a maximum instantaneous wave voltage of 169.7 and a maximum instantaneous wave amperage of 11.37. The maximum instantaneous power would be (169.7 V x 11.37 I = 1929.49 W) 1929.49 W. Part of the time the power is zero and at other times it is somewhere between zero and 1929 W. If the power in the circuit is neither at zero or maximum value it must be somewhere between these two values. The average height of the sine wave is 63.6 percent times the peak value but this is not the desired value. The actual value under the sine wave is determined by trigometric calculations. If you square a sine wave and take the average height, and then take the square root of this average, you will obtain the effective value of the wave. This is also called the root-mean-square (rms) because of the process to obtain the value. The rms is 70.7 percent of the peak. This 70.7 percent factor holds only for sine waves.

Thus, there are instantaneous, maximum and effective values for volts and amperes. Referring to the previous calculations; 169.7 V x .707 = 120 volts and 11.37 I x .707 = 8 ampere. The effective values would be W = V x I or 120 V x 8 I = 960 W. In the calculations of power, loads represented by the unit Watts (W), only the effective values for V and I are used, although the generating plant is producing volts and amperes ranging from zero to a maximum peak value. The important feature is that as the current value increases so does the magnetic field and as the current value decreases so does the magnetic field. These rules of magnetism and principles of electricity are the basis for the operation of electric motors.

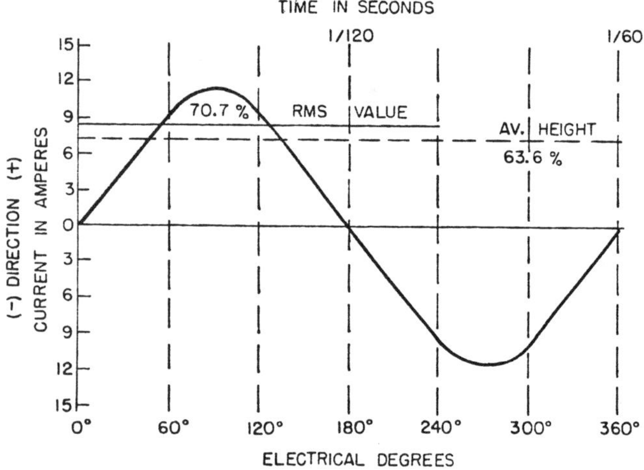

Fig. 5-9. AC Time-Current Graph

MOTOR RUNNING CIRCUIT

As a review, for the parts needed to make an electric motor refer to Figure 5-10. There is first the outside shell or frame made of steel, cast iron or aluminum. It has feet welded to the frame or cast as part of the frame for larger motors. The stationary parts inside the frame are called the stator. There are several electro-magnets (coils of wire) around the outside of the stator with the pole of each magnet facing toward the center of the group. The rotor consists of a cylinder with several groups of electro-magnets around the outside with poles facing toward the stator poles. Using the rule of magnetism that unlike poles attract the remaining task is to progressively change the polarity of the stator poles in such a way that their combined magnetic field rotates, thus the rotor will follow the changing magnetic fields of the stator and turn.

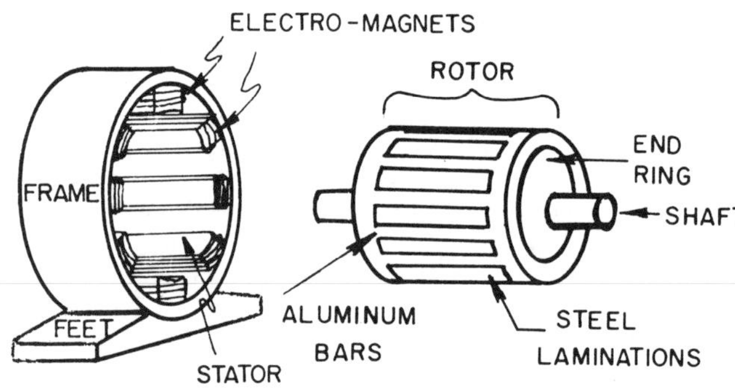

Fig. 5-10. Electrical Components for an AC Motor

The stator in Figure 5-11 has six magnetic poles and the rotor has two poles. At (a) stator pole A-1 is a N-pole and the opposite pole A-2 is a S-pole. The current induced in the rotor produces a N- and S-pole. Assume that the rotor has been given a turn to produce motion. The N-pole of the stator attracts the S-pole of the rotor and when lined up the current, alternates and A-1 becomes a S-pole, A-2 becomes a N-pole and the like poles repel each other turning the rotor. In (b), C-2 becomes a N-pole attracting the S-pole of the rotor and the C-1, S-pole attracts the N-pole of the rotor when the current alternates. At this

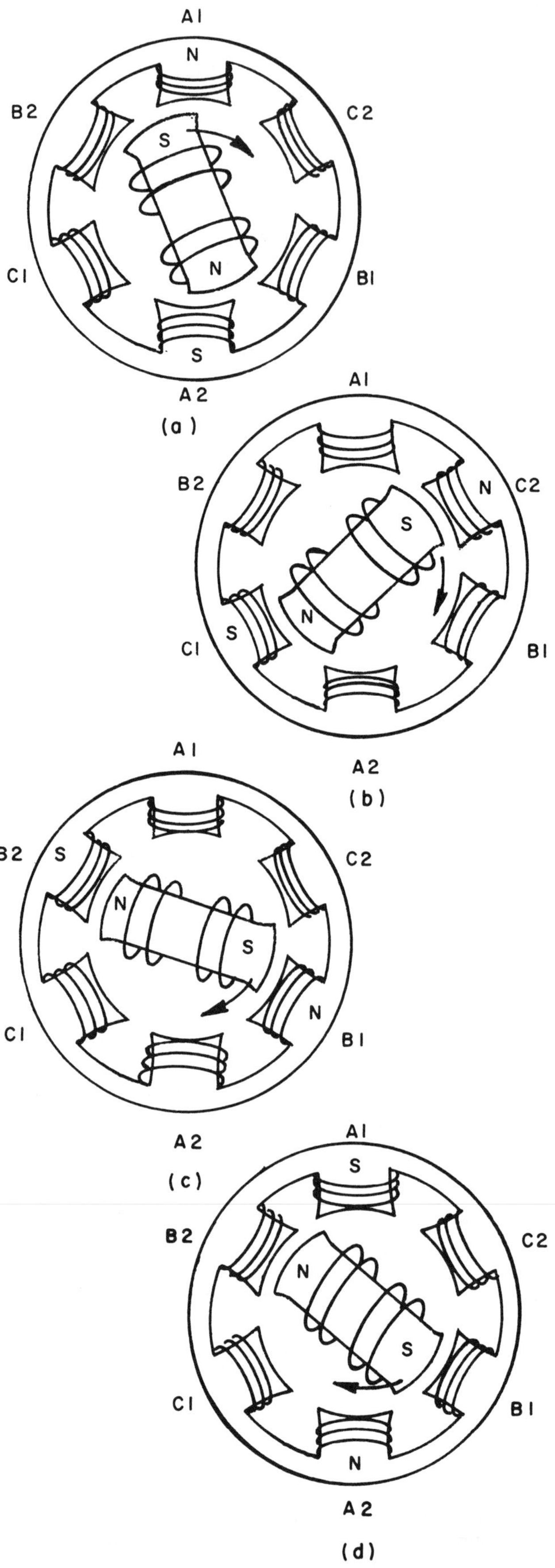

Fig. 5-11. Rotating Magnetic Field of a Six-Pole AC Motor

point one-half of the AC sine wave has been used by the motor. The current alternates again (c) and B-1 becomes a N-pole and B-2 becomes a S-pole again attracting the rotor. At the point when the rotor is in line with B-l and B-2, the rotor has traveled 120 degrees and the second one-half of the AC sine wave has been used to turn the rotor. To complete the 360 degrees travel of the rotor, the previous process must be repeated twice more for use of three sine waves or 1080 electrical degrees. The AC 60 cycles per second cause the changing of the magnetic poles in the stator which induces the current into the rotor creating N- and S-poles. The attraction of unlike poles and repelling action of like poles causes the rotor to be pulled and pushed in the turning of the rotor. Reviewing the synchronous speed formula:

$$\text{Syn. Speed} = \frac{60 \text{ Hz} \times 60 \text{ sec/min}}{\text{Pairs of Stator Poles}}$$

$$\text{Syn. Speed} = \frac{360}{3}$$

$$\text{Syn. Speed} = 1200 \text{ rpm}$$

For the next example, the electric motor has one pair of poles, A-1 and A-2, note Figure 5-12. The A-1 (a), is a N-pole stator that has attracted the S-pole of the rotor and will be the starting position. In view (b) the current alternates, and A-1 becomes a S-pole which repels the S-pole on the rotor and the A-2 stator pole becomes a N-pole attracting the S-pole In views (a) and (b) the rotor has turned 180 degrees. The current alternates again and (c) the A-2 stator becomes a S-pole repelling the rotor S-pole and the A-1 stator pole becomes a N-pole attracting the S-pole of the rotor. The rotor (d) has almost returned to the point of origin as in (a). There has been 360 degrees of rotor travel and 360 degrees of electrical degrees of travel. Determine the synchronous speed by the formula:

$$\text{Syn. Speed} = \frac{60 \text{ Hz} \times 60 \text{ Sec/Min}}{1 \text{ Pair of Poles}}$$

$$\text{Syn. Speed} = \frac{3600}{1}$$

$$\text{Syn. Speed} = 3600 \text{ rpm}$$

One pair of poles has a synchronous speed of 3600 rpm and by adding pairs of poles (3 pair equal 1200 rpm) the rpm is reduced. The addition of poles reduces the rpm of the motor but increases the torque of the motor, note Figure 5-13. There will be extra production costs but it is a satisfactory technique for manufacturing motors with different operational speeds.

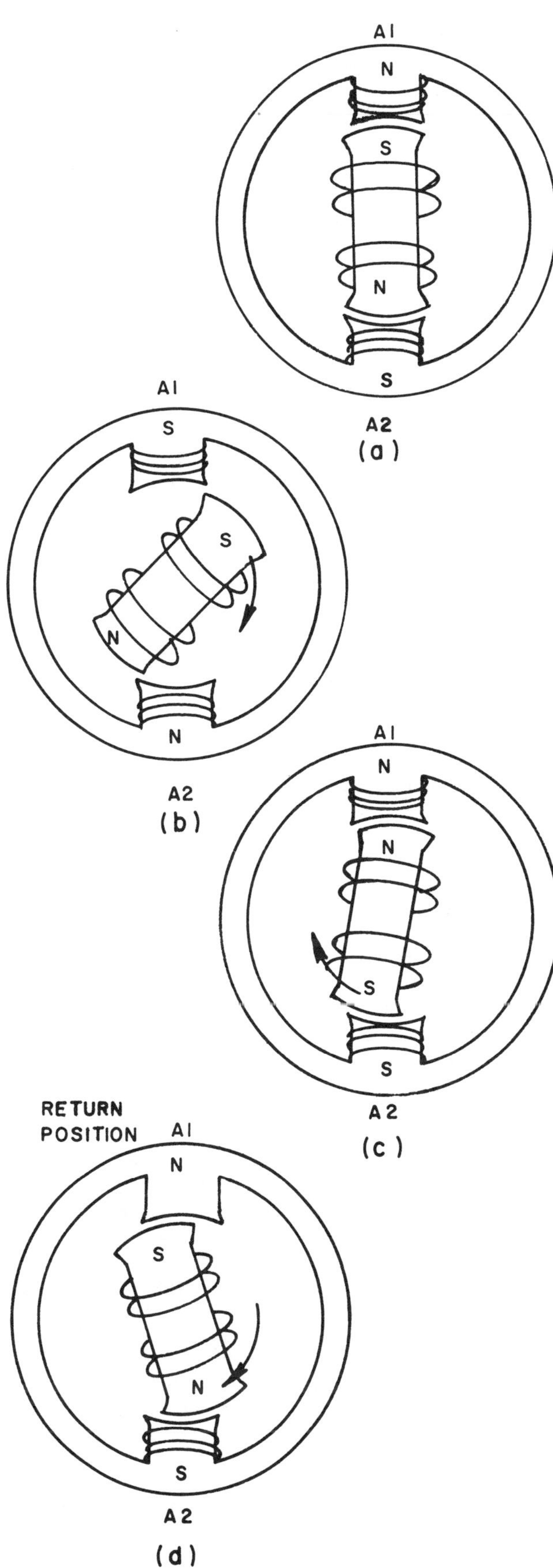

Fig. 5-12. Rotating Magnetic Field of a Two-Pole AC Motor

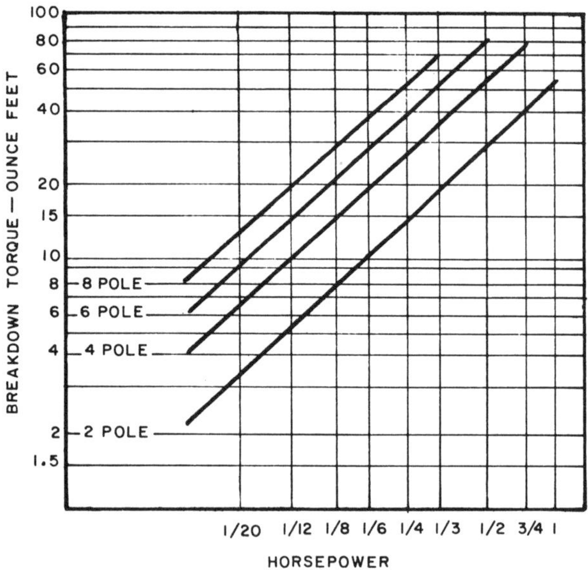

Supplied by Westinghouse Electric Corporation

Fig. 5-13. Breakdown Torque by Horsepower and Number of Poles

The AC motor mechanical and electrical rotation for pairs of stator poles and degrees of travel per sine wave is detailed in Table 5-1.

In summary, as AC is applied to the stator windings, a rotating magnetic field is generated. This rotating field cuts the bars of the rotor and induces current into the rotor. This induced current generates a magnetic field around the conductors of the rotor which try to line up with the stator field. Since the stator field is rotating continuously, the rotor cannot line up with it but must always follow along behind it. It's impossible for the rotor of the induction run motor to turn at the same speed as the rotating magnetic field. If the speeds were the same no relative motion would exist between the two, no lines of flux would be cut and no induced current would result in the rotor. Without induced current (cutting lines of flux), a turning force would not be exerted on the rotor, therefore, the rotor must turn at a speed less than that of the rotating magnetic field. The percentage difference between the two speeds of the rotating stator field and rotor speed is called slip. The lower percent slip the closer the rotor speed approaches the stator field speed. Slip is in reality the governor for the AC induction run motor. When the motor slows down, more lines of flux are cut, more current is induced and the motor speeds up. With an increase of rpm, there is less magnetic field current produced and a reduction in rpm is realized. An analogy to the small gasoline engine is the air-vane governor and its linkage to the carburetor, for example with less rpm and air flow the governor opens the carburetor but with more fuel and more rpm's, the air vane reduces the throttle opening. The governor action in the small gasoline engine can be seen and is easier to understand. The turning of the rotor during the operation of the AC induction run motor is the result of the running windings in the stator. In the explanation of the motor operation in both Figures 5-11 and 5-12, the rotor was always in motion. All motors have a starting system and as listed in Figure 5-1, squirrel cage motors can have a split phase-start, capacitor-start or shaded pole-starting system.

Table 5-1. AC Motor Mechanical and Electrical Rotation for Pairs of Stator Poles

PAIRS OF POLES	ROTOR TRAVEL / ELECTRICAL CYCLE	DEGREES OF TRAVEL SINE WAVE	ROTOR TRAVEL RPM*
1	360°		3600
2	180°		1800
3	120°		1200
4	90°		900
5	72°		720
6	60°		600

*THIS REPRESENTS SYNCHRONOUS SPEED. IN ONE MINUTE OF ROTATION THERE ARE A TOTAL OF 1,296,000 ELECTRICAL DEGREES OF TRAVEL.

MOTOR LIFE

The ideal motor is one that is economical, can be abused and operated forever. The operational life of a motor is dependent primarily on the physical and electrical features, (Unit III), when the human element has been eliminated. The statement that you get what you pay for applies in the anticipated motor life but the added value costs money. The AC Omega XL integral HP production motor by Reliance Electric, Cleveland, Ohio, carries a 10 year warranty and guaranteed efficiencies up to 96.2 percent. To design such a motor the largest drive end bearings possible were selected to handle both the axial and radial loads. The bearings are protected by a non-contact bearing isolator and there is not any loss of performance because there is not any rubbing friction between surfaces. The bearing has a grease fitting but also an over-greasing protection system. The frame and end bells are cast iron, machined to close tolerances and has a more rigid frame to improve rotor alignment and longer bearing life.There is a sealed lead entry point and a terminal block is provided in the conduit box. Inside the motor, the rotor and stator have an epoxy coating. An epoxy vacuum sealed resin material is used on the stator windings which eliminates the voids among the wires and loop ends. The sealing of the stator eliminates vibration, prevents contamination and provides better heat dissipation. The metallic cooling fan is non-sparking and the exterior of the motor is coated with epoxy zinc chromate paint. For a display of confidence in the ten year warranty of the motor when it has been installed properly the company will repair or replace and pay the direct cost of removal, transportation and reinstallation of the motor. Thus, with new products and technology, motor life can be extended. Do not only depend on the manufacturer but as a consumer, follow all service and maintenance recommendations provided with the purchased product.

ENERGY EFFICIENT MOTORS

With every world fuel crisis, interest is generated for energy efficiency. There are not established efficiency standards for electric motors. Motors can be designed to be more efficient and there are a number of possibilities for improved efficiency. In the design of an electric motor, there are two power losses, stator and rotor. When improving the motor performance, there is an interdependent relationship of the efficiency and power factor. As efficiency is increased, the power factor decreases. The improved efficiency may be improved by one percent, but there is a decrease in power loss and the larger the horsepower motor, the greater the power loss. Thus, the "efficiency factor" in a sense becomes academic and might be determined by the price of energy and the application, The power companies have responded by developing product lines of energy efficient motors. In reviewing the sales literature, new product terms are noted. Reliance Electric Company uses XE; Century Electric Inc. uses E-Plus; and General Electric has the Energy Saver. To accomplish greater adjustable speed, controllers have designed for electric motors. These controllers are inverters, being used on 208 or 230 VAC and for either single or three-phase power and can be used on integral or fractional HP motors.

The family of electric motors has been placed in categories and the use of magnetism and electricity to make motors operate has been discussed. The techniques for starting electric motors will be covered in the next unit.

DEFINITION OF TERMS

FLUX: Magnetic lines which create a pattern between opposite magnetic poles. The density of the flux lines is a measure of the strength of the magnetic field.

INDUCED CURRENT: Current that flows as the result of an induced voltage. Current produced by moving a coil in a magnetic field or a magnetic field past a coil.

LEFT HAND RULE: Point the thumb of your left hand in the direction of the electron flow in the conductor and your curled fingers will point the direction of the magnetic lines of flux. The electron theory is that flow is from minus to plus.

MAGNETISM, MAN-MADE: Magnetic attraction created by an electromagnet.

MAGNETS, PERMANENT: Magnet which has and retains its magnetic field.

MAGNETIC FIELD: The total flux lines established around an energized conductor or permanent magnet. These lines form closed loops from the N-pole and enter the S-Pole.

MAGNETISM, NATURAL: The power of attraction of the iron ore called magnetite.

RECTIFIED CURRENTS: Current produced by a rectifier.

RIGHT HAND RULE: Point the thumb of your right hand in the direction of the current flow in the conductor and your curled fingers will point the direction of the magnetic lines of force. The common current flow is said to be from plus to minus.

ROOT MEAN SQUARE: Method for calculating the area under a sine wave to obtain the effective value of the sine wave.

ROTOR: The rotating part of an electric motor (See Definition of Terms, Unit II).

ROTATING MAGNETIC FIELD: The force created by the stator once power is applied and the rotor begins to turn.

SINE WAVE: Curves which represent AC current. A graphical representation of a wave whose strength is proportional to the sine of an angle that is a linear function of time or distance.

SLIP: The difference between the speed of the rotating stator magnetic field and the rotor is known as slip and is expressed as a percentage of a synchronous speed. Slip generally increases with an increase in load.

SQUIRREL CAGE ROTOR: A rotor made by securing laminated iron core sections on a shaft. The core has slots that hold metal bars which fasten to rings at both ends of the iron core.

STATOR: The part of an induction-run motor's magnetic structure that does not rotate (See Definition of Terms, Unit II).

SYNCHRONOUS SPEED: The speed of the rotating field set-up by the stator windings of an induction-run motor (See Definition of Terms, Unit II).

WOUND ROTOR: A rotor made by securing laminated iron core sections on a shaft. The core has slots into which copper conductors are wound forming coils or loops which come out and are fastened to a commutator.

NOTES

CLASSROOM EXERCISE V-A

Motor Classification and Operation

1. Terminology used in naming motors is related to:

    a. ______________________________

    b. ______________________________

    c. ______________________________

2. Single-phase motors are the most common and can be synchronous or induction run. True or False

3. The name induction refers to the current being induced fron the ______________ to the __________.

4. The motor stator gets its name because it remains __________________ while the rotor gets its name because it ________________________.

5. List the two common rotors and a brief description of their design.

    a. ______________________________

    ____

    ____

    b. ______________________________

    ____

    ____

6. Squirrel-cage rotor type motors can have one of three starting systems:

    a. ______________________________
    b. ______________________________
    c. ______________________________

7. List the five (5) common derived names of squirrel-cage type motors:

    a. ______________________________
    b. ______________________________
    c. ______________________________
    d. ______________________________
    e. ______________________________

8. Common wound rotor motors include: repulsion start, repulsion-induction and repulsion. True or False

9. A __________ __________ has a north and south pole and between these poles ______________ ___ __________ represent the magnetic field.

10. A magnetic field is present around a conductor when ______________ flows through the conductor.

11. List four (4) methods of increasing the strength of a magnetic field:

    a. ______________________________

    b. ______________________________

    c. ______________________________

    d. ______________________________

12. Coiled conductors have permanent magnetism or lines of force while permanent magnets have lines of force only when connected to a power source. True or False

13. Discuss cycles or Hertz in respect to AC generation of electricity:

    ____

    ____

14. The basic rule of magnetism is that __________ poles attract, the polarity of the __________ _____ change so the magnetic field rotates causing the ______________________ to turn.

15. Briefly describe the operation of a two-pole AC Motor: (no starting windings)

    ____

    ____

    ____

    ____

    ____

    ____

    ____

16. Compute the synchronous speed in rpm of a four-pole, 60 hertz motor.

17. For an eight-pole (four pairs of poles) motor the degrees of electrical rotor travel per cycle is_______, the electrical degrees of sine wave travel is ____________ and rotor travel in rpm is ____________.

18. The induction motor rotor turns the same exact speed as the rotating magnetic field. True or False

19. __________ is the difference between the speed of the rotor and the speed of the magnetic field.

20. The running windings of the stator causes the motor to operate once it starts but the initial turning motion is caused by some type of __________ mechanism.

LABORATORY EXERCISE V-A

Motors on Machines

GO TO THE LABORATORY, YOUR HOME, FARM, OR A LOCAL STORE TO COMPLETE THIS EXERCISE

MACHINE WITH MOTOR	HP	PHASE	VOLTS	AMPS	RPM	TYPE OF MOTOR (Refer to Figure 5-1, could use several terms)
Example: Drill Press	3/4	1	208	3.96	1725	AC Motor, Single-Phase Induction-Run, Squirrel Cage rotor, Capacitor-Start Motor
A.						
B.						
C.						

UNIT II AS AN ASSIST FOR ANSWERS, IF NEEDED	MOUNTING POSITION V, F, W, C	LUBRICATION SYSTEM (OIL -- GREASE)	END BELL STYLE (OPEN OR TOTALLY ENCLOSED)	OPEN ENCLOSURE (DRIP OR SPLASH PROOF)
Example: Drill Press	Vertical	Oil	Open	Drip Proof
A.				
B.				
C.				

STARTING SYSTEMS AND CIRCUITS

The starting of AC electric motors is a major problem. The addition of special windings, switches, capacitors, special configuration of poles and other design considerations is done to make a motor start by creating magnetic lines of force at an angle to dead center. When a rotor moves off dead center, it will attempt to catch up with the rotating sine wave and run at a constant speed. The problem is to get it started in the correct rotation. The three-phase AC electric motor is truly a self-starter. Imagine a three paddle water wheel. Water being directed from above will cause the wheel to turn because the angle of each paddle is 120 degrees apart and rotation continues. A single-phase AC motor has two paddles and water directed from above would never achieve a rotating action. Therefore, special starting equipment or pole configuration used for starting motors is nothing more, in concept, than adding more paddles.

AC-DC universal motors are self-starters and self-runners. They are the exception to motors having special starting mechanisms, because they either do not have a starting system or do not have a running system. The AC-DC universal motor is non-synchronous in speed. Therefore, it would be difficult to distinguish the starting from running windings as separate enitities. All other motors are constant speed motors and something special must be done to cause the motor to start, gain speed and rotate at a nearly constant or synchronous speed. These starting systems and their circuits will be discussed in this unit.

ALTERNATING CURRENT INDUCTION-RUN MOTORS

SPLIT PHASE-START MOTORS

The sine wave in Figure 3-3 is a composite of the electrical current. The sine wave has two components; the amperes and the volts. The type of circuit involved will determine the volt and ampere reaction. Sketches in Figure 6-1 illustrate the condition in circuits with: (a) only resistance; (b) only inductance; and (c) only capacitance. Many circuits have combinations of these electrical conditions resulting in variations to the standard circuit pattern, for example, a circuit with both resistance and inductance, Figure 6-1, (d) and a circuit with both resistance and capacitance, Figure 6-1, (e). These terms, plus others will be used to describe the starting circuits for electric motors.

The stator starting windings in a split phase motor are identified in Figure 3-7. These starting windings usually have smaller conductors and fewer turns than the running windings. The starting windings, in a motor with one pair of running windings, will be positioned at right angles to the running windings. The electrical phase shift between the current in the two sets of windings is due to unequal impedance. The running windings have high inductance and low resistance and the current lags the voltage by a large angle, note Figure 6-1, (b). The starting windings have a comparatively low inductance and a high resistance and the current lags the voltage by a smaller angle, Figure 6-1 (d). To obtain maximum starting torque a 90 degree angular phase difference is desired. This phase difference, between the starting and running windings, generates a rotating field ahead of the running windings and supplies enough

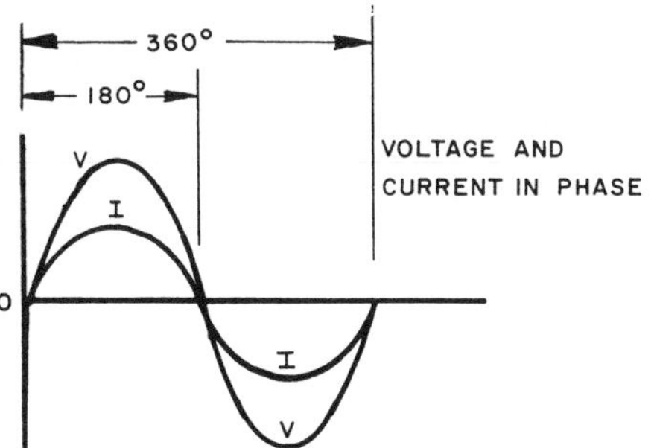

(a) CIRCUIT WITH RESISTANCE ONLY

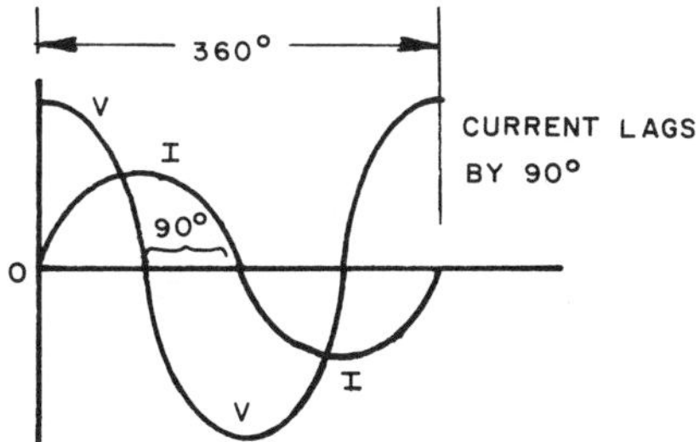

(b) CIRCUIT WITH INDUCTANCE ONLY

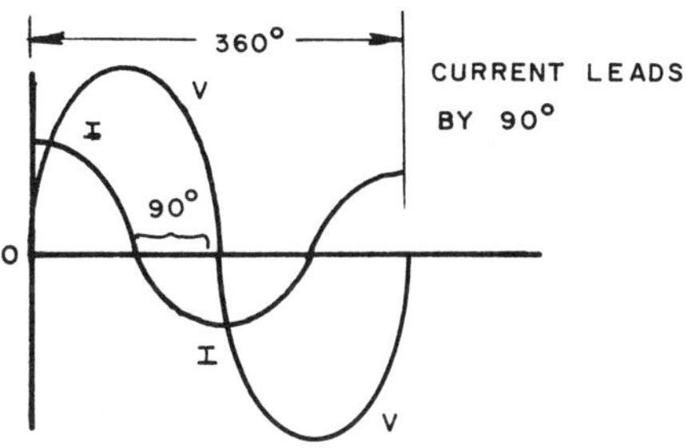

(c) CIRCUIT WITH CAPACITANCE ONLY

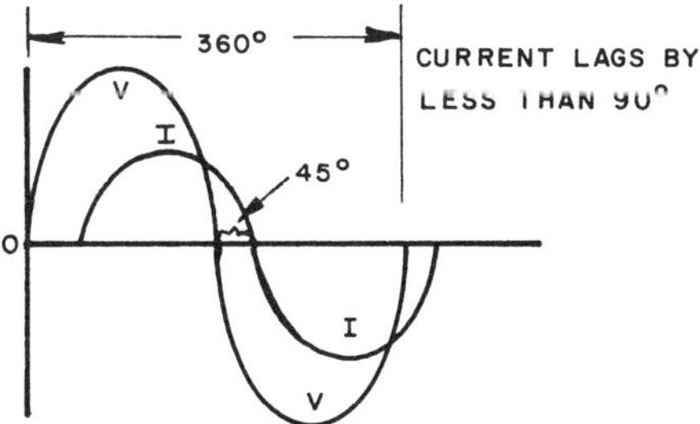

(d) CIRCUIT WITH BOTH RESISTANCE AND INDUCTANCE

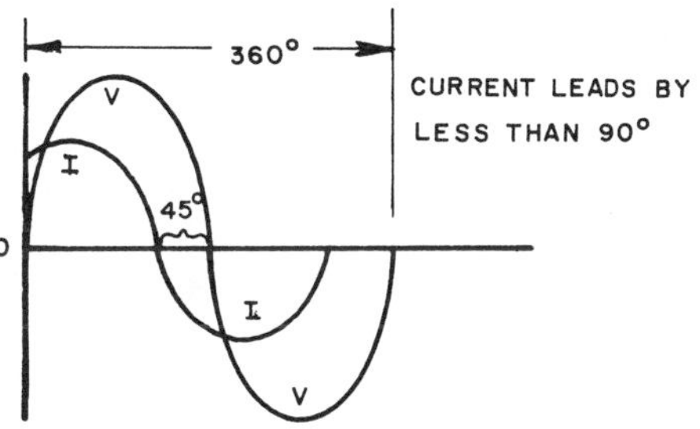

(e) CIRCUIT WITH BOTH RESISTANCE AND CAPACITANCE

Fig. 6-1. Circuits with Resistance, Inductance and Capacitance

torque to turn the rotor. The starting windings in Figure 3-7 are not at a right angle to the running windings but are at a 45 degree angle because the stator has four poles or two sets of running windings. These starting windings help start the motor and are disconnected by means of a centrifugal switch at about 75 percent of rated rpm, see schematic Figure 6-2. Instead of a centrifugal switch some split phase motors have a control relay that disconnects the starting windings.

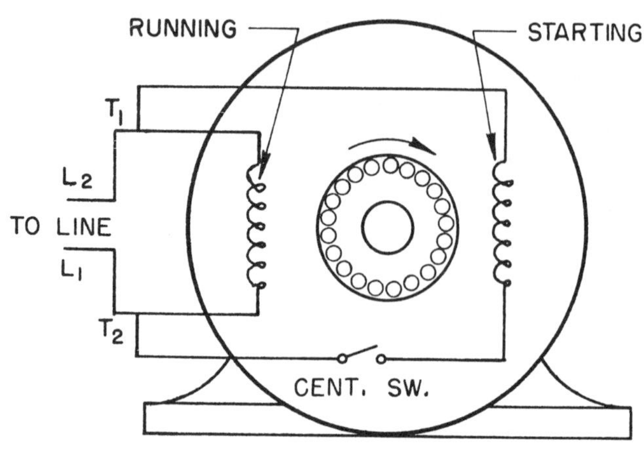

Single Phase, Single Voltage

Fig. 6-2. Split Phase-Start Motor

Impedance is the total resistance to the flow of AC as a result of both reactance and resistance. Reactance is the opposition to AC as a result of either inductance or capacitance. It is beyond the scope and purpose of this manual to become involved with the mathematics of these new terms. Rather, it is desired that the student learn to associate the words with the generation of electrical current, transforming electrical current and use in the operation of electric motors, note definitions and symbols in Table 6-1.

Table 6-1. Definitions for Selected Electrical Terms and Symbols for Identification

TERMS	DEFINITION	MEASUREMENT	SYMBOL
Resistance	Internal friction that opposes the flow of current through a conductor.	ohms	Ω
Inductance (L)	Property of an electrical circuit that opposes a change in current flow. The characteristics of a coil of wire to cause the current to lag the voltage in time phase. The current reaches its peak after the voltage does.	henry	H
Capacitance (C)	Property of electrical circuit that opposes the change in voltage flow. Characteristic that causes the current to lead the voltage in time phase. The current reaches its peak before the voltage does.	micro-farads	MFD
Reactance	The opposition to AC as a result of either inductance or capacitance.	ohms	X
Impedance	Total resistance to flow in an AC circuit as a result of reactance and resistance.	ohms	Z

CAPACITOR-START MOTORS

It is not easy to distinguish between the capacitor-start and split phase- start motors when inspecting the stator and windings except for the presence of the capacitor, Figure 6-3. The capacitor is readily detected as the round cylindrical object on top of the frame. Sometimes it might be at the bottom of the frame between the legs of the motor and in that case, probably in a flattened oval shaped container. It can also be placed away from the motor on the machine. The capacitor is wired in series with the starting windings. The capacitor in the starting winding circuit can cause the current to lead the line voltage by approximately 45 degrees,as illustrated in Figure 6-1 (e). The resistance in the running winding circuit can cause the current to lag the line voltage by 45 degrees, note Figure 6-1 (d). The currents in the starting and running winding circuits are 90 degrees out-of-phase as are their respective magnetic fields. The generated rotating field again supplies enough torque to turn the rotor. Splitting the phase or placing a capacitor in the starting windings accomplishes the same purpose of providing a starting technique to turn the rotor. The advantage of the capacitor-start system is that it can start a more difficult load with a smaller amperage requirement than can the split phase-start system. The split phase motor may draw five to seven times its operating ampere rating when starting a load, while the capacitor-start motor draws three to six times the operating ampere rating.

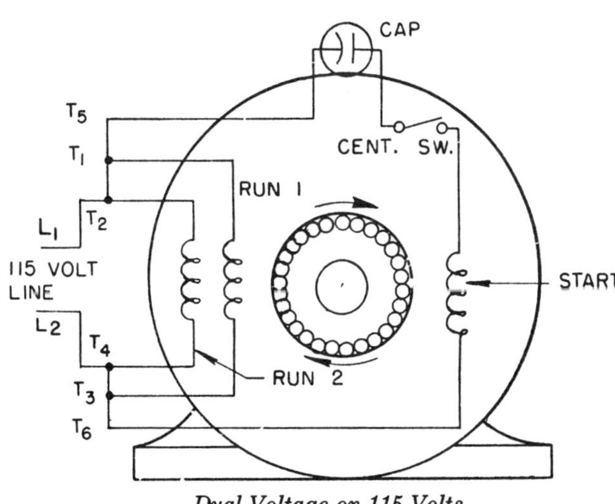

Dual Voltage on 115 Volts

Fig. 6-3. Capacitor-Start Motor

A comparison of the split phase-start and capacitor-start motor is: (1) the split phase motor is started because the phases of the starting windings and running windings in the stator are electrically split whereby the rotating magnet of the rotor will follow from one pole of the running winding to an adjacent pole of the starting winding to complete the 360 degrees of rotation; and (2) the capacitor start motor is basically a split phase-start motor but has a capacitor in the starting circuit which causes a higher starting torque characteristic with less ampere draw.

Starting windings in motors may have less than, equal to or more turns than the running windings and conductor size is smaller than or equal to the running windings. These variables are dependent on the motor type and design characteristics.

SHADED POLE-START MOTORS

The schematic of the shaded pole motor, Figure 6-4, discloses a different stator configuration and windings. The shading technique is used to start the motor. The entire pole piece has the coil of wire around it. One corner of the pole has a closed copper loop made from a strap of round copper conductor and this is called the shaded pole.

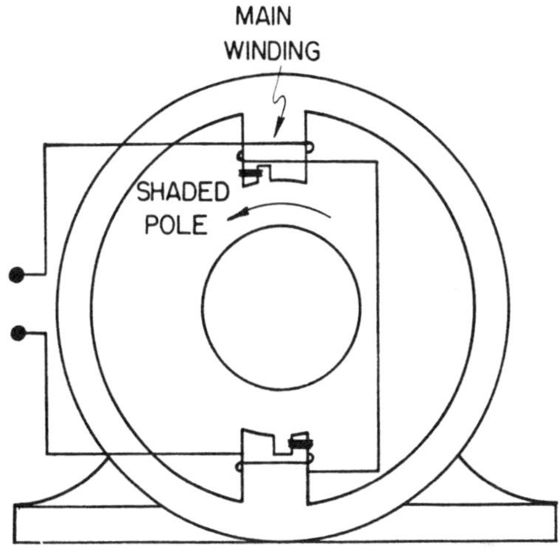

Fig. 6-4. Shaded Pole Motor

When the current increases in the main windings poles, a current is induced in the shading coil. This current opposes the magnetic field build-up in the part of the pole pieces surrounded by the shading coil. This produces the condition shown in Figure 6-5, where the flux is moved away from the pole piece surrounded by the shading coil.

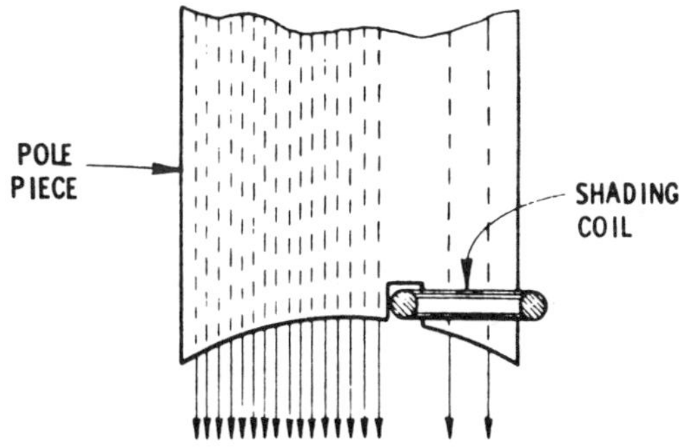

Fig. 6-5. Magnetic Field in a Shaded Pole Motor

When the winding pole current decreases, the current in the shading coil also decreases until the pole pieces are uniformly magnetized. As the winding pole current and the magnetic flux of the pole piece continue to decrease, the current in the shading coil reverses and tends to maintain the flux in part of the pole pieces. When the winding pole current drops to zero, current still flows in the shading coils to give the magnetic effect which causes the coils to produce a rotating or magnetic field that starts the motor. The rotor will turn in the direction from the unshaded to the shaded portion, counter clockwise in Figure 6-4. Because of the shading coil design feature the shaded pole motors are non-reversible. Shaded pole motors do not have separate starting windings or a centrifugal starting switch mechanism as does the split phase and capacitor-start motors. Therefore, it is a less expensive motor to manufacture, but can be used for low starting torque applications only such as a fan.

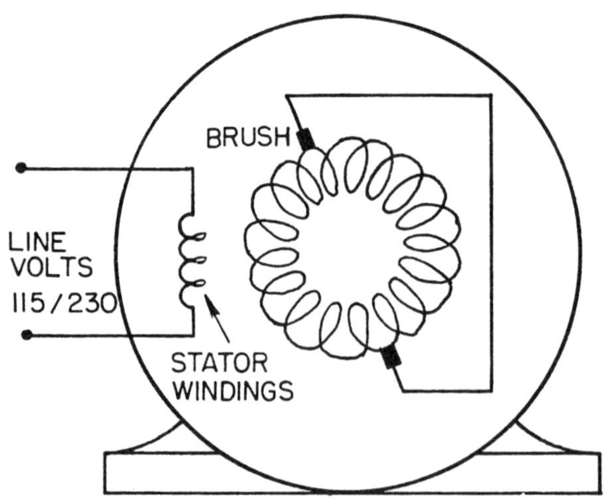

[a] *Repulsion Motor*

REPULSION-START MOTORS

The repulsion-start induction run electric motor does not have a traditional squirrel cage rotor, but a wound rotor. The wound rotor has wire coil windings instead of a series of connecting bars in the rotor. This motor was designed for very hard starting loads and it requires low starting current, usually two to four times the operating amperes. Because of the size of the motor, the wound rotor and brushes, the production cost is high and manufacture is limited.

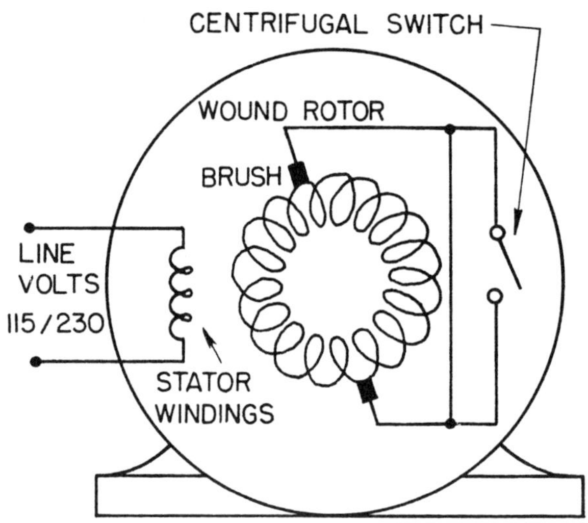

[b] *Repulsion-Start Induction Run Motor*

There are three types of repulsion-start motors and the technical names for each are: repulsion, repulsion-start and repulsion-induction. The schematic for the three types is shown in Figure 6-6.

The repulsion style motor, Figure 6-6 (a), has a wound rotor with commutator ends which are shorted out at 75 percent speed, the brushes continue to ride on the commutator, and the motor operates as an induction run squirrel cage rotor motor. This is a varying-speed motor. The repulsion-start induction run style motor, Figure 6-6 (b), has a wound rotor and brushes and after 75 percent operating speed is attained the ends of the rotor windings are shorted out to achieve the induction run operation. It can be either a brush-lifting or brush-riding motor. This is a constant -speed motor.

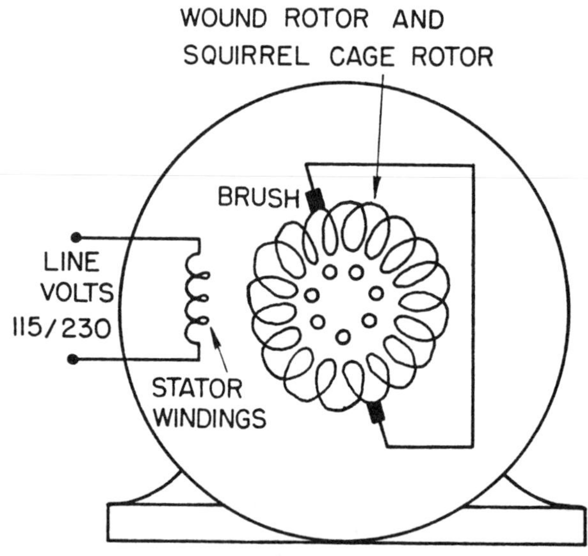

[c] *Repulsion Induction Motor*

Fig. 6-6. Repulsion-Start Type Motors

The third style, repulsion-induction, Figure 6-6 (c) , has a special combination rotor which has both squirrel cage construction and a wound rotor. When starting, the commutated windings supply the torque and the squirrel cage winding is inactive. When armature speed increases the squirrel cage rotor takes over the load and stabilizes the speed. The brushes remain in contact on the commutator. This motor can be either constant- or varying-speed. Many of the repulsion-start motors are being replaced by the capacitor-start, capacitor run single-phase and three-phase motors because of high production costs and extra maintenance required of the brushes and centrifugal switching system.

The starting systems for induction run motors, Figure 5-1, have now been introduced. Another motor that operates on single-phase AC that is quite common is the synchronous motor.

ALTERNATING CURRENT SINGLE-PHASE SYNCHRONOUS MOTORS

Synchronous fractional horsepower motors have a very small current draw, are non-reversible and operate at a constant speed. They cannot start a significant load and have virtually no starting torque. These motors will be found driving recording equipment, sound projectors, timers and clocks. There are two basic types of synchronous motors, (a) reluctance and (b) hysteresis, note Figure 6-7.

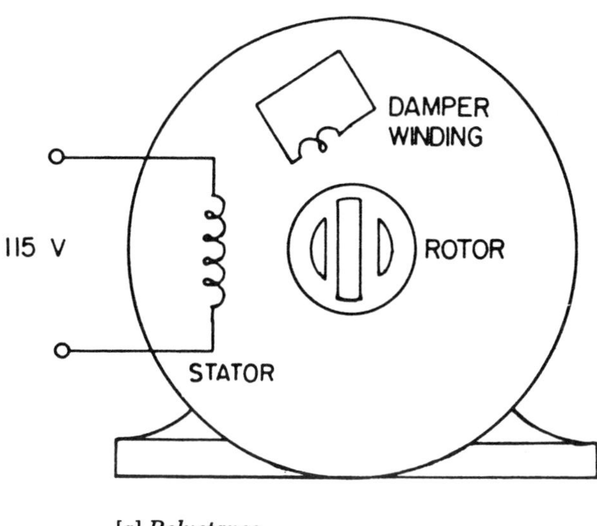

[*a*] *Reluctance*

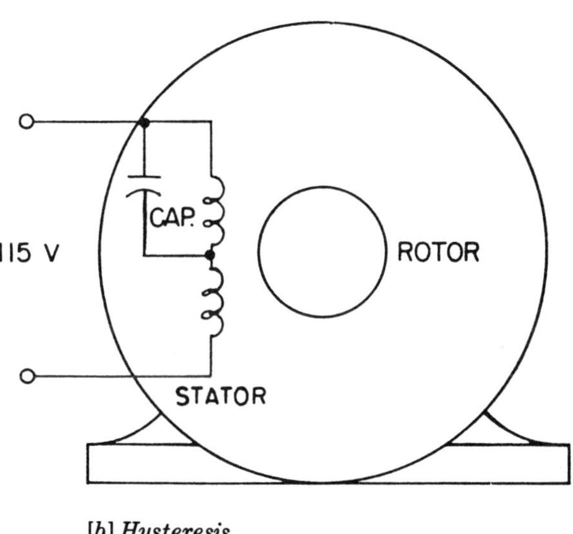

[*b*] *Hysteresis*

Fig. 6-7. Synchronous Motors

The rpm of this motor is at an exact constant speed as determined by the number of poles and Hz. They cannot run as synchronous motors until they run at synchronous speed. The motor needs a starting system and the shaded pole, split phase or capacitor system can be used with the motor torque depending on the starting system. The reluctance synchronous motor, Figure 6-7 (a) has a squirrel cage rotor with one cutout per pole. These cutouts cause magnetic reluctance to be greater between poles than along the pole axis and the motor operation depends on the difference in reluctance. The motor locks into synchronism in less than one cycle of applied voltage. Motor efficiency is low but this is not as important as the low production cost. The hysteresis motor does not have a physical pole, Figure 6-7 (b), arrangement on the rotor. The rotor develops fixed magnetic poles only as it reaches synchronous speed and therefore locks into synchronous speed in any random angular position. The hysteresis motor has less tendency to hum and vibrate around synchronous speeds than the reluctance and are selected for driving the recording instruments requiring precise constant speed motors. Synchronous motors can be designed to operate on single-phase or three-phase AC.

The starting systems for single-phase induction run motors have been covered and also the synchronous motors that can operate on single-phase. There are two other common AC motors, universal and three-phase. The universal motor also operates on single-phase current.

ALTERNATING CURRENT UNIVERSAL MOTORS

The universal motor, Figure 6-8, can be operated on either DC or AC, thus its name is appropriate. Trace the electrical circuit through the motor. Current goes from the power source to the "S" stator pole, to the "N" stator pole to a brush, through the commutator bar and wound rotor,the other commutator brush, and back to the power source. This is a series circuit. First let's follow the operation of a DC motor by reducing the motor to a one-loop armature, refer to Figure 6-9.

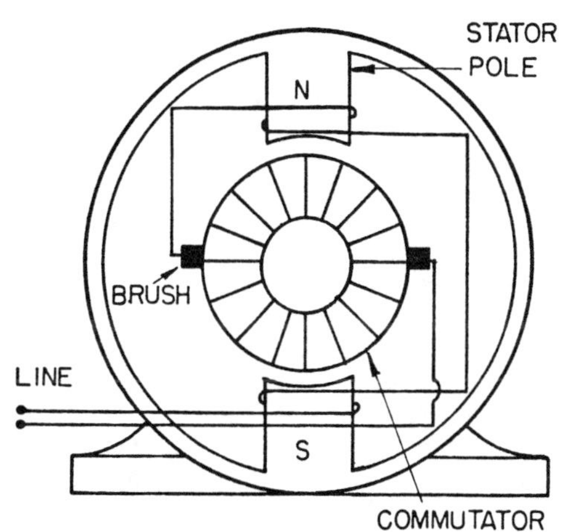

Fig. 6-8. Universal Motor

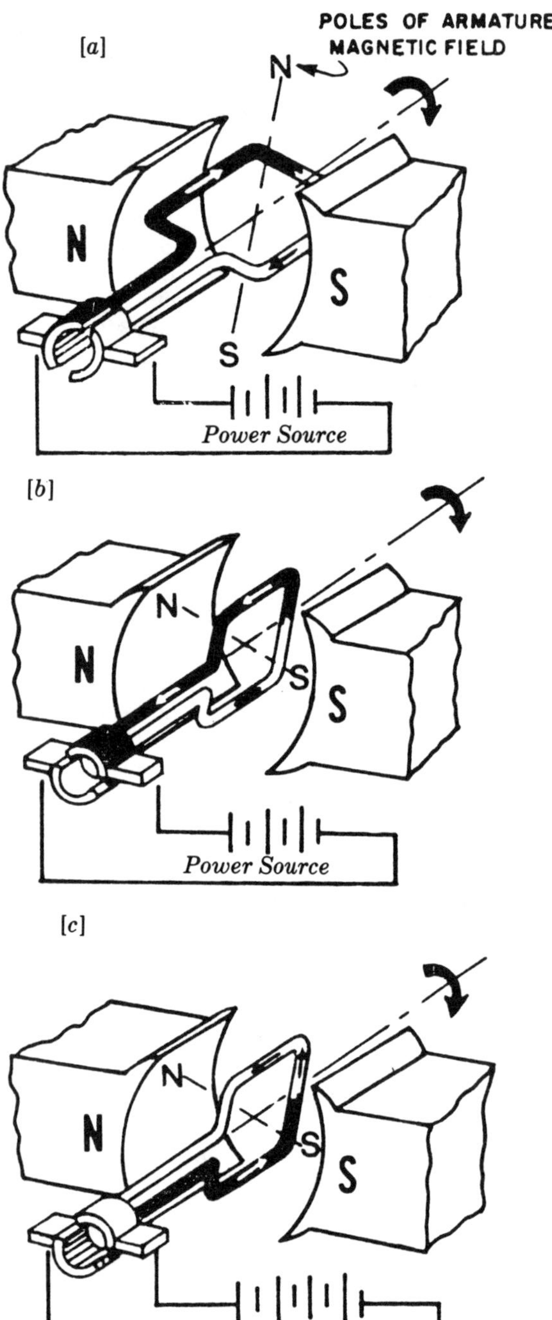

Fig. 6-9. Direct Current Motor Operation

With the armature loop in position (a), the current flowing through the loop makes the top of the loop a north pole and the underside a south pole. The magnetic pole of the loop will be attracted by the corresponding opposite poles of the field coils. The loop will rotate clockwise because unlike poles attract. The loop rotates 90 degrees in (b), commutation takes place and the current through the loop reverses direction. When the loop current reverses, the magnetic field generated by the loop reverses. With pole changes occurring the like poles are facing each other and they repel each other, and the loop continues rotating. Loop position, Figure 6-9 (c) is 180 degrees past Figure 6-9 (b). Commutation takes place again and the loop continues to rotate to repeat the cycle of events. The commutator causes the current through the loops to reverse at the instant unlike poles are facing each other. This causes the reversal of the polarity in the field, repulsion occurs instead of attraction and the loop continues to operate. The more loops there are in the armature the smoother the armature will turn and the better the torque characteristics. The motor in Figure 6-8 is called a series motor and is appropriately named based on the electrical circuit. Two other types of DC motors are the shunt and compound types. A common DC motor is the automobile cranking motor which is powered by a battery.

The universal motor, Figure 6-8, can be operated on AC. When AC is applied to a series motor, the current through the armature and field changes at the same time and the motor will rotate in one direction. The number of field turns in the AC series motor is less than in the DC series motor which decreases the reactance of the field and the required amount of current will flow. Reducing the size of field reduces the motor torque. These series motors are classed as varying speed motors with low speeds for large loads and fast speeds for light loads. The large loads will require large torque and a large current demand and a light load requires a small torque and small current demand. The load demand on the universal motor regulates the speed. The universal motor operates at a lower efficiency than either the special DC series or AC series motor. The motor is designed for fractional horsepower motors and can be found on drills, saws, grinders, sanders, routers and similar electric hand tools. Universal motors do not operate at synchronous speed and will operate to the limit of the power input versus the applied load. The motor could explode due to high centrifugal forces when not on an appropriate load.

DIRECT CURRENT MOTORS

[a] Brush Type

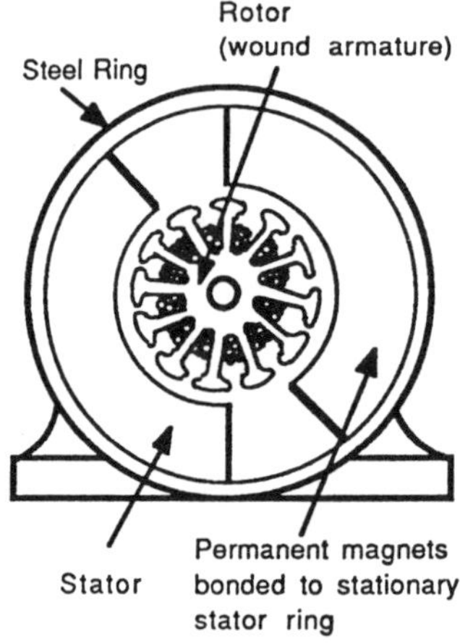

[b] Brushless Type

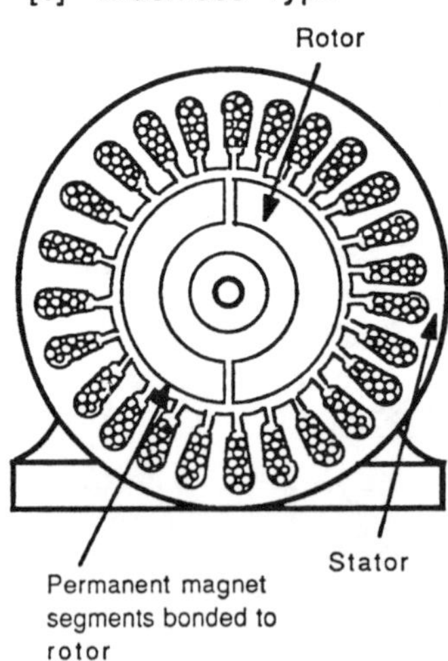

Fig. 6-10. Permanent Magnet DC Motors

Brush and brushless permanent magnet (PM) motors are efficient, have high starting torque and their linear speed-torque curves are ideal for motion control applications where predictable motor performance is essential. A majority of the PM motors in industry are the brush type which are called mechanically commutated PM motors. If down time becomes a factor designers will consider the brushless PM motors as an alternative. The brushless PM motors are called electronically commutated PM motors. Although the brushless PM motors cost approximately twenty-five percent more they operate will less mechanical noise, offer the same or more horsepower in a smaller package, greater starting torque, and the brush maintenance has been reduced. In the brushless motor, Figure 6-10 the placement of the permanent magnets and the windings are usually reversed. Instead of placing the magnets on the stator, the permanent magnets are bonded to the rotating member replacing the wire wound armature. The windings are placed around the perimeter of the stator, see Figure 6-10 (b).

Mechanical commutation is the same as used in shunt motors to perform current switching in the armature windings. Current is carried to the armature windings by a pair of spring-loaded, carbon brushes in mechanical contact with the commutator bars to which windings are terminated. When current flows through the armature windings a magnetic field is created. This armature winding field reacts with the field generated by the permanent magnet on the stator setting up a magnetic field force which causes the armature to rotate. If the armature had a single winding the rotor would only rotate until the fields reach an equilibrium position. However, to create a continuous motion a number of physically separated windings sets are used. Thus, different winding sets are energized and de-energized as the armature rotates, causing the brushes riding on the commutator bars to make and break contact with pairs of commutator bars connected to different armature windings.

The windings are switched on and off sequentially so that the relationship between the armature winding field and the permanent magnet stator field is virtually constant, thereby assuring that there is nearly a constant attractive force and the fields never reach an equilibrium condition. The greater the number of armature windings switched on at separate times by the brushes as they ride past the commutator bar pairs, the more consistent the motor's rotational speed and torque output.

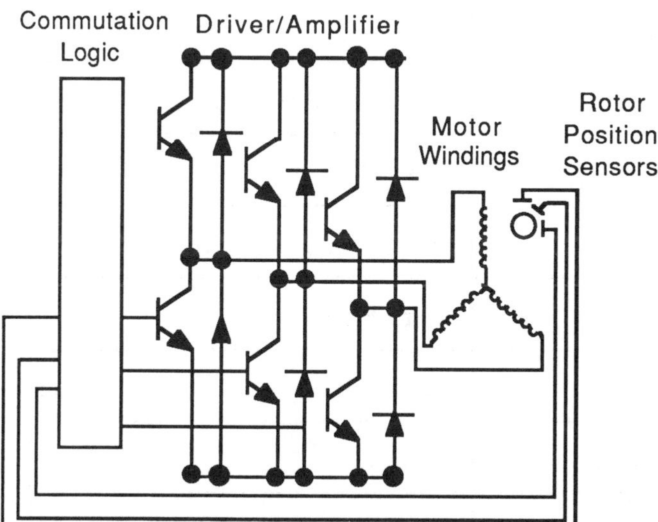

Fig. 6-11. Control Circuitry for Electromagnetic Commutated Brushless PM Motor

Electronic commutation is more complex, Fig. 6-11. The permanent magnet field generated by the rotor magnets is no longer stationary. Electronic control circuitry is needed to determine the rotor's position and to switch current to the windings in an appropriate sequence to cause rotation. The electronic commutation has non-contact electronic sensors to detect rotor shaft position. Devices used to sense rotor position in electronic commutation include Hall effect devices, LED's and photo sensors, resolvers or magneto resistive devices. These electronic sensors monitor the position of the permanent magneto poles of the rotor and control the distribution of current to the field windings. A feedback circuit signals a control amplifier which drives the power amplifier, delivering current sequentially to the stator field coils creating a revolving magnetic field which maintains a virtually constant spatial relationship between the stator winding field and the rotor's permanent magnetic field and thus a constant rotational force (torque).

While the commutation techniques used to generate a constant attractive force in the brushless PM motor is different, the end result is the same. The motor has a linear speed-torque characteristic and consistent torque output similar to the brush PM motor.

ALTERNATING CURRENT THREE-PHASE MOTORS

The Three-phase motor, frequently called polyphase, operates on a three-phase AC current. In Figure 6-12, only three pairs of windings can be observed.

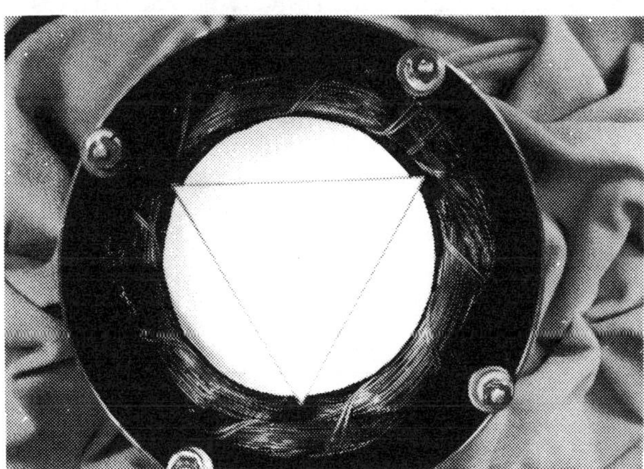

Fig. 6-12. Three-Phase Stator

The three pairs of windings are positioned 120 degrees apart. Since they are 120 degrees apart it eliminates the need for starting windings. The rotor may be either the squirrel cage or the wound rotor type and will not have a centrifugal switch. The three-phase motor, for comparable horsepower ratings, will generally be smaller in size and less expensive than other electric motors. The squirrel cage rotor style is simplest in construction, resulting in lower production costs, less maintenance and increased acceptance.

The three-phase motor may be wired for the "wye" or "star", or the "delta" connections. Combination connections are also possible. It is recommended that the three-phase motor have a wiring connection which matches the electrical service with the same voltage. As illustrated in Figure 6-13. either the wye or delta wiring connections can be used for polyphase motors.

The stator windings in the single-phase motor could be used to readily determine the rpm of the motor if the

[a] *Delta Connection*

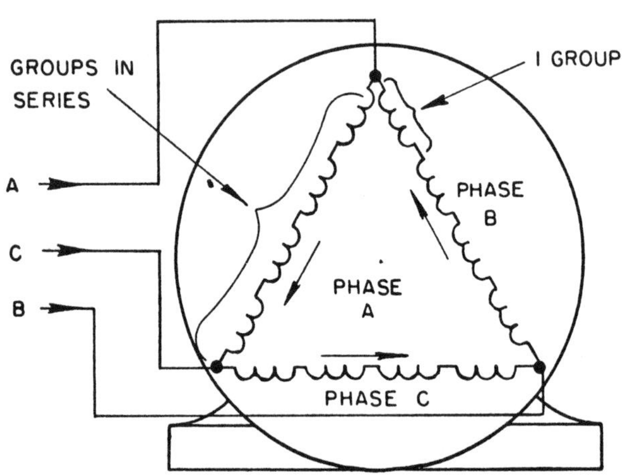

Three-Phase, 4 Pole in Series

[b] *Wye or Star Connection*

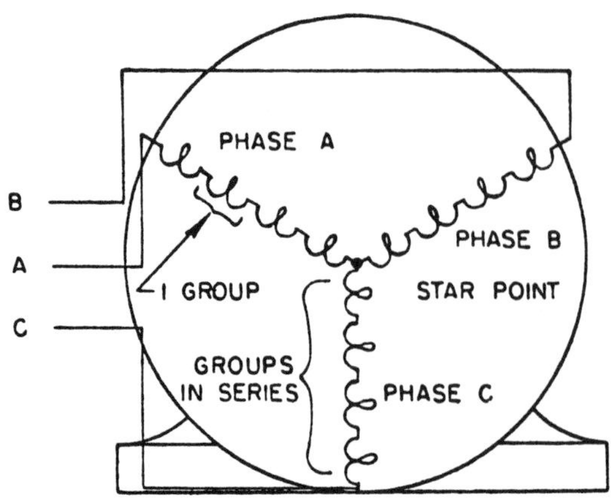

Three-Phase, 4 Pole in Series

Fig. 6-13. Polyphase Motor Connections

nameplate data were lost. The easiest way to determine the rpm of a three-phase motor is to never lose the nameplate data. When viewing the stator, there will be many coils of wire that are close together overlapping each other and covered with insulation. Generally there are as many slots in the stator as there are coils. Assume that the number of slots and coils can be counted and they are equal in numbers, for example 36. Dividing the coil numbers by three phases results in twelve coils per phase. These coils are connected to produce three separate windings called phases. To find the number of coils in each pole, divide the number of coils by the desired number of poles, for example 36 coils divided by four poles equals 9 coils per pole. Each pole must have coils designated for each phase and with 9 coils and three phases, there will be three coils per phase. The three coils per phase will be connected in series. The four pole motor will operate in theory at 1800 rpm synchronous speed. Thus, the rpm of the three-phase motor may be very difficult to determine with the eye due to insulation on the coils. Motor rewind shop employees might have to remove the insulation and make a visual count as the coils are removed from the stator. If the motor is operational, it could be placed on a multi-voltage power panel and a voltmeter and ohmmeter can be used to

Table 6-2. Starting Current and Load Starting Ability of Common Electric Motors

TYPE OF MOTOR	STARTING CURRENT	TYPE OF STARTING LOADS
Split Phase	5-7 times full load	Medium -- 130-170% of full load
Capacitor-Start	3-6 times full load	Hard -- 350-400% of full load
Two-Value Capacitor	3-5 times full load	Hard -- 350-400% of full load
Perm.-Split Capacitor	2-4 times full load	Easy -- 150% of full load
Shaded Pole	Low	Easy
Universal or Series	High	Hard -- 350-400% of full load
Synchronous	Low	Very Easy
Three-Phase	Normal to High	Medium Hard -- 200-300% of full load
Soft-Start	1.5-2 times full load	Easy
Repulsion	2-4 times full load	Very Hard -- 350-400% of full load

determine the optimum operating voltage and rpm. There are some other facts that can be determined by the stator leads, for example: (1) if nine leads are brought out, it's a dual voltage motor, (2) if six leads are brought out the motor is two speed, and (3) if three leads, the motor is for single voltage. Thus the three-phase motor may have a lesser number of moving parts but the eyeball inspection will disclose fewer details than when viewing a single-phase motor.

The initial current and load starting ability of electric motors are dependent upon the motor's starting system. Synchronous and shaded pole motors have virtually no ability to start a load while capacitor-start, capacitor run motors compare to the repulsion-start induction run motors which they are replacing. Note the comparison in Table 6-2, Starting Current and Load Starting Ability. The ability of a motor to operate on more than one voltage and the techniques for reversing the rotational direction will be covered in Unit VII, CHANGING VOLTAGES, REVERSING ROTATION and CHANGING MOTOR SPEED.

DEFINITION OF TERMS

CAPACITANCE: Property of electrical circuit that opposes the change in voltage flow. The current leads the voltage because the current reaches its peak before the voltage.

CENTRIFUGAL SWITCH: Switch activated by weights being moved as the result of turning of the rotor.

COIL: A group of conductors wound into the core slot, electrically insulated from the iron core. These coils produce the magnetic field when the current passes them.

FARAD: Unit of measurement of capacitance. A capacitor has a capacitance of one farad when a charge of one coulomb raises its potential one volt. The formula is $C = Q/V$.

FIELD: Term used to describe the stator member of a motor. The field provides the magnetic field with which the mechanically rotating member (armature or rotor) interacts.

HENRY (H): Unit of measurement of inductance. A circuit is said to have an inductance of one henry when a current change of one ampere per second induces a voltage of one volt.

HYSTERESIS: The resistance offered by materials to becoming magnetized (magnetic orientations or molecular structure) results in energy being expended and corresponding loss. Hysteresis loss in a magnetic circuit is the energy expended to magnetize and demagnetize the core.

IMPEDANCE: The total resistance to flow in an AC circuit as a result of reactance and resistance.

INDUCTANCE: Property of an electrical circuit that opposes a change in current or amperage flow. The amperage lags the voltage because the amperage reaches its peak after the voltage.

LAMINATIONS: Thin sheets of steel used in cores of transformers, motors and generators.

LAGGING ANGLE: Angle current lags voltage in inductive circuits.

LEADING CIRCUIT: Angle current leads voltage in capacitive circuit.

MICRO: Prefix meaning one millionth of. Thus, Microfarad or microhenry.

OHM (R or Ω): A unit of measurement of resistance.

OHMS LAW: Relationship between amperes, voltage and resistance: $V = I \times R$. I represents ampere in equations and formulas but may be represented by A in written text and the metric system.

POLARITY: Property of a circuit or device to have poles, N or S, and positive or negative.

POLYPHASE MOTORS: Motors using three-phase current, have identical windings, currents are balanced and a set of windings for each phase.

REACTANCE: The opposition to AC as a result of either inductance or capacitance.

RELUCTANCE: The characteristics of a magnetic material which resists the flow of magnetic lines of force through it.

RESISTANCE: Internal friction that opposes the flow of current through a conductor.

SALIENT POLES: A motor has salient poles when its stator or field poles are concentrated into confined arcs and the winding is wrapped around them (as opposed to distributing them in a series of slots). The shaded pole motor has salient poles.

SHUNT: Field coils connected in parallel with armature circuit. To connect across or parallel with circuit or component.

SYNCHRONOUS SPEED: The speed of the rotating magnetic field set up by the stator winding of an induction-run motor. In a synchronous motor the rotor locks into step with the rotating magnetic field, and the motor is said to run at synchronous speed.

NOTES

CLASSROOM EXERCISES VI-A

Electric Motor Starting Systems

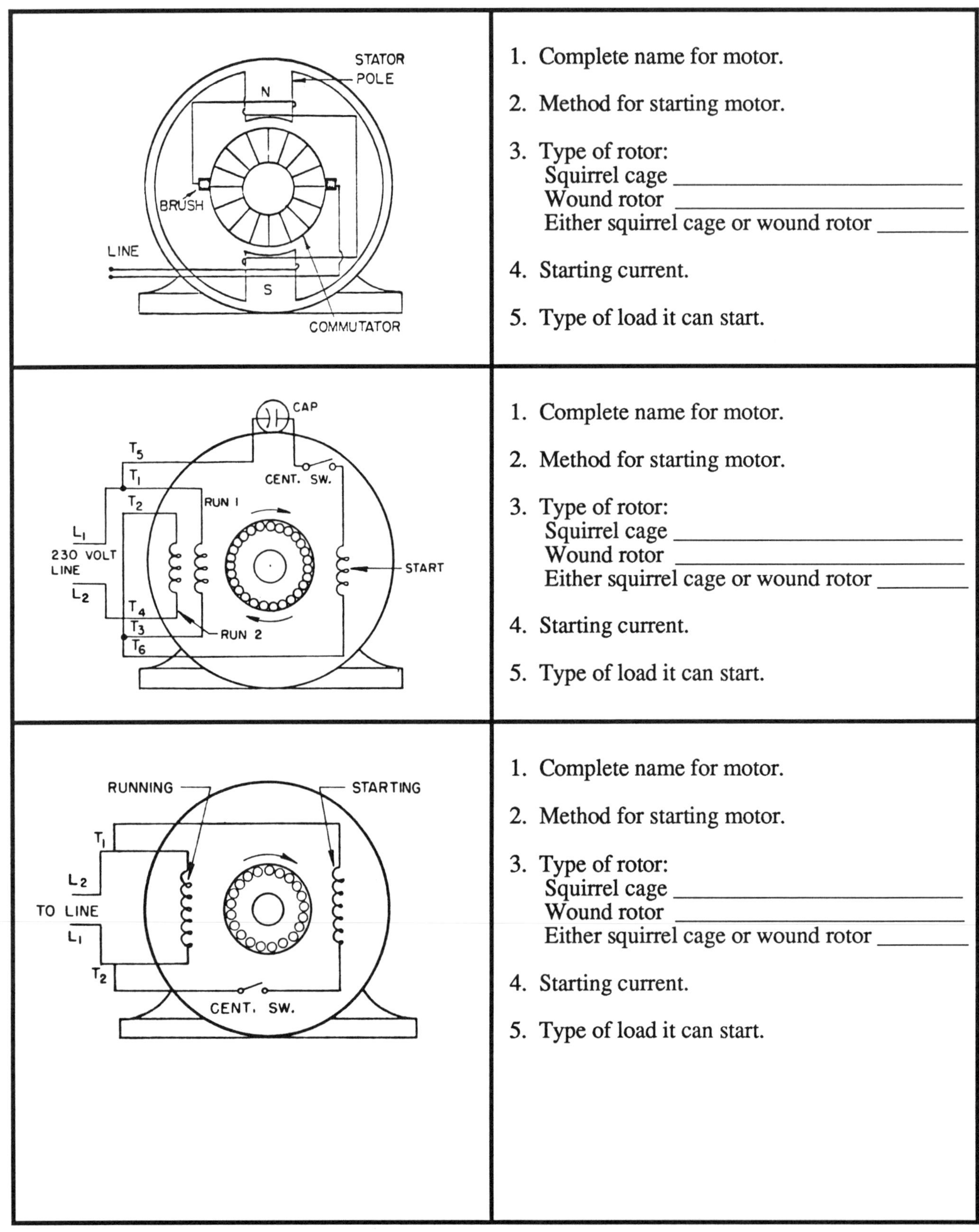

1. Complete name for motor.
2. Method for starting motor.
3. Type of rotor:
   Squirrel cage ______________________
   Wound rotor ______________________
   Either squirrel cage or wound rotor ________
4. Starting current.
5. Type of load it can start.

1. Complete name for motor.
2. Method for starting motor.
3. Type of rotor:
   Squirrel cage ______________________
   Wound rotor ______________________
   Either squirrel cage or wound rotor ________
4. Starting current.
5. Type of load it can start.

1. Complete name for motor.
2. Method for starting motor.
3. Type of rotor:
   Squirrel cage ______________________
   Wound rotor ______________________
   Either squirrel cage or wound rotor ________
4. Starting current.
5. Type of load it can start.

CLASSROOM EXERCISES VI-A (Cont'd)

Electric Motor Starting Systems

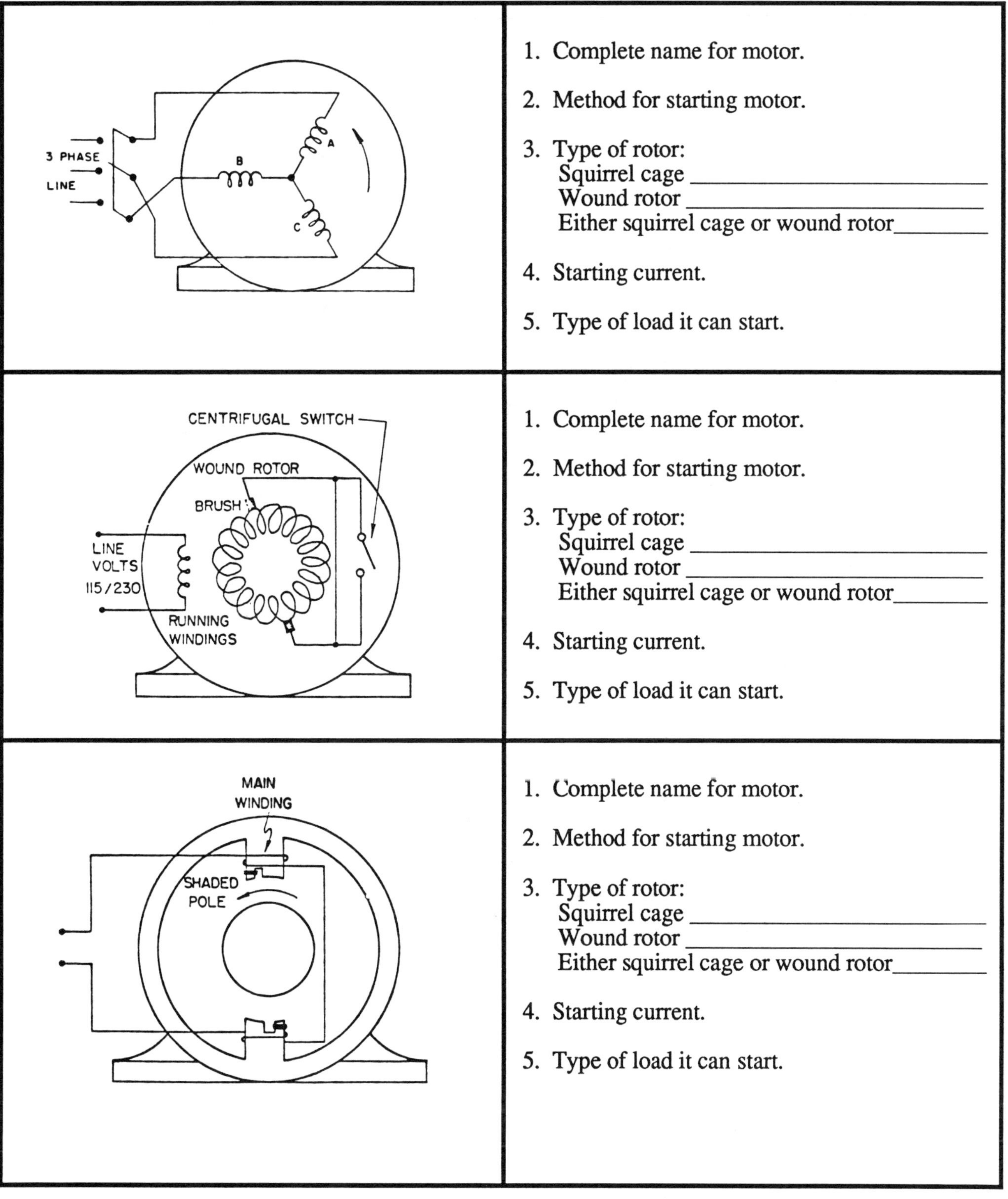

1. Complete name for motor.
2. Method for starting motor.
3. Type of rotor:
   Squirrel cage ____________________
   Wound rotor ____________________
   Either squirrel cage or wound rotor________
4. Starting current.
5. Type of load it can start.

1. Complete name for motor.
2. Method for starting motor.
3. Type of rotor:
   Squirrel cage ____________________
   Wound rotor ____________________
   Either squirrel cage or wound rotor________
4. Starting current.
5. Type of load it can start.

1. Complete name for motor.
2. Method for starting motor.
3. Type of rotor:
   Squirrel cage ____________________
   Wound rotor ____________________
   Either squirrel cage or wound rotor________
4. Starting current.
5. Type of load it can start.

LABORATORY EXERCISE VI-A

Starting Systems for Motors

GO TO THE LABORATORY AND COMPLETE THIS TABLE WITH THE ELECTRIC MOTORS SUPPLIED

ACTIVITY	MOTOR A	MOTOR B
1. Type of Motor		
2. Name of Starting System		
3. Type of Rotor	Squirrel Cage ____________ Wound _________________	Squirrel Cage ____________ Wound _________________
4. Number of Stator Windings	Starting _____ Running _____	Starting _____ Running _____
5. Pairs of Poles		
6. Centrifugal Switch	Yes ________ No ________	Yes ________ No ________
7. Other Switching System		
8. Phase	1 __________ 3 __________	1 __________ 3 __________
9. Outline how Starting System operates		

CHANGING VOLTAGE, REVERSING ROTATION AND CHANGING MOTOR SPEED

Electric motors can be designed to operate on a single voltage or dual voltage. Dual voltage is the term used to describe those that can be used on two voltages, for example, 115 or 230 volts. The dual voltage feature improves the versatility of a motor and can easily be changed from one voltage to another.

CHANGING VOLTAGE SERVICE

The induction run electric motor that operates on 115 volts will have the running and starting windings in parallel across the line voltage at L1 and L2. Dual voltage (115/230) motors operating on 230 volts will have the running windings in series and the starting winding in parallel with one running winding and in series with the other running winding. Trace these circuits on the dual voltage capacitor-start motor schematic in Figure 7-1 (a) which is wired for 115 volts. The leads are identified by T2, T1, T5 and T4, T3 and T6. To change the motor to 230 volts, Figure 7-1 (b) T1 and T5 are connected to power L1. Lead T4 is connected to power L2 and lead T2 is tied together with T3 and T6 to connect the running windings in series, and the starting winding in parallel with the number one running winding and in series with the number two running winding. It is possible to change the voltage connections and not understand the motor circuits. If it is a dual voltage motor, the instructions for changing the connections will be illustrated on the nameplate or inside the coverplate where the connections are made, note examples in Figure 7-2. There is a non-conducting terminal plate inside the motor where electrical connections are made. The number of leads visible to the outside of this terminal plate is limited to those going to the power source, and those essential for dual voltage and reversing. Leads behind the plate go to centrifugal switches, stator windings, thermal overloads and capacitors if included in the motor. Common bars are found behind this terminal plate where several connections are made. This can cause confusion to the person looking at the schematic and then at the actual motor connections. An understanding of the motor circuits helps clarify what's going on behind the terminal plate. If the motor is listed as dual voltage these changes can be made. If it is not a dual voltage motor, it is impractical to attempt to wire the motor for dual voltage.

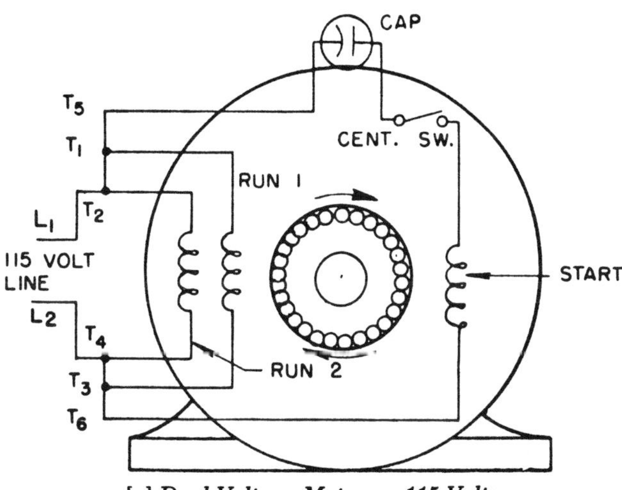

[a] Dual Voltage Motor on 115 Volts

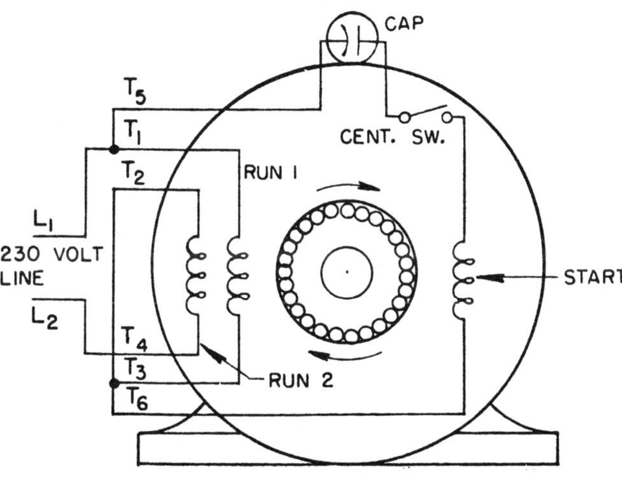

[b] Dual Voltage Motor on 230 Volts

Fig. 7-1 Dual Voltage Capacitor-Start Motor

Motor Connections

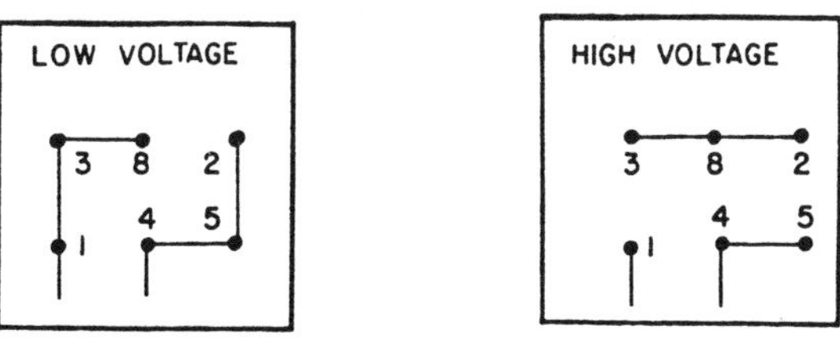

To reverse rotation interchange W5 and W8.

[a] Voltage and Reversing Instructions

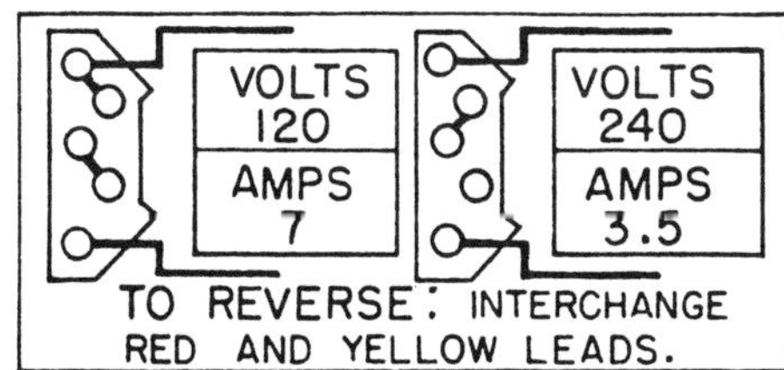

[b] Voltage and Reversing Instructions

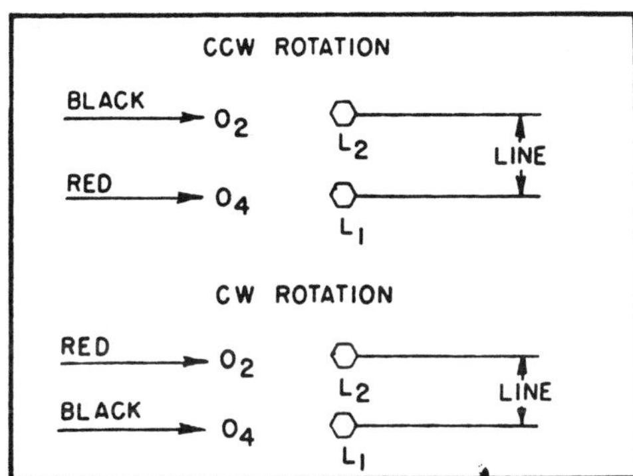

Connect L2 to grounded line when the motor has thermal protection.

[c] Reversing Lead Connections

Fig. 7-2 Selected Dual Voltage Motor Connections and Reversing Instructions

REVERSING MOTOR ROTATION

What direction the motor is turning is a common question. The choices are clockwise (cw) and counterclockwise (ccw), but the motor has two ends. The ends will be called the drive end and opposite the drive end. When facing the end of the motor opposite the drive end, the standard NEMA operating direction is ccw for DC motors, AC single-phase, synchronous and universal motors. Standard motors will be connected for this direction of rotation unless specified otherwise. If the direction of standard rotation as viewed opposite the drive end is ccw, it will be cw as viewed looking at the drive end. To avoid confusion when reporting the rotation direction of a motor always identify the viewing position. This rule does not apply to three-phase motors because their direction of rotation can be readily changed and the phase sequence of the power line is not standardized. AC and DC generators will have cw rotation.

If an electric motor does not have a starting cirucit or the starting circuit is not operating, it could be operated cw by providing that direction of rotation manually as the switch is turned on. Likewise, to obtain the ccw direction of rotation turn it manually in that direction. For changing directions, the induction run motor needs to have the current flow reversed in the stator starting windings. This direction of flow determines the direction of the magnetic field and turning direction of the rotor. For DC motors, changing the direction of rotation is achieved by changing the direction of current flow in the armature or field connections depending upon the leads reversed. This reverses the motor rotational direction. Actual motor information on the nameplate and cover plate can be found in Figure 7-2. Sometimes the reversing instructions identifies colors of conductors to switch and in other cases changing positions of conductors by numbers. Rotation of the motor can be changed by following directions but knowledge of the circuits removes some of the mystery of why the motor changes its direction of rotation. To change rotational direction of the split phase and capacitor-start motors the rule is to change the relationship of the starting winding leads to the running windings leads.

Rotation reversal for the capacitor-start motor, Figure 7-3 (a) is done by interchanging T5 and T6. For the split phase motor, Figure 7-3 (b) interchange the starting winding lead at T1 with the one at T2.

The reversal of the three-phase motor, Figure 7-4, is achieved by interchanging connections between any two of the three phases. In this illustration phases "B" and "C" were interchanged. The interchange may be done at the terminal plate of the motor, at the power cord plug cap or at service feed lines.

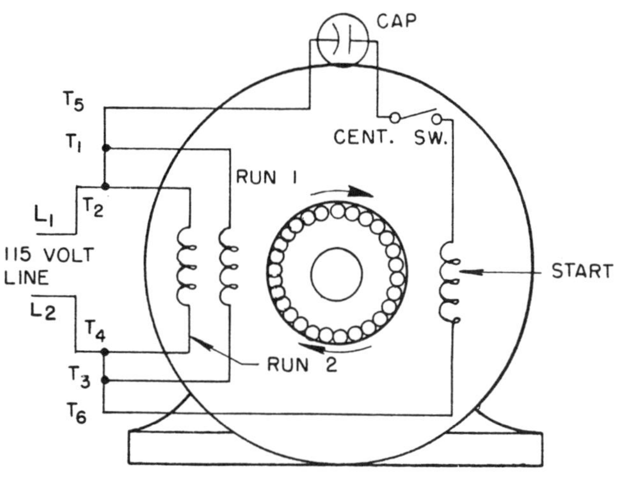

[a] *Dual Voltage Capacitor-Start Motor*

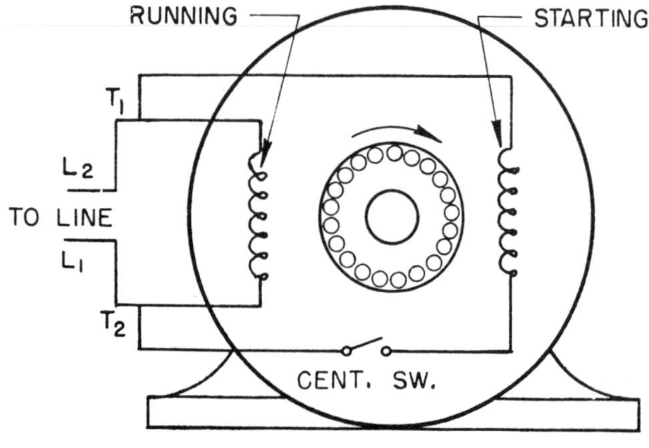

[b] *Single Voltage Split Phase-Start Motor*

Fig. 7-3. Reversal of Split Phase and Capacitor-Start Motors

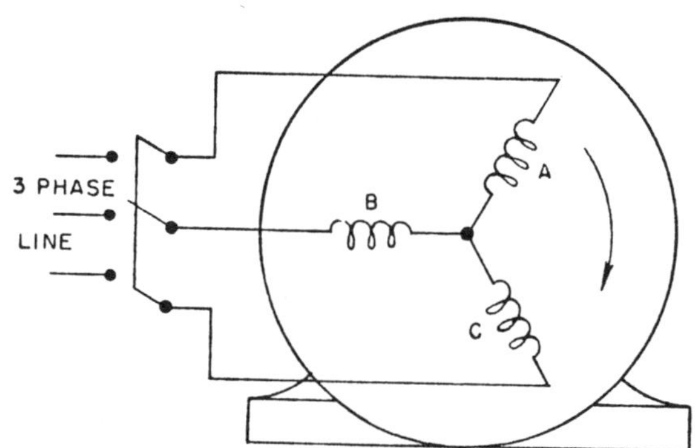

Polyphase Motor

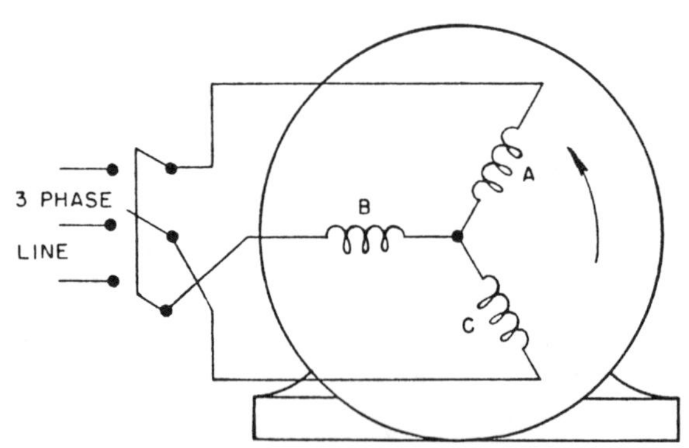

Interchange any Two Motor Leads

Fig. 7-4. Reversal of the Three-Phase Motor

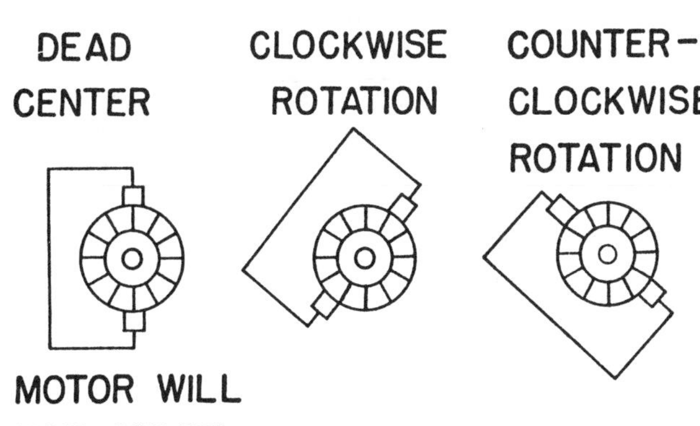

Fig. 7-5. Reversal of the Repulsion-Start Motor

Changing direction of rotation on induction run type motors involves reversing the starting winding leads. The easiest motor to reverse direction is the repulsion-start induction run. Opposite the drive end on the motor end bell is a screw that holds the brush ring in a fixed position on the commutator. Loosen the screw and turn the brush unit a few degrees in the desired direction of rotation as described in Figure 7-5. Because the "best" brush angle is sometimes difficult to obtain, an ammeter is used to detect the starting amperage to obtain a brush position yielding the lowest possible amperage flow and good starting characteristics. Although easier to change, this type of motor is not common because of cost of production. Shaded pole and synchronous motors are not reversible. AC series and universal motors can be designed for operating in either direction. Direction of travel can be changed in impact tools and electric drills by changing brush position or with switching devices which changes polarity of the armature in relationship to the field windings. Note Table 7-1, Method of Reversing Electric Motors.

CHANGING MOTOR SPEED

Induction run motors have their basic rotational speed determined by the synchronous speed and the number of stator poles, refer to Unit III, NAMEPLATE INFORMATION. Speeds less than synchronous can be contributed primarily to slip. Slip is determined by the engineering design of the motor. Friction from the bearings and the fan on the rotor would contribute to small speed losses in some motors. There are five induction run motors which are non-synchronous motors. The shaded pole, permanent-split capacitor and polyphase motors can have speed changes made in their design which changes the percent slip. Split phase-start and capacitor-start motors are not designed to alter the slip unless the rotational speed doesn't go low enough to engage the starting switch. If the motor is running at reduced speeds with the starting switch closed, the starting windings will burn out. The possibility for a speed change can also be achieved by a change in frequency, for example, from 60 Hz to 50 Hz but this would be an impractical option. Wound rotor motors have their speeds determined by engineering design and the attached load. The speed adjustment options have been limited and many times it would be advantageous to have a variety of rotational options.

Table 7-1. Method of Reversing Electric Motors

TYPE OF MOTOR	YES/NO	METHOD
Split-Phase	Yes	Reverse starting winding leads
Capacitor-Start	Yes	Reverse starting winding leads
Two Value Capacitor	Yes	Reverse starting winding leads
Permanent-Split Capacitor	Yes	Reverse starting winding leads
Shaded Pole	No	---
Universal or Series	Yes/No	Adjustment of brush rings or change current flow directions
Synchronous	No	---
3-Phase	Yes	Reverse two of the three leads
Soft-Start	Yes	Reverse starting winding leads
Repulsion	Yes	Shifting of commutator brushes

SPLIT PHASE MOTOR,TWO SPEED

There are several ways to change the speed of the split phase-start induction run motor but each technique requires changes in the number of stator poles being used. The schematic for a two speed split phase motor is illustrated in Figure 7-6. This motor has a low and high speed switch and a double contact centrifugal switch. The centrifugal switch is normally closed in the high speed position. In your mind flip the speed control switch to the low position.

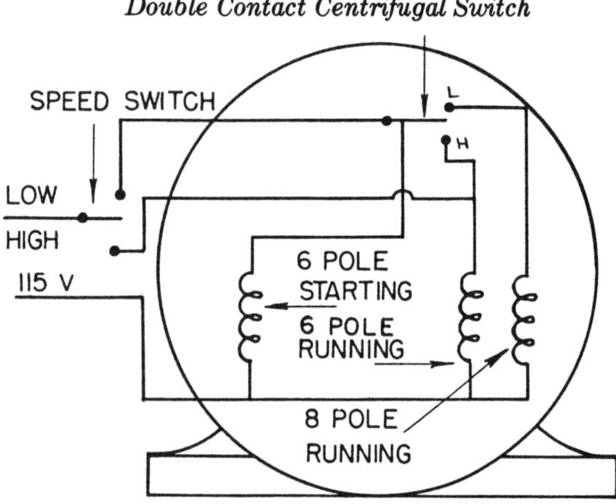

Fig. 7-6. Two Speed Split Phase Motor

The current flows through the speed switch to the 6 pole running windings. The current also flows through the 6 pole starting windings. After the motor reaches its predetermined speed the centrifugal switch disconnects the 6 pole running windings and changes the current flow to the 8 pole running windings which is the slower speed. When the motor stops the centrifugal switch returns to the high position. For the high speed, mentally flip the speed switch to the high position and remember the centrifugal switch is normally closed (N/C) in the high position. The current flow can go directly to the 6 pole running windings. For current to flow to the starting windings it goes through the centrifugal switch to the 6 pole starting windings. After the predetermined speed is attained the centrifugal switch opens and disconnects the starting windings. The speed is governed by the running windings with the 6 poles providing the higher speed and the 8 pole running windings providing the low speed. There are other wiring techniques for changing speed of rotation.

PERMANENT-SPLIT CAPACITOR

Permanent-split capacitor motors can be wired for a variety of conditions. Figure 7-7 shows the schematic for low voltage and low speed operation and there is a switch for the low and high speed operation. The running windings (1 and 2) and auxiliary windings (3 and 4), Figure 7-7, are in series across the line. The line voltage must be divided between the two windings and only a portion is applied to the running windings. The lowered voltage decreases the

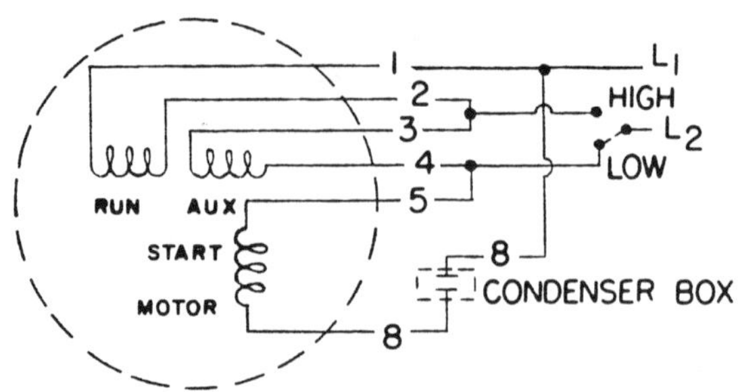

Fig. 7-7. Permanent-Split Capacitor Motor, Low Volts and Low Speed

field strength of the running windings and causes a decrease in speed. The starting winding (5 and 8) for low speed operation is in series with the capacitor and across the line (L1 and L2). Trace the current flow with the switch in the low position, Figure 7-7, to confirm that the running and auxiliary windings are in series. The same motor with the switch in the high position is illustrated in Figure 7-8.

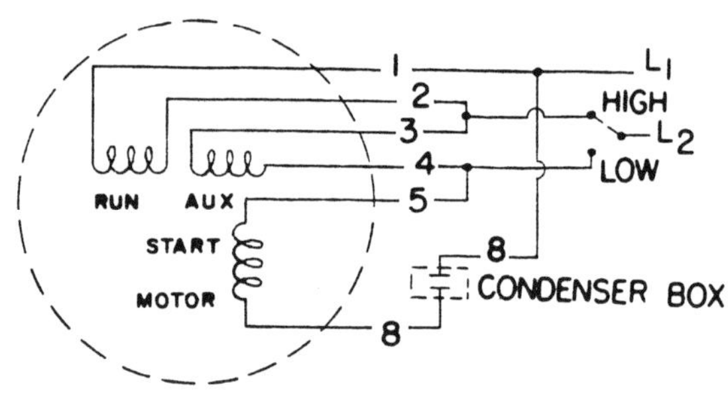

Fig. 7-8. Permanent-Split Capacitor Motor, Low Volts and High Speed.

The auxiliary winding may be wound with a different size conductor than the running winding but it is always placed in the same stator slots as the main winding. The running windings are placed in the slots first, next the auxiliary windings and then the starting windings which are 90 electrical degrees from the other windings. Insulation must be placed between all windings. To reverse this motor, reverse the starting winding leads (5 and 8).

The high voltage and high speed permanent-split capacitor motor is illustrated in the schematic in Figure 7-9. The running winding (1 and 2) and auxiliary winding (3 and 4) are in series, Figure 7-9, which is essential for the high voltage. The starting winding (5 and 8) is in parallel with the auxiliary winding (3 and 4) . Trace the current flow in this motor.

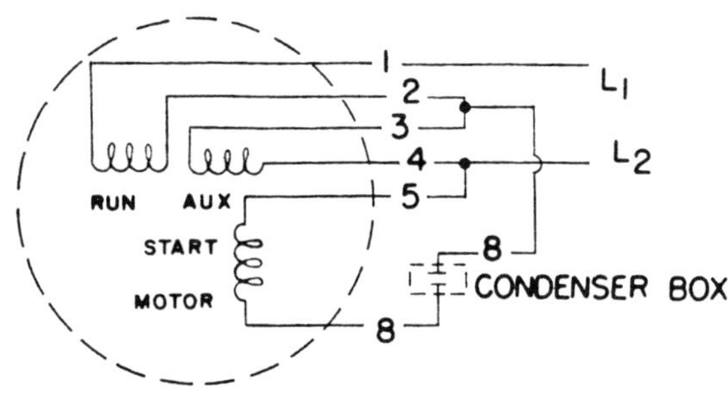

Fig. 7-9. Permanent-Split Capacitor Motor, High Volts and High Speed

A triple pole, double throw switch (TPDT) has been added to the permanent-split capacitor motor, Figure 7-10, to illustrate high and low speed at low volts. The switch flipped to the right (low speed) places the running winding (1 and 2) and auxiliary winding (3 and 4) circuits in series. For high speed, flip the switch to the left and all circuits are in parallel. Trace the current flow in Figure 7-10.

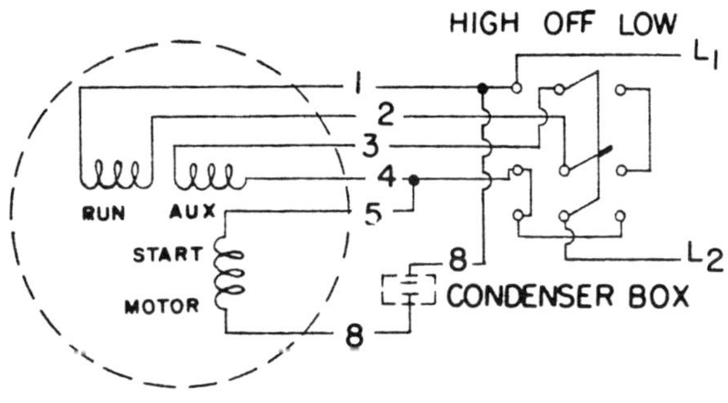

Fig. 7-10. Permanent-Split Capacitor Motor, Low Volts and Two Speeds.

The permanent-split capacitor motor can also be adapted for three speed operation, note Figure 7-11. This motor differs from the previously discussed motors because it has a center tap on the auxiliary winding. On high speed the running winding is across the line (L1 and L2) and the starting winding and both auxiliary circuits (1 and 2) are in series across the line (L1 and L2). For medium speed the running winding and one-half the auxiliary winding (2) are connected in series with the starting winding circuit. The auxiliary winding (1) is in series with the running winding. For low speed, the running is connected in series with both auxiliary circuits (1 and 2) across the Line L1 and L2. The starting circuit is also across the line. The capacitor is always in the starting circuit. Two auxiliary circuits were made by center tapping the auxiliary winding. The permanent-split capacitor motor is usually a small horsepower unit of 1/6, 1/4, or 1/3 hp. Its primary use is on low torque fans especially for blowers found on humidifiers, small air conditioners and small heat exchangers. Fans for greenhouses and livestock confinement housing will be larger, 1/2 and 3/4 hp.

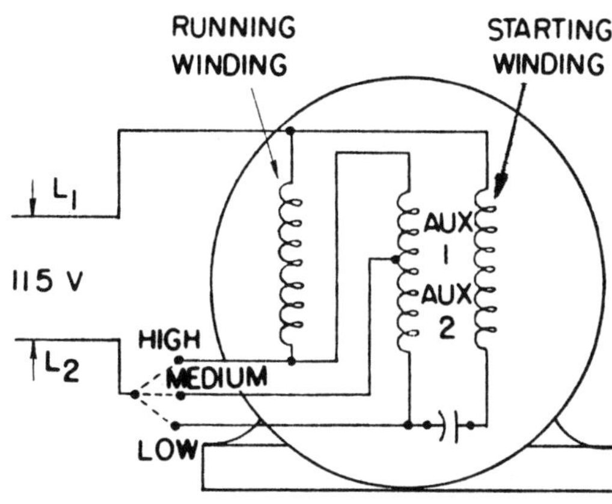

Fig. 7-11. Three Speed Permanent-Split Capacitor Induction-Run Motor

An outstanding feature of the AC induction run motor is its ability to maintain constant speed under normal voltage and load variations. The previous methods for changing rotational speed have been (1) change of frequency, (2) changing rotor slip, and (3) change in the number of stator poles. The series wound AC or DC (Universal) motor has greater flexibility for changing rotational speeds by changing the voltage either in the field or armature or a combination of both. This voltage change is accomplished by having a series rheostat control, a shunt rheostat control, a variable voltage control or a solid state control. Solid state control devices can also be used to vary the speed on the permanent-split capacitor motors and shaded pole motors. Polyphase motors are not generally made in fractional sizes and solid state control devices would not be applied to them.

In previous units, the generation of electrical current, the starting technique for motors, the running systems, the changing of voltage service, the reversing of motors and changing of motor speeds have been explained. In Unit VIII there will be a discussion of the motors most frequently used.

DEFINITION OF TERMS

BRUSH: A piece of current conducting material (usually carbon or graphite) which rides directly on the commutator of a commutated motor and conducts current from the power supply to the armature windings.

CAPACITOR: Unit providing capacitance in a circuit. Electrical component that stores electrical charges. A device which, when connected in an alternating current circuit, causes the current to lead the voltage in time phase. The peak of the current wave is reached ahead of the voltage wave. This is the result of the successive storage and discharge of electric energy.

CLOCKWISE: Operated the direction that a clock rotates.

COMMUTATOR: A cylindrical device mounted on the armature shaft and consisting of a number of wedge-shaped copper segments arranged around the shaft (insulated from it and each other). The motor brushes ride on the periphery of the commutator and electrically connect and switch the armature coils to the power source.

COUNTERCLOCKWISE: Operates the opposite direction that a clock operates.

DUAL-VOLTAGE MOTORS: Motors which can be operated on one voltage and a second voltage usually twice the value of the first voltage.

REVERSING ROTATION: Changing from clockwise (CW) to counterclockwise (CCW) or vice versa.

SINGLE-VOLTAGE MOTOR: Motor which can operate on only one voltage.

UNIVERSAL MOTOR: Motor which will operate either on AC or DC.

WINDING, AUXILIARY: A special set of windings in the permanent-split capacitor motor. The auxiliary windings makes possible speed and voltage changes.

WINDING, RUNNING: The circuit in the stator which provides the rotational force when the rotor is turning at its rated revolutions per minute.

WINDING, STARTING: The circuit in the stator which provides the initial rotational force to start the rotor.

NOTES

CLASSROOM EXERCISES VII-A

Wiring Motor Circuits

(Schematics in Units III and VII are to be used to complete this assignment)

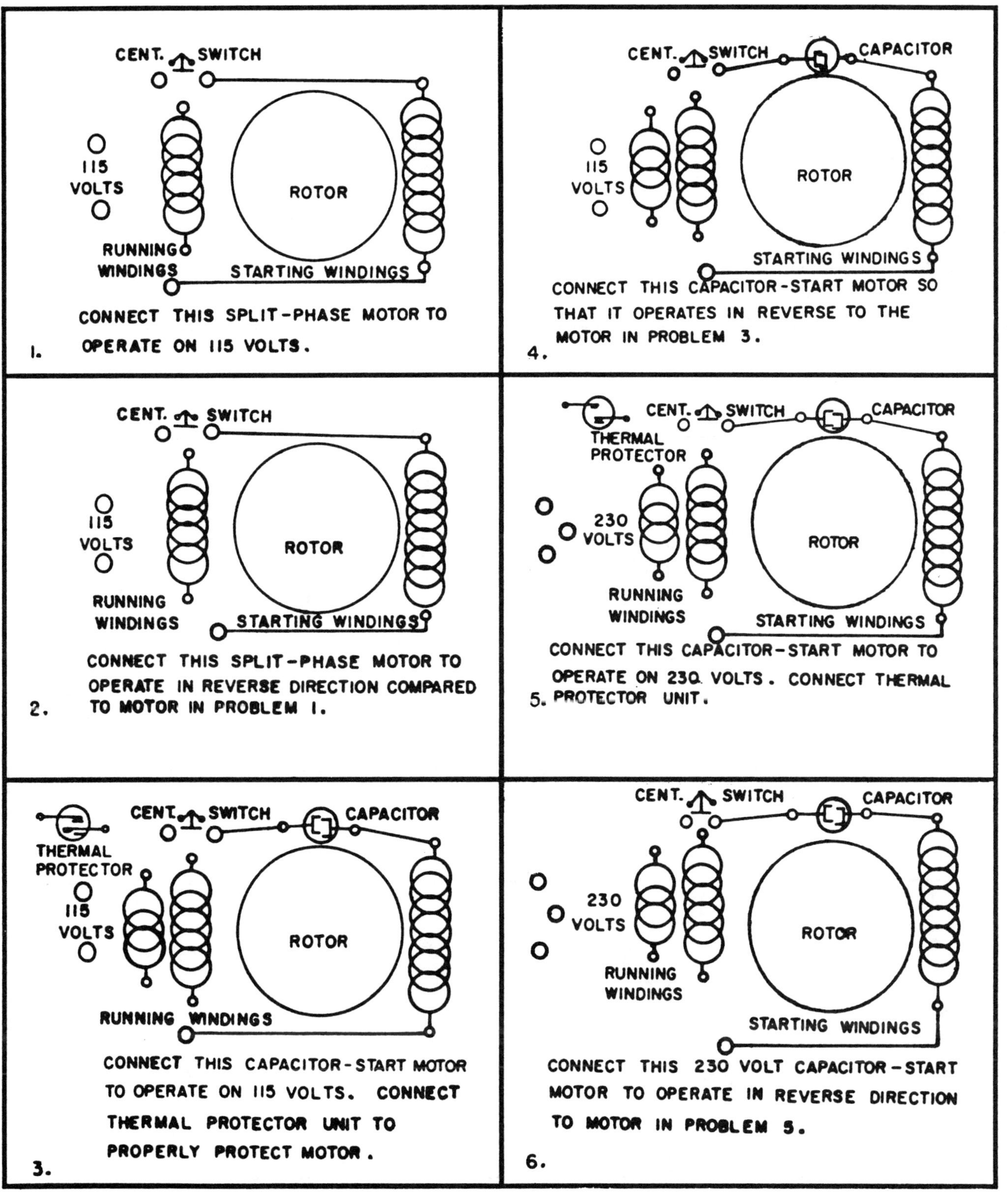

LABORATORY EXERCISE VII-A

Dual Voltage, Direction of Rotation, and Speed Control

GO TO THE LABORATORY AND COMPLETE THIS TABLE WITH THE ELECTRIC MOTORS SUPPLIED		
ACTIVITY	MOTOR A	MOTOR B
1. Type of Motor		
2. Present Operating Voltage		
3. Dual -- Voltage	YES NO	YES NO
4. Method to Change Voltage		
5. Rotation Direction	CW CCW	CW CCW
6. Method to Change Direction of Rotation		
7. Speed Control	YES NO	YES NO
8. If yes, how?		

THE MOTORS WE USE

There are many types of electric motors and a multitude of applications for each type. The hp range, starting ability, percentage of full load torque, starting current and operating characteristics will be high lighted in the tables in this unit.

INDUCTION RUN TYPE MOTORS

The most common electric motor is the split phase-start induction-run. It can be found on air conditioning fans, belted blowers, oil burners, poultry feeders, attic fans, small belt driven tools, lathes, centrifugal pumps, office machines and other applications.

The characteristics of a split phase motor are listed in Table 8-1. This motor is generally designed for single voltage.

*Table 8-1. Split Phase Electric Motor Features**

ITEM	VALUE
Horsepower Range	1/6 to 3/4
Load Starting	Medium
Full Load Torque	130-170%
Starting Current	5-7 x Rated
Reversible	Yes
Speed	Vary Load, Speed Nearly Constant

**Schematics found in Units Units VI and VII.*

Shaded pole motors can be found on small fans and blowers, space heaters, humidifiers, seed cleaners, freezer blowers vent fans, small exhaust fans, floor fans, portable evaporation coolers and similar load applications. The characteristics of the shaded pole motor are listed in Table 8-2. The shaded pole motor is generally a single voltage unit and is low in efficiency.

*Table 8-2. Shaded Pole Electric Motor Features**

ITEM	VALUE
Horsepower Range	1/250 to 1/4
Load Starting	Easy
Full Load Torque	Limited
Starting Current	Low
Reversible	No
Speed	Constant

**Schematics found in Unit VI.*

There are three types of capacitor motors; (1) capacitor-start, induction-run, (2) capacitor-start, capacitor-run, and (3) permanent-split capacitor. The first capacitor motor developed was the capacitor-start, induction-run. It has been a workhorse for hard starting loads and can be found on refrigeration compressors, conveyors, pumps, augers, ventilating fans, barn cleaners, vacuum pumps, manure pumps, milk coolers, drying fans, aeration fans, elevators and feeders. This motor is usually a dual voltage unit and the other characteristics of the motor are listed in Table 8-3.

*Table 8-3. Capacitor-Start Induction-Run Electric Motor Features**

ITEM	VALUE
Horsepower Range	1/8 to 10
Load Starting	Hard
Full Load Torque	350 to 400%
Starting Current	3-6 x Rated
Reversible	Yes
Speed	Vary Load, Speed Nearly Constant

**Schematics found in Units III, VI and VII.*

The schematics for the capacitor-start, capacitor-run motor, (a) 115 volts and (b) 230 volts, are illustrated in Figure 8-1.

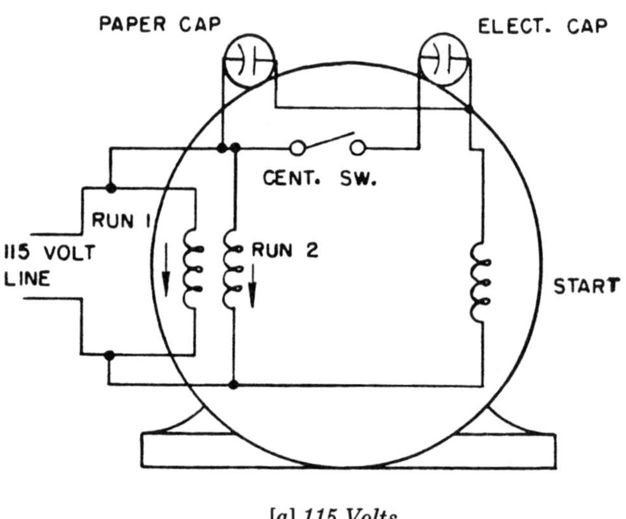

[a] 115 Volts

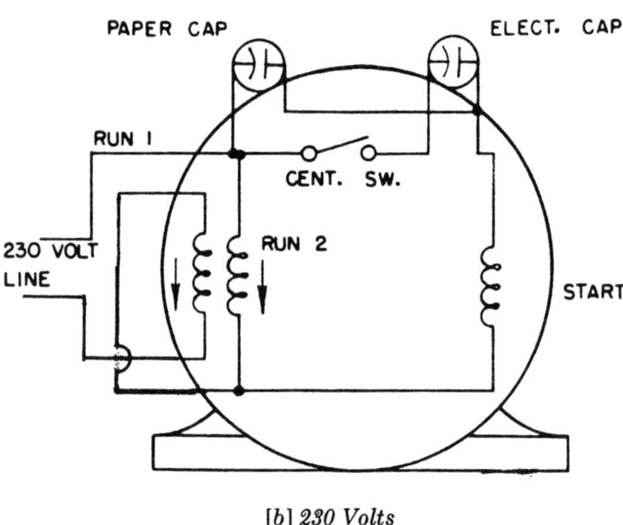

[b] 230 Volts

Fig. 8-1. Capacitor-Start, Capacitor-Run Electric Motors

The two-value capacitor electric motor is frequently called the capacitor-start, capacitor-run motor. The addition of the capacitor in the running circuit greatly improves the motor efficiency, power factor and reduces the load current. This capacitor may have been added just to improve efficiency or to match motor losses caused by the motor's cooling system. Capacitor motors larger than three hp are typically capacitor-run type. These motors are found on silo unloaders, feed mills and large bucket elevators and similar applications where the repulsion-start, induction-run motors were previously used, The shift to the capacitor-run motor is a reflection of economics because both motors are capable of handling hard starting loads. This is usually a single voltage motor, either 115 or 230. The features of this motor are listed in Table 8-4, Two Value Capacitor.

*Table 8-4. Two-Value Capacitor Electric Motor Features**

ITEM	VALUE
Horsepower Range	2 to 20
Load Starting	Hard
Full Load Torque	350 to 400%
Starting Current	3-5 x Rated
Reversible	Yes
Speed	Vary Load, Speed Nearly Constant

**Schmatics found in Unit VIII.*

In the development of electric motors, the permanent-split capacitor is a relative new comer. It is used on shaft mount fans and blowers, room and central air conditioners, furnace blowers, unit heaters, fans for confinement housing buildings for poultry, hogs and cattle, and for window ventilating fans. The facts on the permanent-split capacitor motors are listed in Table 8-5. The permanent-split capacitor is designed for jobs handled by the shaded pole motor. The motor can handle the loads with higher efficiency and power factor. It can be custom designed for special applications and adapted easily to solid state speed control. It is a single voltage motor, either 120, 208 or 240. The schematic, Figure 8-2, is for the permanent-split capacitor. The permanent-split capacitor motor has the same value of capacitance in both the starting and running windings. The motor differs from the capacitor-start induction-run in that the capacitor and starting windings are in the motor circuit at all times. The capacitor is generally the oil impregnated type and there is not a centrifugal switch or other disconnecting mechanism. The low value of the capacitor results in the low starting

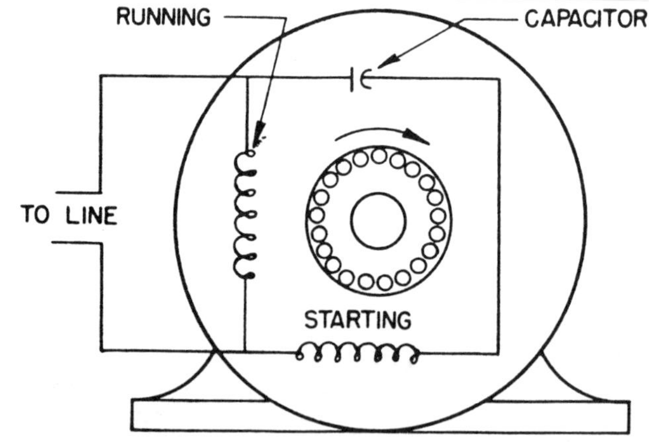

Fig. 8-2. Permanent-Split Capacitor Electric Motor

torque. To reverse the motor, reverse the leads of the starting windings in respect to the running windings. If four leads are brought out to the terminal board the change is easy; if not, it will be necessary to remove the motor end bracket to make the change.

*Table 8-5. Permanent-Split Capacitor Electric Motor Features**

ITEM	VALUE
Horsepower Range	1/20 to 1
Load Starting	Easy
Full Load Torque	150%
Starting Current	2-4 x Rated
Reversible	Yes
Speed	Lower Voltage, Reduce Speed

**Schematics found in Units VII and VIII.*

Another common electric motor is the three-phase. Its use is dependent upon the availability of three-phase electrical service or the use of phase converters. Features of the three-phase motor are listed in Table 8-6. The three-phase motor does not need a starting system because the three electrical phases each being 120 degrees apart handle that function with the running windings. The range for three-phase motors varies from 1/12 to 200 hp. The availability of type includes; (1) the totally enclosed fan-cooled, (2) open drip proof, and (3) explosion proof. Although available in smaller sizes, the three-phase motor is recommended for agricultural applications requiring two hp and larger. The motor requires little maintenance or repair because of the absence of the starting circuit

*Table 8-6. Three-Phase Electric Motor Features**

ITEM	VALUE
Horsepower Range	1/12 to 200
Load Starting	Medium to Hard
Full Load Torque	200-300%
Starting Current	Normal to High
Reversible	Yes
Speed	Vary Load, Speed Nearly Constant

**Schematics found in Units VI and VII.*

centrifugal switch and the brushes or slip rings which are essential on repulsion and universal motors. Because of the absence of arcing which results when a motor has brushes; the three-phase motor is utilized in chemical plants, flour and lumber mills and other hazardous locations. Data in Table 8-6, three-phase motors, list a wide range of operating conditions. This is possible because the three-phase motor can have three designs of squirrel-cage rotors which alter the performance. Note Table 8-7, for physical and electrical features of the three-phase motor.

The three-phase motor can have either a Delta or Wye internal wiring system.

The absence of three-phase service and the large amperage draw on large capacitor-start, capacitor-run motors prompted a need for a new motor design, called Soft-Start. In reality it is a two value capacitor motor with certain electrical modifications. Agricultural applications for the soft-start motors are crop drier fans and forage blowers which operate on single-phase service. This motor is also used by light industries with similar problems. The features of the soft-start motors are listed in Table 8-8.

Table 8-7. Three-Phase Electric Motor Squirrel-Cage Rotor Performance

ROTOR BAR SIZE	CODE LETTER	STARTING TORQUE	STARTING CURRENT	ROTOR FEATURES	APPLICATION
Small, Round o	A	High	Low	High Resistance	Shears Punch Press
Deep, Oval	B-E	Fair	Low	High Reactance, Low Resistance	Pumps, Fans, Blowers
Medium Rectangle	F-V	Poor	High	Low Reactance, Low Resistance	Pumps, Fans, Blowers

*Table 8-8. Soft-Start Electric Motor Features**

ITEM	VALUE
Horsepower Range	10 to 75
Load Starting	Easy
Full Load Torque	50-100%
Starting Current	2-2.25 x Rated
Reversible	Yes
Speed	Constant

**Schematics found in Unit VIII.*

The soft-start motor design permits a starting current 2.0 to 2.25 times the full-load current so that the starting surge does not affect other loads on the line. This reduced starting current also reduces the starting torque to 50 to 100 percent of the full-load torque.

Soft-start motor schematics are illustrated in Figure 8-3. In schematic (a) Figure 8-3, the soft-start motor is in the starting position. The running windings are in series which reduces the starting current to one-fourth the value as if the windings had been in parallel. This permits the large

[a] Starting

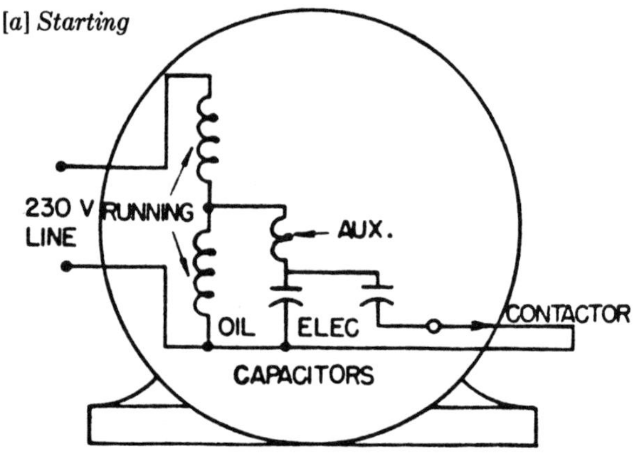

[b] Running

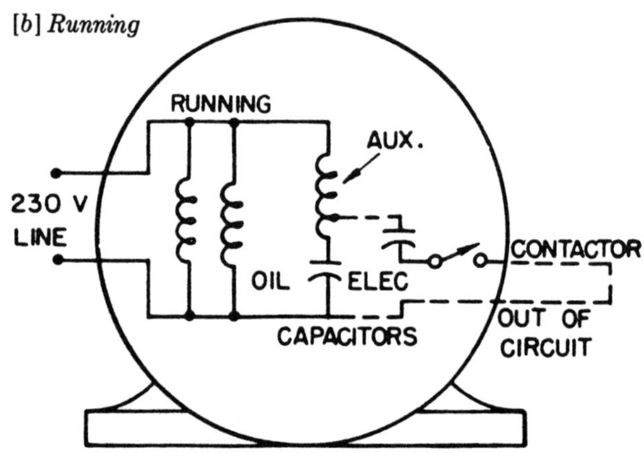

Fig. 8-3. Soft-Start Electric Motors

hp, single-phase electric motors to be used without placing extra load on the transformer. The auxiliary winding and two capacitors complete the starting circuit controlled by a centrifugal switch and a magnetic contactor. When the motor has reached two-thirds of its rated speed, the centrifugal switch activates a magnetic contactor which has the ability to handle the heavy current, and disconnects the starting circuit. The running windings, Figure 8-3 (b), are now in parallel and the current surge does not adversely affect the transfoprmer because of the motor's speed. In Figure 8-3, the centrifugal switch and other wiring connections for this motor have been omitted. The starting mode (a) has been illustrated along with the running mode (b). The reader must accept the change from series to parallel in the running windings even though it can not be traced on the schematics

WOUND ROTOR MOTORS

The universal or series motor has adaptability from the electric hand drill to power units for industrial trucks, cranes and hoists. Hand powered tools have universal motors and can be operated on either AC or DC. The universal motor does not operate at a constant speed but operates as fast as the load permits. If not loaded, these motors will overspeed and could damage or ruin the motor. Study the features of the universal or series motor in Table 8-9.

*Table 8-9 Universal or Series Electric Motor Features**

ITEM	VALUE
Horsepower Range	1/160 to 2
Load Starting	Hard
Full Load Torque	350 to 400%
Starting Current	High
Reversible	May Be
Speed	Vary Load, Speed Changes

**Schematics found in Unit VI.*

Repulsion type motors can be used for barn cleaners, grinders, feeders and similar hard starting loads. The details for this motor are listed in Table 8-10. This motor is reversible by the brush-ring adjustment method or can be reversed electrically by the use of the proper type of switch. The wound rotor, brushes and switching technique contribute to the high production cost for this motor and increase the maintenance and repair costs. The production of this motor has been dropped by many firms, however, it was still being produced by the Baldor Electric Company in the eighties with sizes ranging from 1,1/2 to 15 hp.

*Table 8-10. Repulsion [Wound-Rotor] Electric Motor Features**

ITEM	VALUE
Horsepower Range	1/6 to 15
Load Starting	Very Hard
Full Load Torque	350-400%
Starting Current	2-4 x Rated
Reversible	Brush Ring Adj.
Speed	Vary Load, Speed Nearly Constant

* *Schematics found in Unit VI.*

Electric motors have a rated rpm and if the driven machine requires a different rpm the change can be made by use of pulleys and belts or gears and chains. These combinations may require a lot of space, odd arrangements and achievement of low speeds may be a problem. Speed can be changed by the use of a combination of gears. If the gear arrangement to convert the basic motor speed to the rated output speed is assembled as a unit it is called the gearhead. Attach this gearhead permanently to the motor and the combined package is called a gearmotor.

[*a*] *In-Line Shaft Gearmotor*

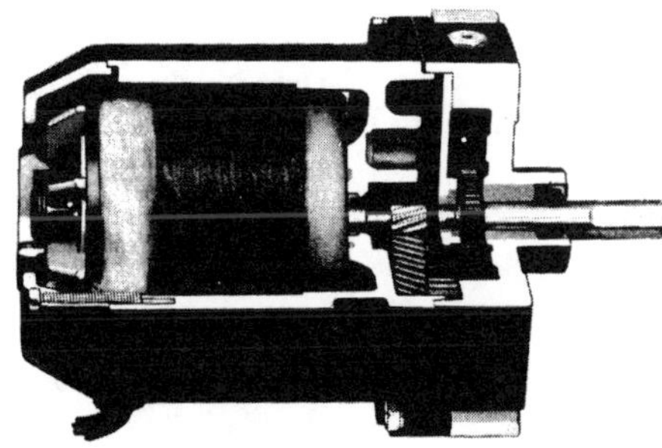

[*b*] *Parallel Shaft Gearmotor*

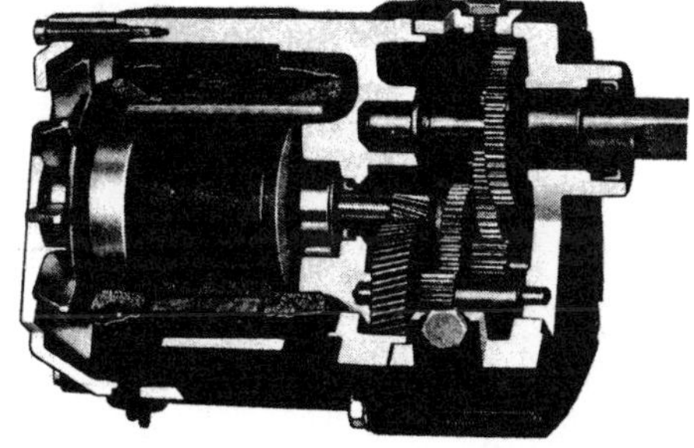

Courtesy of the Bodine Electric Company, Chicago.

Fig. 8-4. In-Line and Parallel Shaft Gearmotor

The gearhead is compact, efficient and when attached to the fractional hp electric motor provides speed-reducing and torque-multiplying with excellent reliability. The term "parallel shaft" is applied to gearmotors having their output shaft either in-line (concentric) with or offset from (parallel to) the motor shaft axis, see Figure 8-4. The offset output shaft arrangement is generally more compact than the inline design. The offset shaft can be located at the 12, 3, 6, or 9 o'clock position which provides versatility in mounting to machines. If a "parallel shaft" design is not satisfactory a right angle (90 degree) gearmotor may meet the driven machine requirement. The right angle gearmotor is desirable when a vertical output shaft is required but is also used when a horizontal shaft is needed, note Figure 8-5. Refer to Figures 8-4 and 8-5 to determine the types of gears needed to make the gearhead work. Spur gears are straight, helical gears are straight and angled and the worm gear has curved teeth. Combinations of the types of gears, size and number of teeth, makes possible many combinations of speed and torque output.

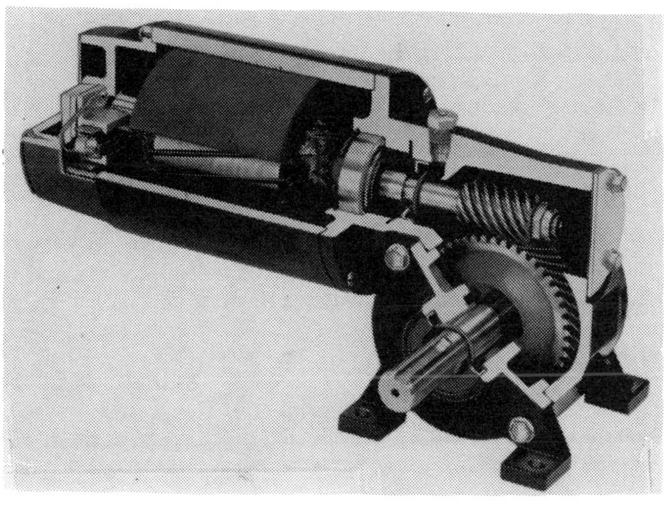

Courtesy of the Bodine Electric Company, Chicago.

Fig. 8-5. Single Reduction Right Angle Gearmotor

The electric motor applications discussed in this unit can be found in the home, at work, on the farm and in industrial production plants.

The factors to consider when selecting electric motors for specific jobs and how to test the performance of electric motors will be covered in UNIT IX, ELECTRIC MOTOR SELECTION AND PERFORMANCE TESTING.

DEFINITION OF TERMS

CAPACITOR: Unit providing capacitance in a circuit.

CAPACITOR, TWO-VALUE: Electric motor with two or more capacitors; located in both starting and running windings.

CONTACTOR: A device operated other than by hand. A contactor employs an electromagnet for opening or closing a circuit under normal operating conditions.

FULL LOAD TORQUE: Torque necessary to produce its rated horsepower (hp) at full -load speed.

GEARHEAD: The portion of a gearmotor which contains the actual gearing which converts the basic motor speed to the rated output speed.

GEARMOTOR: A gearhead and motor combination to reduce the speed of the motor to obtain the desired rpm's.

MAGNETIC STARTER: A device for starting motors utilizing a contactor for making electrical line connections.

PHASE CONVERTER: A device which can be used on single-phase electrical service to make a conversion which will permit the use of a three-phase motor.

POLYPHASE MOTOR: Two or three phase induction motors have their windings, one for each phase, evenly divided by the same number of electrical degrees.

POWER FACTOR: A measurement of the time phase difference between voltage and current in an AC circuit.

RATED or FULL CURRENT: The current drawn from the line when the motor is operated at full-load torque and full-load speed at rated frequency and voltage.

REACTANCE (INDUCTIVE): The characteristic of a coil, when connected to alternating current, which causes the current to lag and voltage in time phase. The current wave reaches its peak later than the voltage wave reaches its peak.

RESISTANCE: The degree of obstacle presented by a material to the flow of electric current is known as resistance and is measured in ohms.

SOFT-START: An electric motor of larger hp that can be operated on single-phase service when three-phase service is not available. Starting current of the motor is one to two times full-load current and it has a starting torque of 50-90 percent of the full-load torque.

STARTING CURRENT: Amount of current drawn when a motor is energized. In most cases is higher than that required for running.

SWITCH, CENTRIFUGAL: Switch operated by forces caused by rotary movement.

TORQUE: Force delivered by a motor and expressed in lb-ft or oz-ft.

TORQUE, STARTING: Turning force delivered when the rotor is not allowed to move. Current reading is taken at rated voltage and frequency. Also called locked rotor torque.

VOLTAGE, DUAL: Some motors can operate on two voltages depending upon how they are built and connected. The voltages are either multiples of two (120/240) or the square root of 3 of one another. (Square root of 3 = 1.73 x 120 = 208)

VOLTAGE, SINGLE: A motor built and connected for only one voltage.

WOUND ROTOR: A rotor made by securing laminated iron core sections on a shaft. The core has slots into which copper conductors are wound forming coils or loops which come out and are fastened to a commutator.

NOTES

CLASSROOM EXERCISE VIII-A

Identification of Motor Types

Study the schematics and identify the motor types and related features. Schematics found in Units VI and VIII are to be used to complete this assignment.

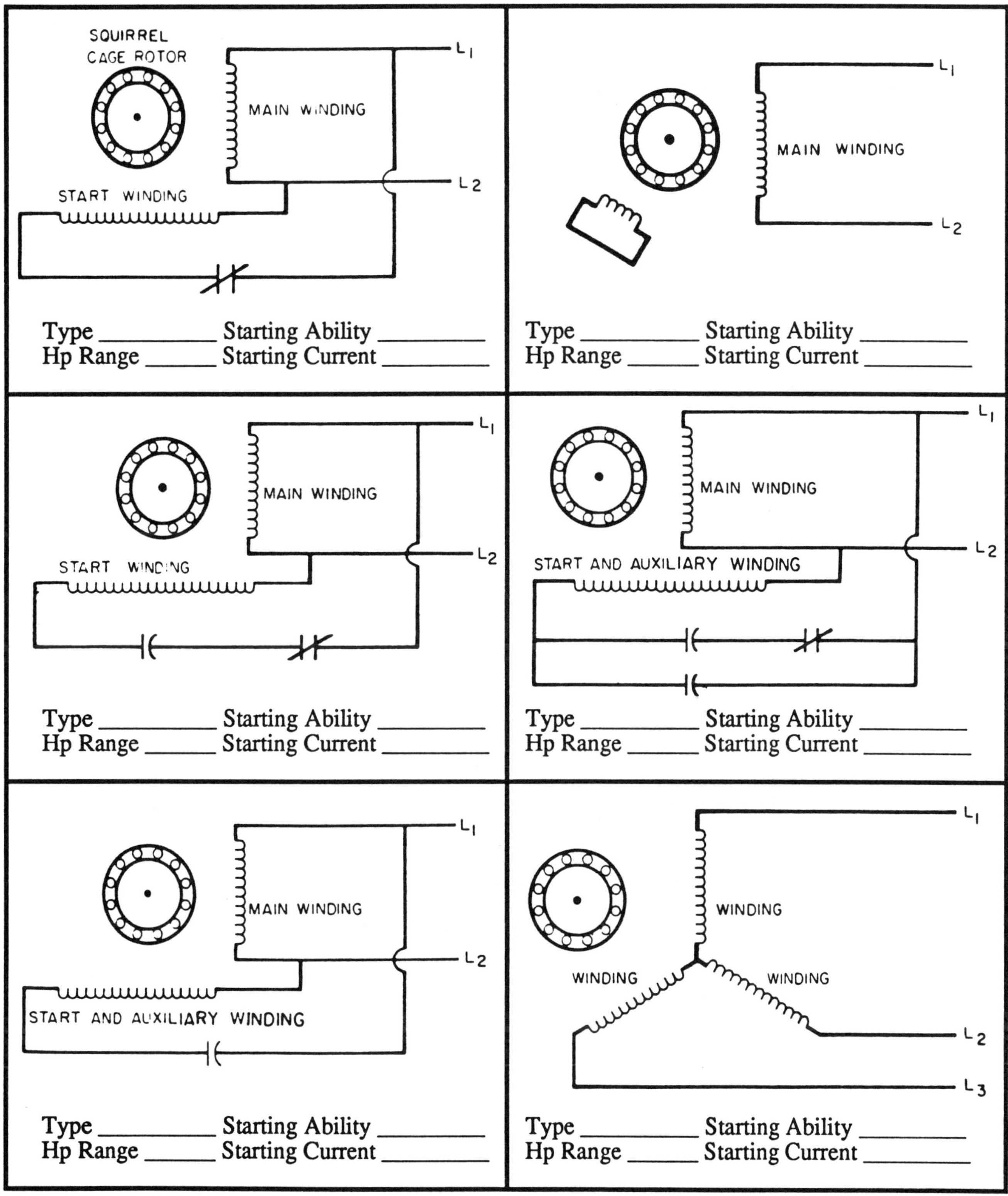

LABORATORY EXERCISE VIII-A

Electric Motor Part Identification

Select an electric motor, disassemble and identify component parts and complete the following questions. It may be necessary for you to refer to Units II, III, and VII for assistance with this assignment.

1. Complete the following nameplate data: Serial No. ____________________

   a. H.P. ____________ e. Cycles ____________

   b. R.P.M. ____________ f. Service Factor ____________

   c. Volts ____________ g. Temperature Rise ____________

   d. Amperes ____________ h. Locked Rotor ____________

2. Disassemble the motor and study its component parts. Follow recommended instructions (Unit XII) for disassembly to avoid damage to the motor.

   a. Does the motor have starting windings? ____________ Is the conductor size larger or smaller than the running windings? ____________________

   b. How many pairs of stator poles are there? ____________________

   c. What type of rotor does the motor have? ____________________

   d. Does the rotor have a commutator? ____________ Brushes? ____________

   e. Does the motor have a capacitor? ____________ Where located? ____________

   f. Does the motor have a centrifugal switch? ____________________

   g. Does the motor have built-in overload protection? ____________________
   If so, is it manual or the automatic reset type? ____________________

   h. What type of bearings are in the motor? ____________________
   How many? ____________ Where located? ____________

   i. How are the bearings lubricated? ____________________

   j. Can the motor be changed from one voltage to another? ____________________
   How? ____________________

   k. Can the direction of the motor be changed? ____________________
   How? ____________________

   l. What positions can this motor be mounted? ____________________

   m. What type motor is this? ____________________

ELECTRIC MOTOR SELECTION AND PERFORMANCE TESTING

TYPES OF LOADS FOR MOTORS

There are several types and sizes of electric motors. You may have purchased a piece of equipment driven by an electric motor, purchased a replacement electric motor or purchased an electric motor for a new machine. These three conditions for electric motor selection will be discussed briefly.

When a commercial firm selects a motor for a piece of equipment, it may require a large number of motors. This will result in an engineer testing and matching the operating characteristics and load requirements of both the motor and the equipment. Special electrical and mechanical designs can be utilized under these conditions. The engineer has been trained to make these decisions of motor specification for the machine. These are called original equipment motors, (OEM). The purchaser of equipment is protected to a certain extent when buying a package unit.

When a replacement motor is needed the cause of its failure should be studied. The nameplate data are analyzed and the replacement motor should match this rating or exceed the prior rating. Replacement motors from the manufacturer's stock are generally called standard, stock or catalog motors. Standard, stock and catalog motors have been built to mechanical and electrical guidelines established by the National Electrical Manufacturers Association (NEMA). Whether the motor is an OEM or a standard, stock or catalog motor, it will be easier to understand load types, speed and torque characteristics, and horsepower (hp) rating if the motor is classified according to operating conditions as; (1) continuous-running steady loads, (2) continuous-running intermittent loads, (3) adjustable speed loads, and (4) cyclical loads.

CONTINUOUS-RUNNING STEADY LOADS

Motors used on pumps, fans and conveyers are examples of constant speed and constant load. The hp requirement can be determined on the basis of the average load of the machine being driven. Consideration should also be given to cold starting conditions, and any overload, if possible.

CONTINUOUS-RUNNING INTERMITTENT LOADS

Motors used on punch-presses, shears and rolling mills are examples of continuous-running and intermittent loads. They may be running continuously but the load is applied only when the machine is doing its assigned job. In this application, the maximum load must be determined, not the average, because it's impossible to determine how long the maximum load will be required. Frequently when determining the hp rating, the motor's breakdown torque will be increased by 30 percent above the anticipated load torque. Margins of safety make it possible for the motor to carry the load without overheating even under low voltage and reduced frequency. Motors are normally designed for the rated voltage, plus or minus 10 percent and rated frequency, plus or minus 5 percent.

ADJUSTABLE SPEED LOADS

Motors used on adjustable speed loads operate over a wide speed range with a load that is either constant or variable with speed. Examples are conveyers, hoists and grinders. With this type of load, it is essential to determine how the load varies with speed, and which combination of torque and speed produces the maximum hp output required by the driven equipment. The three common types of adjustable speed loads are constant torque, variable torque and constant hp, note Figure 9-1, for constant torque and hp load application curves.

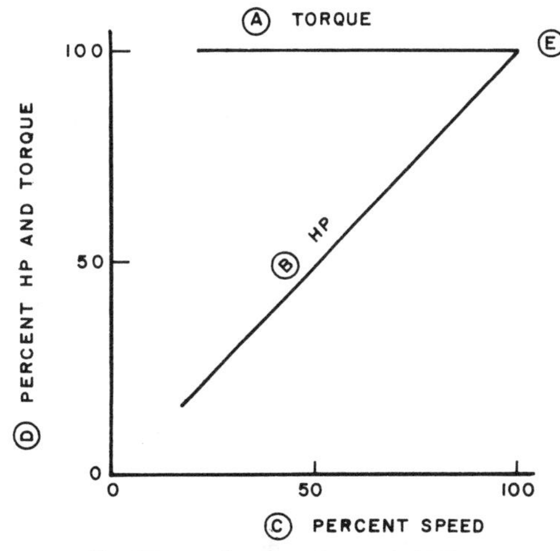

The Electrification Council, NY.

Fig. 9-1. Constant Torque and Horsepower Load

Constant torque loads are the easiest to analyze and the most common. The torque requirement, (A), is the same in Figure 9-1, whether the speed is low or high. The torque remains constant throughout the speed range and the diagonal hp curve, (B), increases in direct proportion to the speed, (C). The maximum hp is at the maximum speed point, (E), and the amount of work done increases with an increase in speed. The hp rating for the motor to drive a constant load can be the same as the hp required to drive the load at top speed.

Fans, blowers and centrifugal pumps are examples of variable torque loads which require much lower torque at low speeds than at high speeds. The variable torque and hp load curves are shown in Figure 9-2. The hp varies approximately as the cube of the speed and torque varies approximately as the square of the speed. Thus, only one-fourth as much torque is required to drive a fan at half speed as at full speed. If the motor selected with a hp rating (B) that will drive the load at top speed, both torque (A) and speed will be adequate, refer to Figure 9-2, point (E).

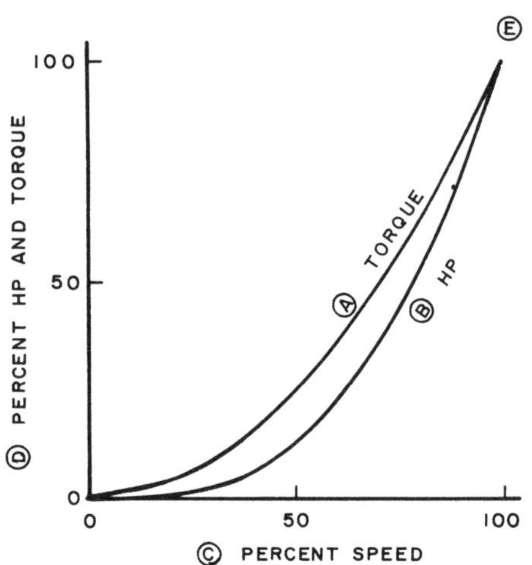

Fig. 9-2. Variable Torque and Horsepower Load

For variable torque and constant hp load conditions refer to Figure 9-3. Motors on large lathes, boring mills and coiling machines in the metals industry are examples of constant hp loads which require high torque at low speeds and low torque at high speeds. When the lathe, Figure 9-3, has the greatest torque (A), the speed is low (E) and at 100 percent speed, the torque is the lowest but at all times, the hp (B) requirement is constant.

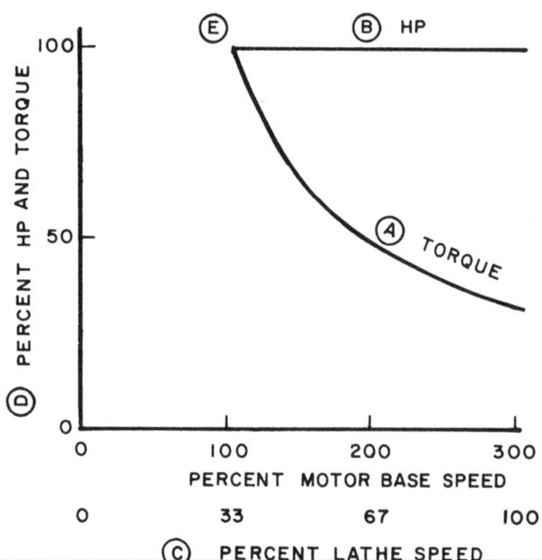

The Electrification Council, NY.

Fig. 9-3 Variable Torque and Constant Horsepower Load

CYCLICAL LOADS

The fourth category is cyclical loads and hoist and cranes are examples of this type of application. The machine must be started, run for a period of time, decelerate, stop, remain idle and then start again. Refer to Figure 9-4, for the cycle of a motor operating a hoist. Each of these operation conditions requires a different hp requirement from the motor. It is possible to select a motor with a smaller hp rating than the peak value dictated even though it will become overheated because the motor can cool down during nonpeak periods.

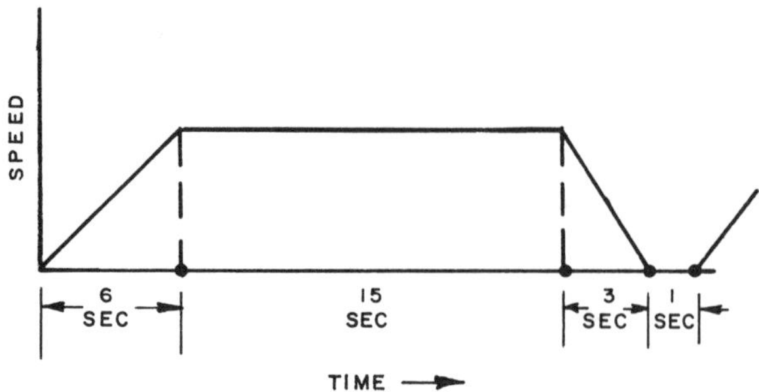

The Electrification Council, NY.

Fig. 9-4. Duty Cycle For A Typical Hoist Drive

MOTOR CHARACTERISTICS

Regardless of the type of load the motor must drive there are common terms; torque, motor speed, and hp. The main selection feature for an induction run motor is the ability of the motor to start a load. This turning force is called torque and is measured in ounce-inches for fractional hp motor and pound-inches for integral hp motors. Values based on size of motor and rpm are listed in a Motor Torque Table in the Appendix. Note Figure 9-5, Torque Calculation. Torque can exist with or without rotation and is determined by the formula:

Torque = Force x Distance

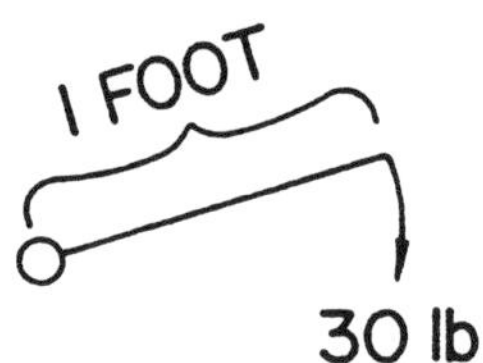

TORQUE = FORCE. x DISTANCE

TORQUE = 30 lb • FT

EXISTS WITH OR WITHOUT ROTATION

Fig. 9-5. Torque Calculation

Motor manufacturers provide motor performance data plotted as curves for the motors they produce. The first step for illustrating the electric motor torque and speed characteristics is to develop a graph. The vertical line, ordinate, is to represent speed and the horizontal base line, absicca, represents torque, note Figure 9-6. The torque values which will be added to the graph are; starting, pull-up, breakdown and full-load torque.

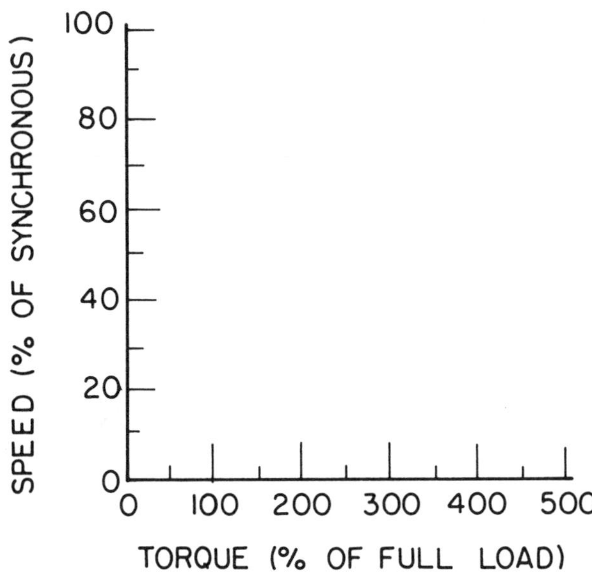

Fig. 9-6. Motor Performance Graph

The starting torque is the amount of torque the motor has available at zero speed. This is also called the locked rotor torque and is represented by a code letter on the nameplate. The letter represents kilovolt-ampere (KVA) and is determined by the following formula:

$$KVA = \frac{\text{Line Voltage x Amperage}}{1000}$$

This value is determined by holding the rotor to prevent turning and measuring the amperage flow and voltage. This represents the power required to start the motor. From Figure 9-7 a sequence of graphs will be developed to demonstrate torque characteristics for a capacitor-start motor, note the locked-rotor torque point.

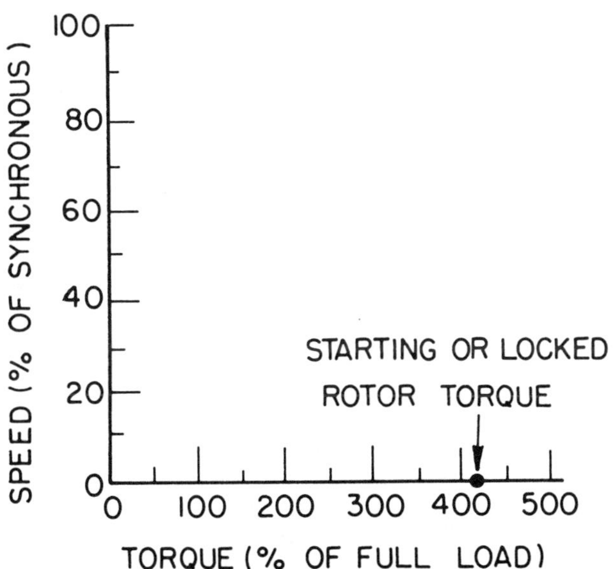

Fig. 9-7. Capacitor-Start Motor Locked-Rotor Torque

The next torque value, Figure 9-8, is pull-up torque. This curve represents the minimum torque developed between zero speed and breakdown torque. The pull-up torque for the induction run motor must be adequate to accelerate a load to its operating speed. Do not confuse pull-up with pull-in torque. Pull-in torque is the value that brings the synchronous motor's load to synchronous speed.

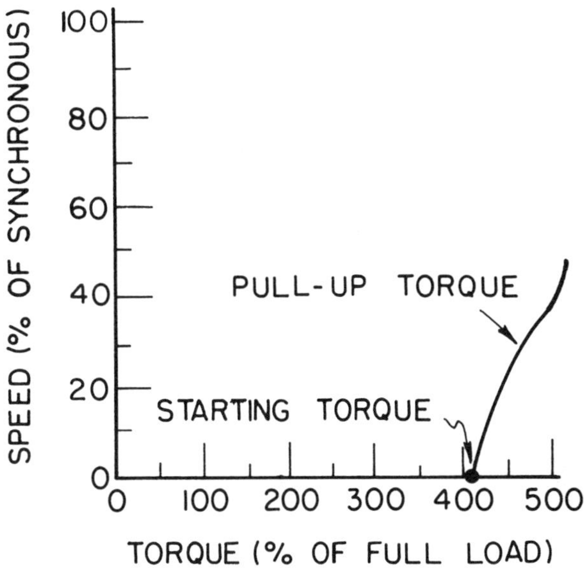

Fig. 9-8. Capacitor-Start Motor Pull-Up Torque

The breakdown torque, Figure 9-9, is the maximum torque that a motor develops at rated voltage without an abrupt drop in speed or without stalling. There is a break in the curve to indicate there are two segments to this part of the torque curve. Both the starting and running windings are operational when the motor starts, and this curve represents the combined efforts of the starting and running windings. The starting windings are, however, providing about three to four times as much torque as the running

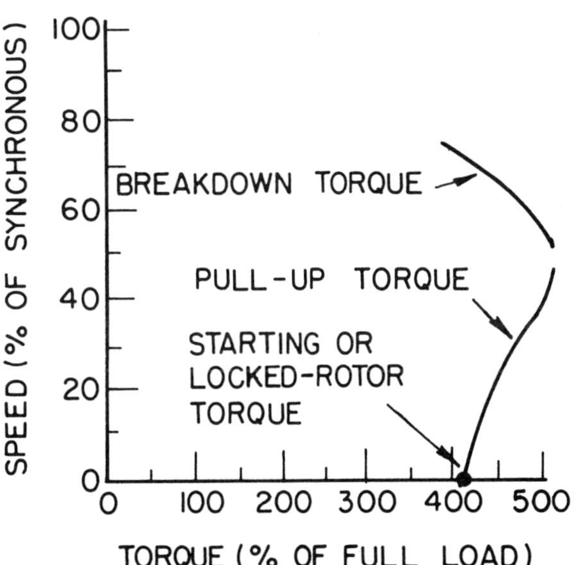

Fig. 9-9. Capacitor-Start Motor Breakdown Torque

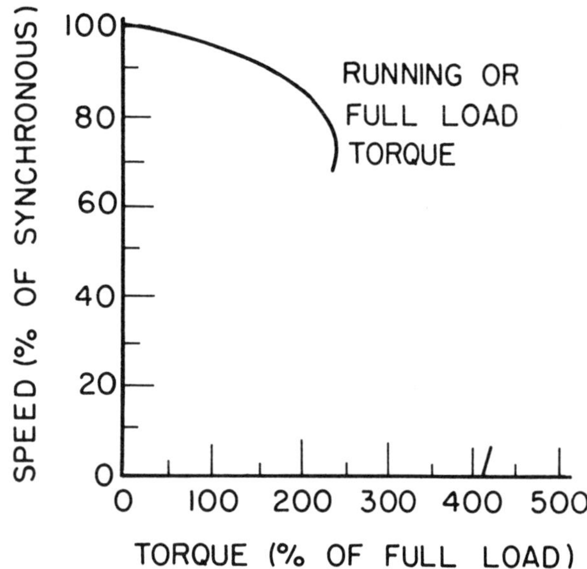

Figure 9-10. Capacitor-Start Motor Full-Load Torque

windings. Full-load torque, Figure 9-10, is the force that the motor will deliver continuously at rated voltage and rpm without exceeding its temperature rating. This full-load torque usually determines the basic rating and size of motor that must be selected for a load. Maximum speed is achieved at no load and the full-load point is at the rated operating torque and speed. All four torque curves for the capacitor-start motor are in one graph, Figure 9-11. The pull-up torque (A) increases to a value greater than 400 percent because both the starting and running windings are functioning. This torque value, pull-up and breakdown curve is greater than the maximum full-torque value (B). As the motor speed increases (approximately 75 percent of rated rpm) the motor's centrifugal switch opens and this is represented by the open space between (B) and (C). If the breakdown torque curve (C) were extended it would cross the full-load torque curve and go below it. These curves are used when selecting electric motors for specific job requirements. The torque supplied by the motor must be greater than the torque required by the driven load;

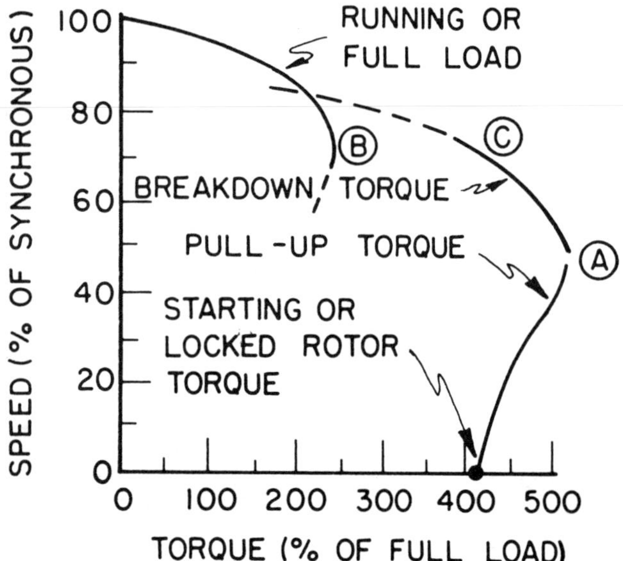

Figure 9-11. Capacitor-Start Motor Torque Curve

pull-up torque when starting the load and full-load torque when operating the machine at rated rpm. If the motor has excess torque, acceleration will be greater than desired. If the motor does not have enough torque, there will be motor overheating because of the extra load in the starting winding circuit. Not all capacitor-start motors will have torque curves exactly as illustrated in Figures 9-7 through 9-11, therefore expect variations depending upon motor manufacturer and design features.

The torque curve for a split phase-start motor, Figure 9-12, illustrates a more vertical curve than for the capacitor-start motor. The total torque, split phase-start, approaches 300 percent as compared to 500 percent for the capacitor-start motor. The capacitor in the starting circuit accounts for this difference. The split phase-start motor must be used on easier starting loads. The shape of curve and values are essential when selecting an electric motor for a specific load.

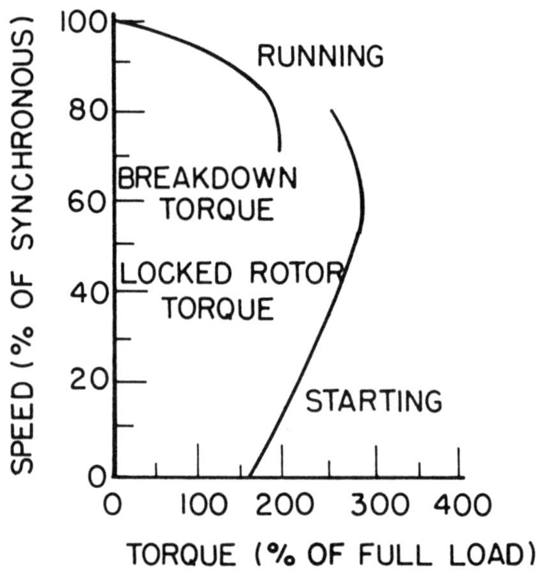

Fig. 9-12. Split Phase-Start Motor Torque Curve

In Figure 9-13, there are torque curves for three permanent-split capacitor motors; a high torque (A), medium torque (B) and low torque motor (C). These motors have low starting torque characteristics, 5 to 45 percent of the full-load torque, and will be used on light starting loads such as direct connected fans and blowers. This motor runs quietly because it does not have a starting switch. The capacitor is permanently wired in series with starting windings and in parallel with the running windings. The dotted curve in Figure 9-13 represents a specific fan load (D). Motors (A) and (B) have enough pull-up torque to start the load but motor (C) would be questionable. Only one motor (A) would have enough full-load torque to drive the fan at 95 percent of synchronous speed.

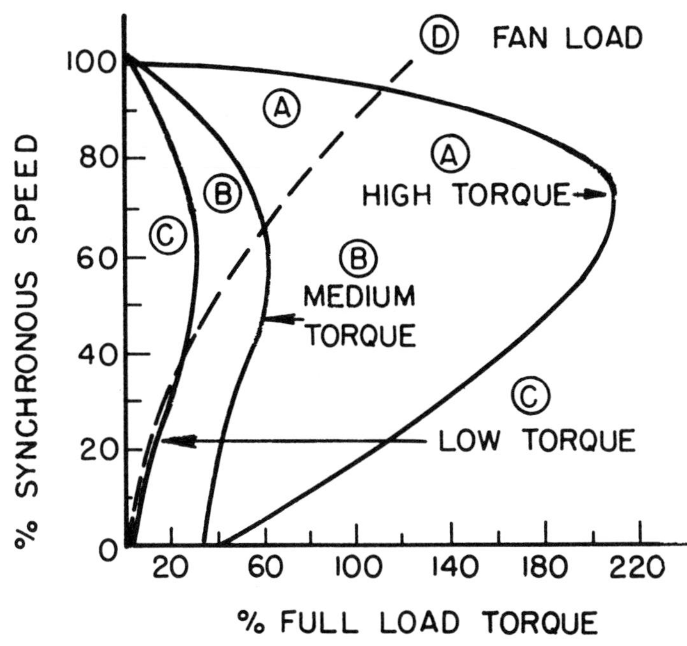

Fig. 9-13. Three Permanent-Split Capacitor Motor Torque Curves

The continuous torque curve in Figure 9-14 represents a three-phase motor. There is no break in this curve because the three sets of windings (one for each phase) serve the function of both the starting and running windings. Three-phase motors are noted for bringing driven loads smoothly to operating speeds as compared to motors with special starting mechanisms.

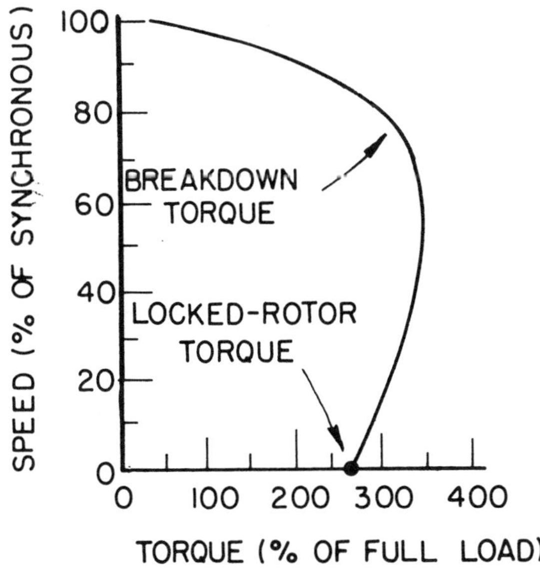

Fig. 9-14. Three-Phase Motor Torque Curve

Typical motor and load torque characteristics are illustrated in Figure 9-15. Note, the load requirement curve is exceeded by the electric motor's torque curve over most of the entire range of speed. The full-load torque point is where the curves cross and the rated rpm and the rated torque values are projected and crossed for the motor being evaluated. This motor could be selected to drive this machine load if the intersect is 95 percent of synchronous speed.

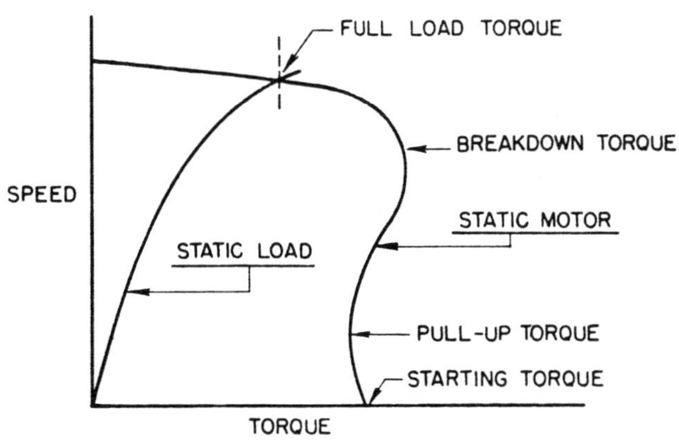

Fig. 9-15. Motor and Load Torque Characteristics

PERFORMANCE TESTING

EQUIPMENT AND TEST METERS

Completing a laboratory activity to test the performance of an electric motor will help you understand the nameplate terms, the starting and running circuits, electrical terms and the use of electrical test meters. There are excellent quality electric motor dynamometers on the market which the authors endorse and recommend. There is a question whether a secondary education class will have the budget for such a piece of test equipment. Post-secondary classes, vocational and technical, technical colleges and universities will probably have the more accurate dynamometers. The same basic principles can be taught (with a lesser degree of accuracy) using the following test equipment, note Figure 9-16. Motor Loading Assembly.

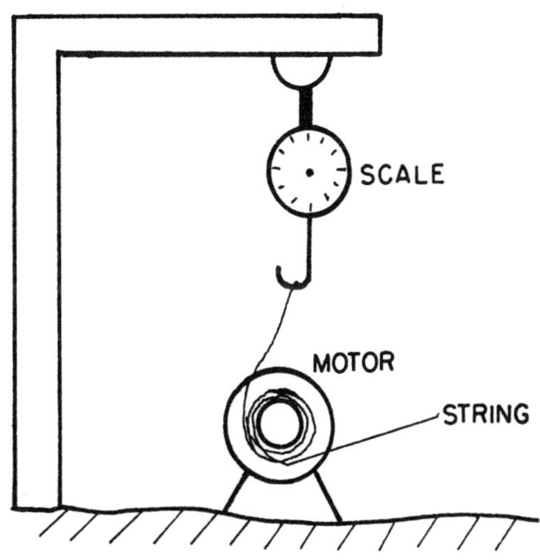

Fig. 9-16. Motor Loading Assembly

The necessary components are an electric motor with a flat or v-belt pulley, a hanging spring scale, chalk line and a platform base for mounting the motor. From this platform there should be an inverted "L" for suspending the scale above the motor's driving pulley. The pulley should be two to three inches in diameter for successfully demonstrating with motors of 1/4 to 1/2 hp.

Test meters needed include: tachometer (the more accurate the better); meter to measure line voltage, starting and running amperes; and wattmeter or kilowatt-hour meter. The ammeter can be the panel or the clamp-on type. The voltmeter will be used with the test leads. Availability of a wattmeter or kilowatt meter may be a problem. Without the wattmeter reading it would be impossible to determine power factor and efficiency. However, if testing a 1/4 hp split phase-start electric motor, refer to Table 9-1, Test Data. Watt values can be substituted which will enable the calculation of power factor and efficiency. Each class will need to fabricate terminal connections for the power supply and make necessary test instrument hookup points.

SAFETY RULES FOR DATA COLLECTION

Always wear industrial quality eyewear in performing laboratory exercises. The electric motor must be mounted securely with the scale and cord directly above the pulley. Use caution when applying force with the cord. Release the tension on the cord between test readings or the cord will break as the result of friction. Vary the number of turns of cord around the pulley to give desired pound readings on the scale. The cord will probably break occasionally, especially in the upper load ranges. The voltmeter is to be connected in parallel, ammeter in series and the wattmeter a combination of both parallel and series because it contains both a voltmeter and ammeter. If an ammeter is connected across parallel service lines it will be ruined. Consult hookup instructions for the meters being used. Double check electrical connections for safety features, both for personnel and instrument protection. The authors recommend the use of a ground fault circuit interrupter (GFCI) extension cord unit for testing all circuits. The use of a GFCI unit is required in school laboratories in some states.

FORMULAS AND METER READINGS

Column 2 of Table 9-1, Test Data, requires a calculation to develop the motor testing laboratory. Use the following formula (2) to determine the force in pounds which must be loaded on the motor by wrapping two or more loops of cord around the pulley to supply the load. Placing tension on the cord by tightening can result in inaccurate values.

$$\text{(2) Force} = \frac{\text{Hp} \times 33{,}000}{2 \times \text{Pi} \times r \text{ (in ft)} \times \text{rpm}}$$

$$= \frac{\text{Hp} \times 5252}{r \text{ (in ft)} \times \text{rpm}}$$

The values are: hp value is to be the rated hp of the motor, for example 1/4; "pi" is to be 3.1416; r is the radius of the pulley in feet, therefore determine the radius in inches and divide by 12 inches; and rpm is the motor's synchronous speed. The pound value is for test # 5 or the motor's 100 percent rated load. Other values in column are determined by reducing or increasing the value of test # 5 by 25 percent, for the # 4 and # 6 value, respectively. Reduce or increase the previous values by 25 percent until the column of values have all been determined. It may be difficult to

Table 9-1. Test Data Collected From a Typical 120 Volt Split Phase-Start 1/4 Horsepower Motor.

(1) TEST NO.	(2) EST. % OF RATED LOAD	(3) EST. LBS.	(4) ACT. LBS.	(5) RPM	(6) AMPS A	(7) VOLTS V	(8) APPARENT POWER V X A	(9) TRUE WATTS	(10) POWER FACTOR	(11) EFFICIENCY	(12) ACTUAL HORSEPOWER	(13) TORQUE OZ FT.	MOTOR INFORMATION (NAMEPLATE)
1	0%	0	0	1791	4.7	123	578.1	140	.242	0%	.0	0	Type:
2	25%	1.35	1.3	1783	4.8	122	585.6	180	.307	31.9%	.060	2.83	Hp
3	50%	2.71	2.7	1771	4.85	121	586.8	230	.392	40.6%	.125	5.93	Amps
4	75%	4.06	4.0	1760	4.90	121	592.9	270	.455	51.3%	.185	8.83	Volts
5	100%	5.41	5.4	1745	5.1	121	617.1	330	.535	56.3%	.249	11.99	Cycle
6	125%	6.76	6.7	1731	5.25	121	635.2	380	.598	60.7%	.309	15.00	Phase
7	150%	8.12	8.1	1715	5.5	120	660.0	435	.659	64.2%	.374	18.33	RPM
8	175%	9.47	9.4	1696	5.8	120	696.0	480	.690	67.4%	.434	21.50	
9	200%	10.80	10.8	1672	6.4	120	768.0	560	.729	66.5%	.499	25.08	
10	225%	12.10	12.1	1647	6.9	120	828.0	640	.773	65.1%	.559	28.52	
11	250%	13.50	13.5	1604	7.8	119	928.0	735	.792	63.2%	.623	32.64	
12	275%	14.80	14.8	1540	9.2	118	1085.0	890	.820	57.3%	.683	37.27	
Locked Rotor				0	29	108	3132	1300	.415				

continue the test to 275 percent of the motors rated load, if so stop at an earlier point. At the 275 rated load this specific motor was approaching its stalling point which is the breakdown torque part of the curve in Figures 9-9 and 9-11.

Column 4 is the actual value that is loaded and read on the scale. Column 5 is determined by reading the speed measuring device at the exact instant the estimated pound value is loaded on the scale. Values for columns 6, 7, and 9 are to be read on the electrical meters at the exact same instant as the rpm. Thus, several students are needed to perform all of these functions instantaneously. Column 8 is determined by multiplying the ampere value of column 6 times the volt value of column 7. Power factor (pf) column 10, is determined by dividing the column 9 value by the column 8 value. Power factor is defined as the ratio of the actual power used in the circuit, measured in watts, to the apparent power delivered by the utility company expressed as volt-amperes. To understand pf consider a soldering iron load and an electric motor load. The soldering iron load is converting the utility company power directly into heat or actual power. The actual power is equal to apparent power, so the pf is 1.0 or 100 percent. The formulas for resistance and motor loads would be:

Resistance Load, $W = V \times I \times 1.0$

$W = V \times I$

Motor Load, $W = V \times I \times pf$

With the pf of 1.0 this AC resistive load would be the same as if it were a DC circuit because the pf of 1.0 does not influence the results of the formula. In the single-phase AC electric motor circuit the actual power is the sum of several components: (1) work performed by the driven unit; (2) heat developed by the power lost in the motor windings; (3) heat developed in the iron of stator and rotor through eddy currents and hysteresis losses; (4) friction losses in bearings; (5) windage losses caused by the turning rotor; and, (6) the result of the inductive load. All of these actual power components are expressed in watts or kilowatts and can be measured with a wattmeter or kwhr meter when used as a wattmeter. Some common pf values are listed in Table 9-2. As the pf is being calculated for column 10 it will be noticed that the inductive electric motor is sensitive to motor loading. Motor loading with pf ratings are listed in Table 9-3 and are typical of modern medium sized motors of 1 to 20 hp.

The actual hp, column 12, needs to be calculated with this formula:

$$(12)\ \text{Hp} = \frac{2 \times 3.1416 \times r \times \text{rpm} \times \text{lbs.}}{33{,}000}$$

$$= \frac{r\ (\text{in feet}) \times \text{rpm} \times \text{lbs.}}{5252}$$

Table 9-2. Power Factor Values of Common Equipment

EQUIPMENT	POWER FACTOR %
Soldering Iron	100
Incandescent Lamp	100
Fluorescent Lamp*	50
Mercury Vapor*	40
Distribution Transformer	Very Low When Unloaded
Induction Motors	30-90
Synchronous Motors	90-100
Arc Welders	50-60
Industrial Heaters	90-100

*Will usually have correct ballasts or other devices added to unit.

Table 9-3, Motor Loading and Resulting Power Factor

LOADING %	POWER FACTOR
0	17
25	55
50	73
75	80
100	84
125	86

In theory there are 746 watts per hp. Efficiency is a ratio of inputs to outputs. Efficiency for column 11 is determined by two calculations:

$$(11a)\ \text{Watt Hp} = \frac{\text{Watt (col. 9)}}{746\ \text{Watt/Hp}}$$

$$\text{Efficiency} = \frac{\text{Actual Hp (col.12)}}{\text{Watt Hp}} \times 100$$

The torque in ounce-feet, column 13, is determined by:

$$(13)\ \underset{(\text{oz-ft})}{\text{Torque}} = \frac{\text{Hp (col 12)} \times 5252 \times 16}{\text{rpm (col 5)}}$$

The locked-rotor data can be collected and recorded at the bottom of the data table. The entire test data table can now be completed. Power and Ohm's Law formulas have been combined into one formula circle in Figure 9-18. This teaching aid may help as other problems are calculated in working with motors and electricity.

ANALYSIS OF DATA

The balance of this laboratory will be to graph the data, note Figure 9-17, and to analyze the data. Typical conclusions might be: (1) as rated load is increased the hp increases , rpm decrease, and efficiency increases; and, (2) as hp output increases, the wattage, amperes and torque values increase. The most significant points on the graph are to compare the 100 percent rated load to the motor nameplate data. The motor hp was 1/4 and data revealed 0.249 and the 1745 rpm is lower than rated rpm listed on the nameplate. The 56.3 percent efficiency is slightly lower than normal but the 0.535 or 53.5 percent pf is normal for the motor being tested. It must be remembered that the testing method described would not be the most accurate, therefore, do not become too ambitious with the conclusions but be content with the opportunity to teach basic motor and electrical principles.

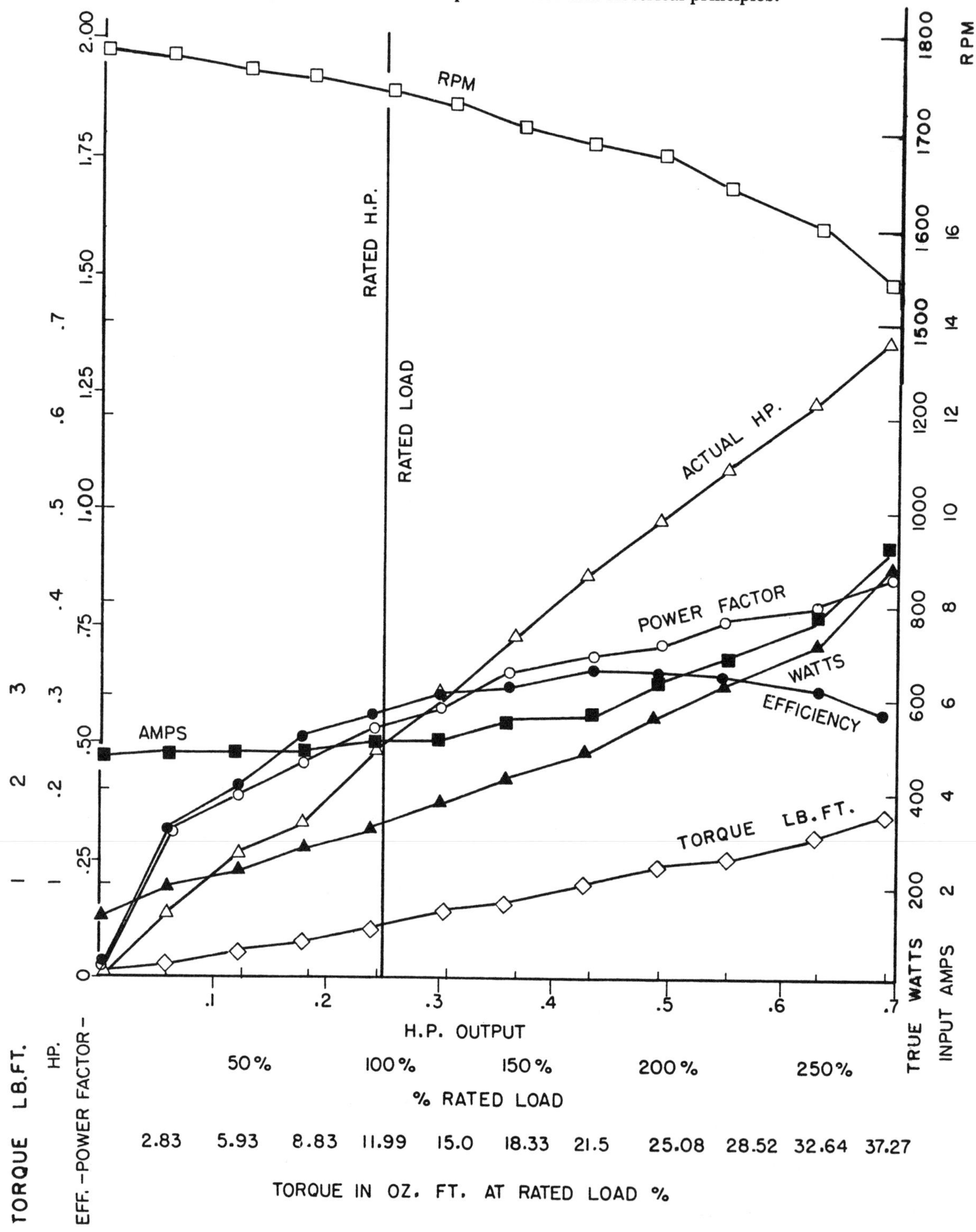

Fig. 9-17. Electric Motor Performance Testing Analysis

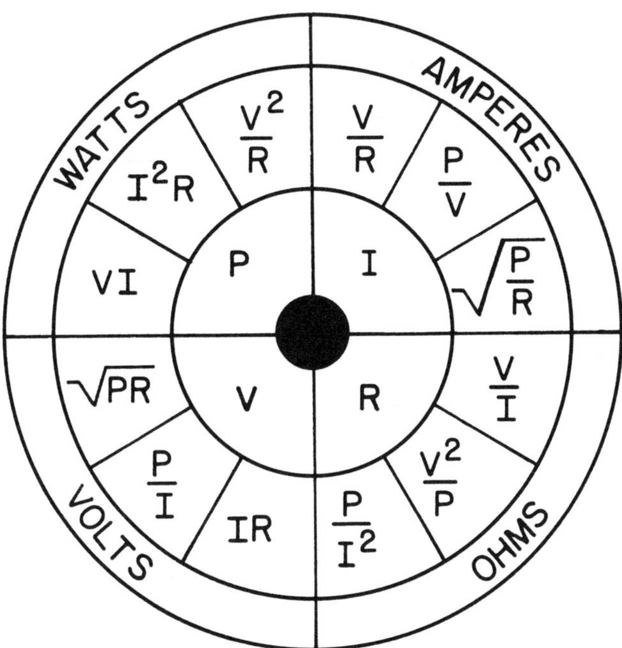

Fig. 9-18. Power and Ohm's Law for DC and AC Resistance Applications

The electric motor has features of speed, torque and hp which provide it with the capability to handle loads. If a specific motor does not meet these requirements a different size and/or type of motor can be selected. The engineer has other options as he studies the machine requirements because of the different type of rotor designs for squirrel cage rotors.

MOTOR PERFORMANCE AND DESIGN

The performance of a motor is determined by its speed, torque and hp characteristics. To manufacture squirrel-cage motors to fit the varying needs, the resistance and/or reactance of the rotor bars is changed. The resistance of a rotor bar can be increased by decreasing its cross-sectional area or by using higher resistivity material such as brass. The reactance of a rotor bar can be increased by placing the conductor deeper in the rotor cylinder or by closing the slot in the air gap. Locating the conductor deeper in the iron positions it further from the air gap between rotor and stator.

An increase in rotor bar resistance will: (1) increase starting torque; (2) lower starting current; (3) lower full-load speed; (4) lower efficiency; and (5) leave the breakdown torque unaffected. If the design engineer desires these characteristics the cross-sectional area of the rotor bars will be decreased or a rotor bar material will be used that has greater resistance.

An increase in rotor bar reactance will tend to: (1) lower starting torque; (2) lower starting current; (3) lower breakdown torque; and, (4) leave the full-load conditions unaffected. If the design engineer desires these characteristics, the rotor bars can be located deeper in the iron rotor cylinder or the slot at the air gap can be closed. The final design could be a compromise between the different possibilities. A design engineer might lower full-load speed by increasing rotor bar resistance and at the same time lower starting current by increasing rotor bar reactance. Three different rotor bar designs are illustrated in Figure 9-19. The results of the different rotor bar designs are listed in Table 9-4.

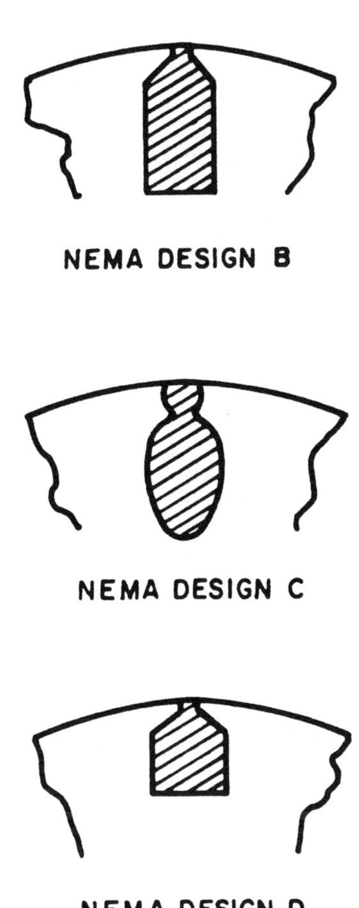

Fig. 9-19. NEMA Rotor Bar Designs

The design engineer can alter the performance of a motor by changing the rotor which results in different torque characteristics, slip, and starting current requirements. When buying a replacement motor it is best to follow all nameplate data and avoid trying to out guess the design engineer. To visualize the results of these designs, it is best to plot the speed-torque curve for the motor. The speed-torque curve example in Figure 9-20, is for a NEMA B design 10 hp squirrel-cage motor at 1200 rpm with a 45 pound-feet full-load torque.

The motor has a starting torque of 67.5 pound-feet which is 150 percent of the full-load torque, point (A). As the motor starts, the curve drops off as the motor picks up speed until about 20 percent of synchronous speed is achieved. The torque starts to increase and continues until about 75

Table 9-4. NEMA Rotor Bar Results

DESIGN	CROSS-SECTION SIZE OF BAR	POSITION IN ROTOR	STARTING TORQUE	STARTING CURRENT
B	Large, Low Resistance	Deep, High Resistance	Normal	Low
C Upper Area C Lower Area	Small Conductor, High Resistance Larger Conductor, Low Resistance	Top Position Lower Position	High	Low
D	Made of Brass, High Resistance	Shallow, Reduces Leakage Reactance	High	Low

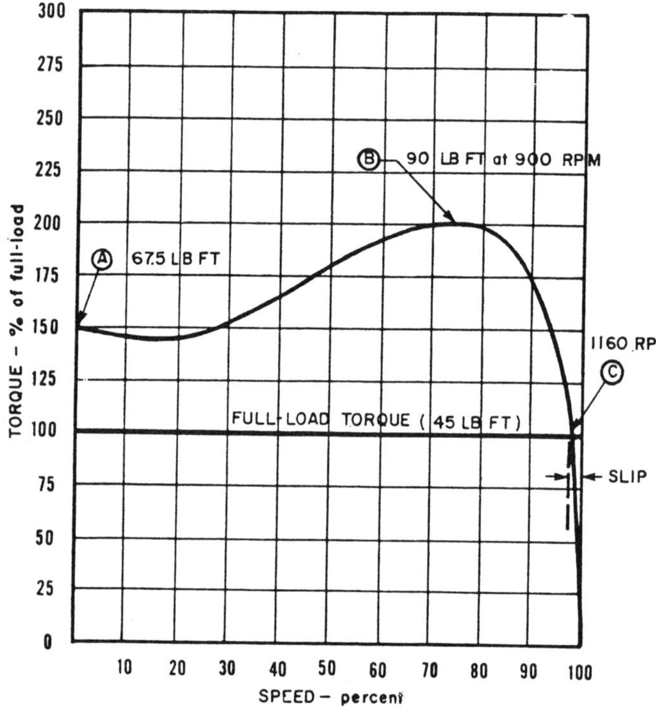

Reliance Electric Company

Fig. 9-20. Speed-Torque Curve 10 HP Squirrel-Cage Rotor 1200 RPM NEMA B Motor

percent of the speed is reached at point (B). The motor has now reached the maximum or breakdown torque. The breakdown torque is 200 percent of full-load torque or 90 pound-feet at a speed of 900 rpm, point (B). Beyond point (B), the torque decreases. The motor reaches the 100 percent torque line which is the full-load torque at 45 pound-feet at the full-load speed or 1160 rpm, note point (C). Now that the motor's speed-torque curve has been explained, the curve can be used to determine if the motor can handle a specific machine. The next step would be to determine the machine demands. Machine demands on the 10 hp, 1200 rpm NEMA B motor are shown in Figure 9-21.

Curves (1) and (2) represent machines and both machines require a full-load torque of 45 pound-feet and the motor has a 45 pound-feet full-load torque capability. Start at the zero speed line and follow curve (1). This machine requires a starting torque of 140 percent which is less than the 150 percent supplied by the motor and the motor will start the machine. Following curve (1) as the speed increases to full-load speed the motor produces torque greater than required by the machine, thus the motor can operate the machine.

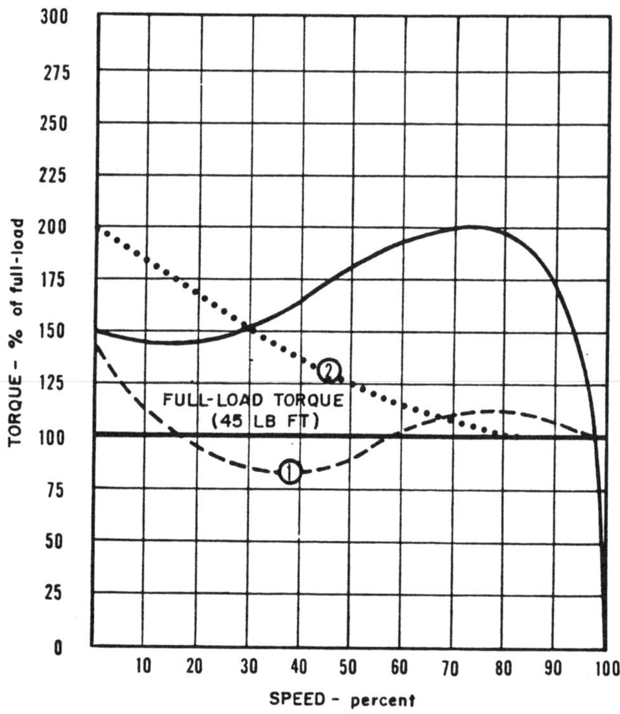

Reliance Electric Company

Fig. 9-21. Machine Demands on the 10 HP, 1200 RPM NEMA B Motor

Refer to a different machine, curve (2), to see if the motor will be satisfactory. Starting again at zero speed, the machine requires a starting torque of 200 percent which is greater than the motor's starting torque of 150 percent, and the motor cannot be used to operate the machine. To handle this load a different motor design is needed. Figure 9-22, has the speed-torque curve for the same size motor but it is a NEMA C design. Both the motor and machine have the full-load torque of 45 pound-feet. At the starting point, machine (2) requires 90 pound-feet starting torque and the motor, design C, produces 102 pound-feet. All of the torque points for machine (2) are less than the torque values produced by the motor and the NEMA design C will be able to meet the requirements of the load. There are five basic NEMA designs for AC motors; A, B, C, D and F and their speed-torque curves are shown in Figure 9-23.

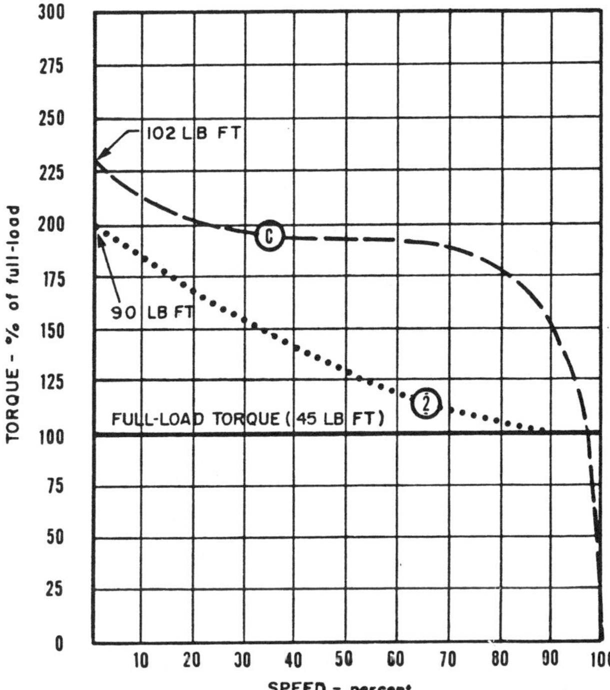

Reliance Electric Company

Fig. 9-22. Speed Torque Curve 10 HP Squirrel-Cage Rotor 1200 RPM NEMA C Motor

Follow this curve analysis in Figure 9-23. Motor designs "A" and "B" are similar but "C" has a higher starting torque than "A" or "B". The breakdown torque for "C" is lower than either "A" or "B", but it is rated 190 percent of full load torque. Design "D" motor has the highest starting torque, 280 percent, of the full-load torque. The curve indicates that the torque decreases all the way to full-load speed, and there is not an official breakdown torque. When curve "D" crosses the 100 percent torque line, the motor's slip or the amount the rotor lags behind the rotating magnetic field is identified; 90 percent of full speed or a 10 percent slip. Motor "D" has a higher slip than the other motors. The high starting torque for the "D" design motor makes it suited for hard-to-start loads. The slope of the speed-torque curve allows the motor to slow down during periods of peak loads, allowing any flywheel energy that has been stored by the load to be released. A machine requiring this type of motor would be a punch press or press brake.

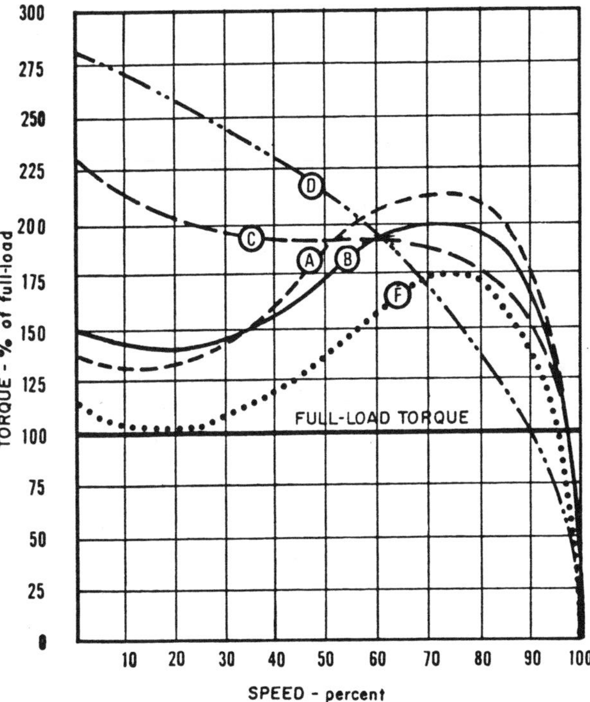

Reliance Electric Company

Fig. 9-23. Speed Torque Curves of NEMA A, B, C, D and F Motors

The NEMA design "F" has the lowest starting torque and the lowest breakdown torque. This motor would be used for easy starting loads such as fans and centrifugal pumps.

The three important torque values are starting, breakdown and full-load. These torque capabilities of AC motors have been standardized by NEMA A, B, C, D, and F. Thus when the machine's torque requirements are known, the motor can be selected to handle the load. As a consumer, this task of matching motor to machine will be taken care of by the design engineer, and you will trust that the motor will handle the load. This may not always be true if a manufacturer selects a marginal motor to reduce the selling cost for a machine. Unless given some reason for not trusting the motor match to the machine, one would purchase an identical replacement motor according to nameplate specifications.

SPEED REPRESENTED AS RPM

The speed-torque curves had the speed represented as a percentage of 100 percent. This curve can also be developed for a motor with speed represented as the actual rpm, note Figure 9-24, speed-torque curve for a 1200 rpm motor.

In Figure 9-24, the speed values are listed in rpm and torque values in percent of full-load torque. Assume that the torque developed at 500 rpm is desired. Find 500 rpm on the bottom scale and project a line upward to the motor's performance curve. Next, extend the line to the vertical axis, torque-percent of full-load, and the torque produced is 160 percent of the full-load torque. The curve shows how the motor performs from start-up to full speed operation. Push the start button and the motor's

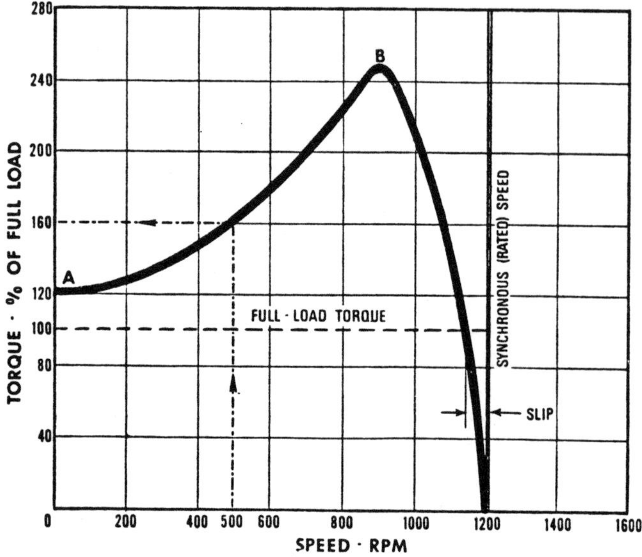

Reliance Electric Company

Fig. 9-24. Speed-Torque for a 1200 RPM Motor

starting-torque (A) is 120 percent of full-load torque. This torque value is what gets the load moving and this torque value can also be called the locked-rotor torque. The motor's torque increases to point (B) which is called the breakdown torque which is 250 percent of the full-load value. The torque decreases all the way to the motor's rated or synchronous speed. Remember, AC motors, (other than synchronous motors), never operate at synchronous speed because the rotor must lag slightly behind the rotating magnetic field by an amount called slip. This motor will operate at 1160 rpm and the percentage slip equals:

$$\text{Percentage slip} = \frac{(1200 - 1160)}{1200} \times 100$$

$$= 3.3 \text{ percent}$$

The selection of an electric motor for driving machines can be based on type, size, speed, torque, and design. Engineers conduct tests to determine the motor for the load application. Consumers will buy replacement motors which match nameplate data from the manufacturers standard motor stock. UNIT X, ELECTRICAL SERVICE AND CONTROL DEVICES will discuss motor electrical service, overload protection, switches and overload protection.

DEFINITION OF TERMS

AMMETER: Measures ampere flow. Meter is hooked up in series.

BREAKDOWN TORQUE: Maximum torque that a motor develops at rated voltage without an abrupt drop in speed or without stalling.

CAPACITANCE: The opposition to a change in voltage flow (See Definition of Terms, Unit VI).

EFFICIENCY: Output or work from a machine based on the energy input. Percent Efficiency = output/input x 100 or Percent Efficiency = Input - losses / input x 100.

FULL-LOAD TORQUE: Torque the motor will deliver continuously at rated voltage and rpm without exceeding its temperature rating.

INDUCTANCE: The opposition to a change in current flow (See Definition of Terms, Unit VI).

LOCKED-ROTOR TORQUE: Torque available at zero speed.

POWER FACTOR (pf): The ratio of the actual power used in the circuit (watts) to the apparent power delivered by the power company (V x I). Pf = Watt / VI.

REACTANCE: The opposition to AC as a result of either inductance or capacitance.

TORQUE: Turning effort and measured in ounce/feet or pound/feet. Torque = Force x Distance.

VOLTMETER: Measures voltage potential between two points. Meters are hooked up in parallel.

NOTES

CLASSROOM EXERCISE IX-A

Electric Motor Performance Testing Analysis

The performance curves for a 15 horsepower, three-phase electric motor are on the graph below. The titles to identify the curves need to be added; Note a, b, c, and d. Complete the questions below the graph.

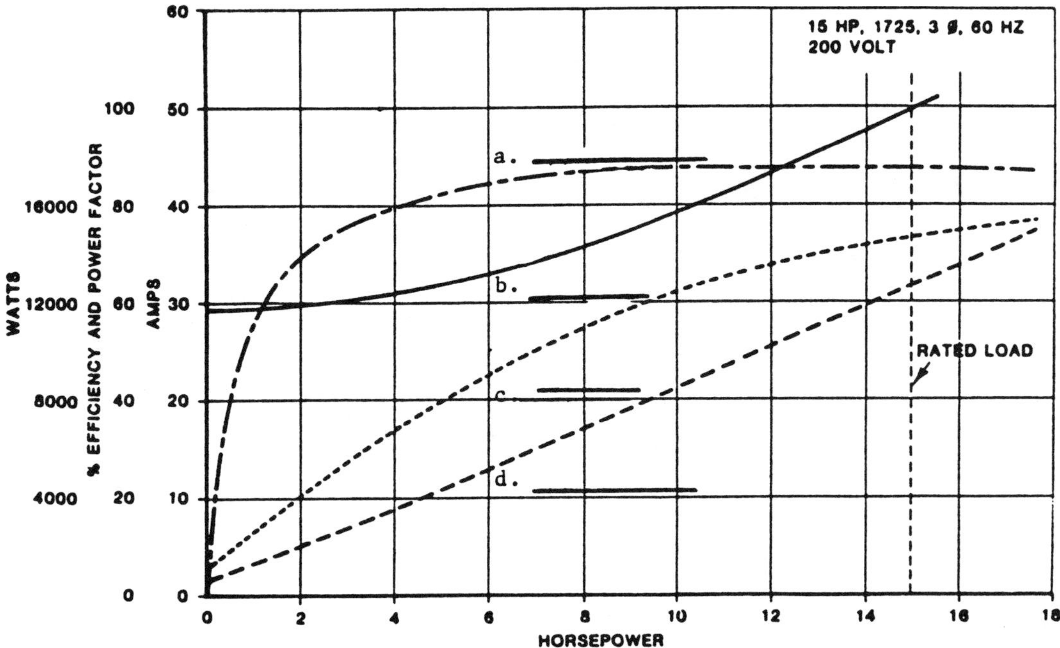

1. Define power factor (Pf):

2. What formula is used to determine power factor?

3. What power factor can be expected from:

   a. Transformer welders? ____________________

   b. Filament light bulbs? ____________________

   c. Electric motors? ____________________

   d. Hot plates? ____________________

4. Does overloading an electric motor inprove its efficiency? ____________________

5. Does overloading an electric motor inprove its power factor? ____________________

6. The efficiency of a motor is relatively constant from _______________ to _______________ percent of rated load.

CLASSROOM EXERCISE IX-B

Motor Characteristics

1. Rank the following electric motors on their load starting ability using easy-E, medium-M, and hard-H.

   a. Shaded Pole __________ d. Capacitor-Start __________ g. Permanent Split ________

   b. Split Phase ___________ e. Two-Value Capacitor _____ h. Three phase ___________

   c. Universal ____________ f. Repulsion ______________ i. Soft Start _____________

2. How many of these electric motors do you have at your farm or home and what are their application?

   a. Shaded Pole ____________________ f. Repulsion ____________________

   b. Split Phase _____________________ g. Permanent-Split ________________

   c. Universal ______________________ h. Three Phase __________________

   d. Capacitor Start __________________ i. Soft Start _____________________

3. What is a single-voltage motor? ____

4. What is a dual-voltage motor? ____

5. The following motor is not reversible ____________, and reversibility of the _____________ depends on the design.

6. Sketch the following gears used in a gearmotor.

   <u>Spur</u> <u>Helical</u> <u>Worm</u>

   Top View:

7. Which motor is easily adapted to solid state speed control? ____________________________
   ____

8. Which motors would have two capacitors? ____________________________________ and
   ____

9. Which electric motor does not have a special starting circuit?__________________________
   ____

10. Which motor is designed to be used for large horsepower loads when only a single-phase electrical service is available? ____

11. The _______________________________________ electric motor can be used on either AC or DC.

12. The ____________________ electric motor requires the greatest starting current and the two motors that require a low-starting current are_________________________and_______________________.

LABORATORY EXERCISES IX-A

Types of Motor Loads

Complete the following table on the types of loads for motors and an application location.

TYPE OF LOAD	MOTOR APPLICATION	TYPE OF MOTOR	HP RATING	PHASE AND VOLTS

1. An electric motor exerts a force of 1 lb. on a 10-inch lever arm. The torque in inch-pounds is ________________ and in inch-ounces, it would be ________________. Another electric motor was found to exert 10 lbs. on a 24-inch lever arm. The torque in inch-pounds is _____________ and in foot-pounds is ____.

2. The locked rotor is indicated by a KVA, which is a letter on the nameplate. The KVA value is listed per horsepower. Data were collected by testing several motors. Complete this table. Refer to Table 3-1 to determine the correct code letters.

MOTOR	HP SIZE	VOLTS	AMPERE	KVA	PER HP	CODE LETTER
1	1/4	110	30			
2	1/2	220	43.2			
3	1/2	115	28			
4	1/2	115	42			
5	1/2	115	28			

LABORATORY EXERCISE IX-B

Voltage Drop

Voltage drop is wasted power and an electric service operating with 90 percent of its rated volts, a motor produces 81 percent of its normal power and a lamp produces only 70 percent of its light. Refer to Fig. 9-16 for a technique for loading an electric motor. Select a 1/4 or 1/3 hp split phase or capacitor-start motor operating on 120 volts. Because of intentional overloading of the motor, a Service Factor (SF) of 1.15 or better is recommended. Use a clamp-on ammeter and voltmeter for obtaining data. For conductors, use 100 feet (min.) to 200 (max.) of AWG #12 and AWG #18. Please remember, motors should never be connected to any wire size less than #12. However, for teaching the concept of voltage drop, an AWG #18 is to be used. CAUTION -- watch carfully for excessive overheating of the conductors when the load is held for a long period of time.

Source Voltage ______________

% MOTOR LOAD, AMPS	LOAD VOLT READING		CALCULATED VOLTAGE DROP**	
	AWG #12	AWG #18	AWG #12	AWG #18
1. *				
2. 100				
3. 125				
4. 150				

*First reading without any mechanical loading. Readings 2, 3, and 4 to be a percentage of the the motor nameplate.

**Percent Voltage Drop = $\frac{\text{Source Voltage - Load Voltage}}{\text{Source Voltage}} \times 100$

1. Voltage drop is a combination of three factors:
   a. ______________ b. ______________ c. ______________
2. What percentage voltage drops are considered permissible?
   a. ______________ b. ______________ c. ______________
3. What voltage drop percentage is the most desirable if electrical energy cost is high? ________%
4. What other cost factors must be considered as plans are made to operate with a lower percentage voltage drop? ______________ ______________
5. Assume you have a 10 amp load on a 125 foot length run with TW copper conductor in conduit. Determine the permissible conductor sizes for the percentage voltage drop.

VOLTAGE DROP	VOLTS	MINIMUM SIZE	SIZE FOR LENGTH	FINAL SIZE
2%	120	________	________	________
3%	120	________	________	________
4%	120	________	________	________
2%	240	________	________	________

LABORATORY EXERCISE IX-C

Using the Kwhr Meter to Measure Energy Consumption

The kilowatt hour meter is a power measuring tool which can be used to determine the watts consumed to calculate voltage drop. In this exercise, use the same conductors (AWG #12 and AWG #18) as you did in Laboratory Exercise IX-B. Use several heater cones to obtain a simulated load of 1000 to 1500 watts for this exercise rather than the electric motor. The kilowatt hour meter will measure the watts consumed by the load: operate the load for 5 to 10 revolutions of the kwhr meter disc in each trial. The equipment set-up is as follows with 120 volt, parallel circuit.

1. SOURCE -> KWHR METER -> CONDUCTOR -> LOAD
2. SOURCE -> CONDUCTOR -> KWHR METER -> LOAD

Disc Constant Kh Value _____ STEPS	AWG #12	AWG #18
1. Meter positioned at power source ahead of conductor and load.		
a. Volt reading at power source	__________	__________
b. Load on the circuit	_____ min. _____ sec.	_____ min. _____ sec.
c. Meter disc revolutions	__________	__________
d. Revolutions per minute	__________	__________
e. Power in watt hours = (RPM x Kh constant x 60 min. per hr.)	__________	__________
2. Meter positioned between conductor and load.		
a. Volt reading at load	__________	__________
b. Load on the circuit	_____ min. _____ sec.	_____ min. _____ sec.
c. Meter disc revolutions	__________	__________
d. Revolutions per minute	__________	__________
e. Power in watt hours = (RPM x Kh constant x 60 min. per hr.)	__________	__________
Watts lost per hour because of wire length 1E - 2E.	__________	__________
Voltage Drop % = $\frac{\text{Source V} - \text{Load V}}{\text{Source V}} \times 100$	__________	__________

LABORATORY EXERCISE IX-D

Determining Power Factor by Using the Kwhr Meter

The power factor (pf) is a unique phenomenon found with induction-run motors. Resistance loads, such as heating and lighting, have a unity power factor of 1.0. With data collected in this exercise, an understanding of pf can be appreciated. Use an induction-run motor of 1/4 or 1/3 hp. Load it to 100 percent ampere rating per nameplate data using the clamp-on style ammeter. Use a loading technique as shown in Fig. 9-16, or any other manner which will maintain a load for the time required for determining watts with the kilowatt hour meter. Voltmeter and ammeter readings need to be taken simultaneously when using the kilowatt hour meter. Remember, inaccurate instruments (equipment error) and inaccurate interpolation of the readings (human error) are major reasons for obtaining results somewhat different than expected.

Kwhr Meter
Disc Constant
Value __________

Power Factor = $\frac{\text{True Watts (Kwhr Meter)}}{\text{Apparent Watts (V X I)}}$

ITEM	INDUCTION-RUN ELECTRIC MOTOR	HEATER CONES
a. Ampere Reading	__________	__________
b. Volt Reading	__________	__________
c. Apparent Watts (a x b)	__________	__________
d. Time of Load on Circuit	_____ Min. _____ Sec.	_____ Min. _____ Sec.
e. Meter Disc Revolutions	__________	__________
f. Revolutions Per Minute = 60 Sec. Per Min. x Sec. (d value) x Disc Revolutions	__________	__________
g. True Watts (Kwhr Meter) = RPM (f value) x Kh Disc Constant x 60 Min. Per Hour	__________	__________
h. Power Factor = True Watts ÷ Apparent Watts	__________	__________

LABORATORY EXERCISE IX-E

Computing Horsepower and Efficiency of a Motor

With a commercial dynamometer or motor loading technique as illustrated in Fig. 9-16 and using the Hp formula on page 127, determine the Hp output of a 1/4 or 1/3 split phase or capacitor-start motor. Operate the motor at 100 percent of rated ampere load per nameplate data when mechanically loading the motor. Make amperage checks using a clamp-on ammeter.

STEP 1. Determine Output Horsepower Ampere

Value of __________ is 100 percent of the nameplate data for the motor.

$$Hp = \frac{r\ (\text{in ft.}) \times RPM \times lbs.}{5252}$$

$$Hp = \frac{\_\_\_\_ \times \_\_\_\_ \times \_\_\_\_}{5252} = \_\_\_\_\_\_\_\_\ \text{Output Hp}$$

STEP 2. Determine Input Horsepower

With the Kwhr meter wired in the service feed, load the motor to draw the same amperage as in Step 1, and calculate the input wattage when using the kilowatt hour meter as a wattmeter to arrive at true wattage consumption.

Watts = RPM of meter disc x Kh Disc Constant x 60
Watts = __________ x __________ x __________ = Input Wattage
Input Wattage of __________ ÷ 746 watts per Hp = __________ Input Hp

STEP 3. Determine the Motor Efficiency

Compare the ratio of the Input Horsepower to the Output Horsepower.

From Step 1. Output Horsepower = __________
From Step 2. Input Horsepower = __________

$$\text{Efficiency} = \frac{\text{Output Hp}}{\text{Input Hp}} \times 100 = \_\_\_\_\_\_\_\_\ \%$$

STEP 4. Determine Motor Efficiency at Other Loads

Determine the eficiency at 90 and 125 percent of rated load. Place all data in the following table and then answer the questions.

STEPS	90%	125%	100% VALUES
Step 1 Hp Output			
Step 2 Hp Input			
Step 3 Efficiency			

QUESTIONS:

1. Is there a correlation between amperes drawn and output Hp? ____________________
2. Did the efficiency of the motor go up or down when the motor load was increased? __________

NOTES

ELECTRICAL SERVICE AND CONTROL DEVICES

Before an electric motor can be used, it must be in an electrical circuit which is called a branch circuit. The branch circuit must have overcurrent protection to protect the conductors against short circuits or ground currents. There must be a disconnecting means within 15 feet of the motor. Overcurrent protection must also be available to prevent overloading the motor under running conditions and a controller device to start and stop the motor. These circuit essentials can be provided individually or all four can be combined into one piece of equipment.

ELECTRIC MOTOR CIRCUIT

SIZE OF CONDUCTOR

Branch circuit conductors to an individual motor should be selected on the basis of carrying 125 percent of the full load current of the motor. The determination of single-phase AC motor currents is listed in Table 10-1 and the current values for three-phase AC motors are listed in Table 10-2.

Table 10-1. Single-Phase AC Motor Currents

	115 Volts		230 Volts	
Motor HP	Full Load (Amps)	125% Full Load (Amps)	Full Load (Amps)	125% Full Load (Amps)
1/6	4.4	5.5	2.2	2.8
1/4	5.8	7.2	2.9	3.6
1/3	7.2	9.0	3.6	4.5
1/2	9.8	12.2	4.9	6.1
3/4	13.8	17.2	6.9	8.6
1	16.0	20.0	8.0	10.0
1-1/2	20.0	25.0	10.0	12.5
2	24.0	30.0	12.0	15.0
3	34.0	42.0	17.0	21.0
5	56.0	70.0	28.0	35.0
7-1/2	80.0	-----	40.0	50.0
10	100.0	-----	50.0	62.0

(Reference Table 430-150. NEC)

If there is more than one motor on the circuit, the conductor size is to be determined by taking 125 percent of the full-load current of the largest motor and 100 percent for all other motors. After the motor current requirement has been determined, the correct size conductor can be selected. The allowable size conductor is determined by six factors: (1) type of conductor; (2) basic load in amperes the conductor can carry; (3) type of insulation on the conductor; (4) whether the conductor is in cable, conduit, the earth or overhead in air; (5) length of run in feet; and (6) acceptable voltage drop percentage. A composite of some of these factors is illustrated in Table 10-3, Recommended Conductor Sizes. In Table 10-3 the increased size of conductor by amperage load and distance is listed. The size of conductor can be reduced as a greater voltage drop is allowed. Aluminum conductors require about one size larger than copper to carry the same ampere load. This is an abbreviated table to illustrate a concept, therefore, consult the appendix for complete tables listing ampere loads and lengths of run values.

RUNNING OVERCURRENT PROTECTION

The motor overcurrent protection is to prevent overloading the motor under running conditions. The selection of overcurrent devices is based on the motor nameplate rating. For motors over one hp, a separate overcurrent protector with a maximum of 115 percent of rated load should be used when the service factor (SF) is under 1.15. If the motor SF is greater than 1.15 and the motor is marked for a 40 degree C temperature rise there should be a separate overcurrent device rated or set at 125 percent of the full load current.

Any motor of one hp or less that is manually started and is in sight from the starter is considered protected by the branch circuit overcurrent protection. If the the motor is out of sight, one hp or less, and is automatically started the overcurrent protection can be one of the following: (1) 125 percent of the full load current for motors marked for 40 degree C rise or SF of 1.15 or more; and (2) overcurrent protection of 115 percent for all other motors. If the one hp or less motor is part of an approved assembly where the motor is not subjected to motor overload (example-safety combustion controls of an oil burner) or if the impedance of the windings is sufficient to prevent overheating (when the motor is stalled) such as clock motors and others of less than 1/20 hp the motor is considered protected by the branch circuit.

When selecting the protective device where the value specified for motor running overcurrent protection does not correspond to standard sizes of protective devices available, use the next higher size but not over 140 percent of full-load current for motors with a temperature rise of 40 degree C, sealed compressor motors, and 130 percent for all others. If a thermal device, in the motor, switch or controller, is selected and it is not capable of opening short circuits, it must be protected by fuses or circuit breakers of correct values. Most manually operated or magnetic type motor controllers making use of a heater device to trip the mechanism on overload do not function fast enough to protect against short circuits.

Table 10-2. Three-Phase AC Motor Currents

	208 VOLTS				230 VOLTS	
	USE CURRENT VALUES BELOW FOR DETERMINING MIN. CONDUCTOR AMPACITY*		USE CURRENT VALUES BELOW FOR SELECTING CONDUCTORS USING 230-240 VOLT TABLES FOR VOLTAGE DROP**			
MOTOR HP	FULL LOAD (AMPS)	125% FULL LOAD (AMPS)	VOLTAGE DROP FULL LOAD (AMPS)	VOLTAGE DROP 125% FULL LOAD (AMPS)	FULL LOAD (AMPS)	125% FULL LOAD (AMPS)
	COL. 1	COL. 2	COL. 3	COL. 4	COL. 5	COL. 6
1/2	2.2	2.8	2.5	3.2	2.0	2.5
3/4	3.1	3.9	3.6	4.4	2.8	3.5
1	4.0	5.0	4.6	5.7	3.6	4.5
1-1/2	5.7	7.1	6.6	8.2	5.2	6.5
2	7.5	9.4	8.6	11.0	6.8	8.5
3	11.0	14.0	12.0	15.0	9.6	12.0
5	17.0	21.0	19.0	24.0	15.2	19.0
7-1/2	24.0	30.0	28.0	35.0	22.0	28.0
10	31.0	39.0	36.0	44.0	28.0	35.0
15	46.0	58.0	53.0	67.0	42.0	52.0
20	59.0	74.0	68.0	86.0	54.0	68.0
25	75.0	94.0	86.0	108.0	68.0	85.0
30	88.0	110.0	102.0	128.0	80.0	100.0
40	114.0	142.0	132.0	165.0	104.0	130.0
50	143.0	179.0	165.0	206.0	130.0	162.0

(Reference Table 430-148, NEC)

* Use these values of current only for determining the minimum conductor ampacity.

** Use these values in Cols. 3 and 4 for selecting conductor size in relation to voltage drop. If motor current exceeds value in Col. 1, determine values for Col. 3 by multiplying motor current by 1.15, and values for Col. 4 by multiplying motor current by 1.44.

*Table 10-3. Recommended Conductor Sizes for Specific Applications *1*

115/120 VOLTS LOAD IN AMPS	TYPES R, T, TW INSUL. MIN. AWG. SIZE CONDUCTOR	50 FT. *3 2% VD		100 FT. 2% VD		50 FT. 3% VD	50 FT. 4% VD
		CU	AL	CU	AL	CU	CU
15 *2	12 *2	10	8	8	6	12	12
20	12	10	8	7	4	12	12
25	10	8	6	7	4	10	12
30	10						
35	8	8	6	4	3	8	10
45	6	6	4	4	2	8	10
60	4	4	3	2	0	6	8
80	2	4	2	1	00	6	6
100	1	3	1	0	000	4	6

ACTUAL SIZE OF COPPER CONDUCTORS: 2/0 1/0 2 4 6 8 10 12 14 16 18

*1 Consult Appendix for complete table.
*2 AWG size #14 conductors will carry 15 amp. For motors No. 12 is the smallest permitted by NEC.
*3 VD = Acceptable voltage drop permitted in percent.

CIRCUIT OVERCURRENT PROTECTION

Branch circuit overcurrent devices are selected with the basis of full-load current values given in Table 10-1 and 10-2. The overcurrent device shall have sufficient current rating or time delay to permit the motor to start and accelerate its load. The National Electrical Code specifies maximum rating of these devices as a percent of the full-load current of the motor. If a time-delay fuse is used, the size can be reduced. If two or more motors are on one branch circuit, each less than 1 hp and each has a full-load current not over 6 ampere (A), they may be on a circuit protected at not over 20 A at 125 volt (V) or less or 15 A at 600 V or less. Two or more motors of any rating, each with individual overcurrent protection, may be connected to one branch circuit if each motor running overcurrent device and controller is approved for group installation. The branch circuit must be protected by fuses with a maximum rating as required by the NEC for the largest motor plus 100 percent of the full-load current of other motors but not exceeding 400 percent of the rating of the smallest motor in the group. The branch circuit overcurrent protection and the motor running overcurrent protection may be combined in one device if all protection rules have been met.

OVERCURRENT DEVICES

Different types of overcurrent devices are illustrated in Figure 10-1. Overcurrent devices can be plug or cartridge fuses, standard circuit breakers or circuit breakers with a ground fault circuit interrupter (GFCI), note Figure 10-1. There are two types of plug fuses, time-delay and non-time delay. The non-time delay, single element, plug fuse is A-1, and the time-delay plug fuses are A-2 and A-3 in Figure 10-1. There are two types of time-delay plug fuses, Fusetron (TM) A-2 and Fustat (TM) A-3. The Fusetron fits in the regular plug fuse receptacle (Edison Base) but the Fustat has a special non-tamperable base which is sized to fit a maximum ampere rating of the base. The cartridge fuses can be either the single element or the time-delay type, B-1 and B-2. The cartridge fuse, B-3 with the knife type connections is for large electrical loads. The regular circuit breaker is C-1 and the GFCI style circuit breaker is illustrated in C-2. Overcurrent device characteristics can be reviewed in Table 10-4. The single element fuse is not designed to provide overload protection. The basic function it performs is protection against short circuits, also called overcurrents. An overload is a damaging flow of current which occurs over a period of time. Since a motor draws a

[a] *Plug Fuses*

[b] *Cartridge Fuses*

[c] *Circuit Breakers*

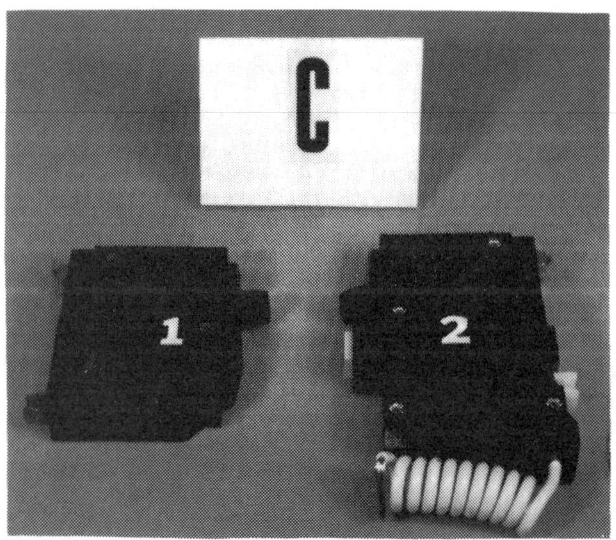

Fig. 10-1. Types of Overcurrent Devices

high amount of current in starting, 2 to 7 times the full-load current, a single element fuse to handle this load might not protect the motor against an overload. Dual element or time delay fuses and circuit breakers provide for the high starting current and may also provide motor overload protection; however, in most applications separate motor overload protection devices are needed.

As indicated earlier, overcurrent protection devices are designed to protect the circuit and its components including equipment operated in the circuit. The question could be asked in regard to how well an overcurrent device protects a circuit. The major factor is the time to trip or open the circuit. Data in Table 10-4 reveal the time it takes to trip an overcurrent device based on the percent rated current. Further, a comparison is shown between the typical time-delay and non-time delay devices. For example, if a circuit is designed and fused for a 15 A load and a load of 15 A is applied to the circuit, neither the time-delay or the non-time delay device will trip. A 15 A load on a 15 A circuit would load the circuit to 100 percent of its rated load. Using Table 10-4, locate 100 percent on the base line and follow the line up the graph. Note that the line does not cross either of the fuse lines. Solve this example. The previous 15 A circuit has a load of 45 A applied to the circuit. Forty-five A is 300 percent rated current for a 15 A circuit; therefore, locate 300 on the base line and follow it up to the curved line. Trace across to the vertical line at the left and read the seconds to open the circuit. If fused with a standard fuse, non-time delay, the fuse would trip in approximately 0.2 seconds. If fused with a time-delay type fuse, the circuit would open in 50 seconds.

A short circuit might be considered to load a circuit at 10 times its rated ampacity. If a load of 150 A were applied to a circuit having a rated ampacity of 15 A (150 A divided by 15 A times 100 equals 1000 percent rated current), the typical non-time delay fuse would blow in 0.011 seconds. A time-delay device would trip or open the circuit in 0.4 seconds under the same short circuit.

NONAUTOMATIC DEVICES FOR CONTROLLING MOTORS

SWITCHES WITHOUT OVERLOAD PROTECTION

The simplest control device is an attachment plug on the end of the conductor which is attached to the motor. This technique for motor control is impractical. Switches can be very small or large and may or may not provide overload protection but each will have a volt and ampacity rating which must be matched to starting loads. Figure 10-2 illustrates some of the pushbutton and toggle switches available.

Table 10-4. Overcurrent Device Characteristics

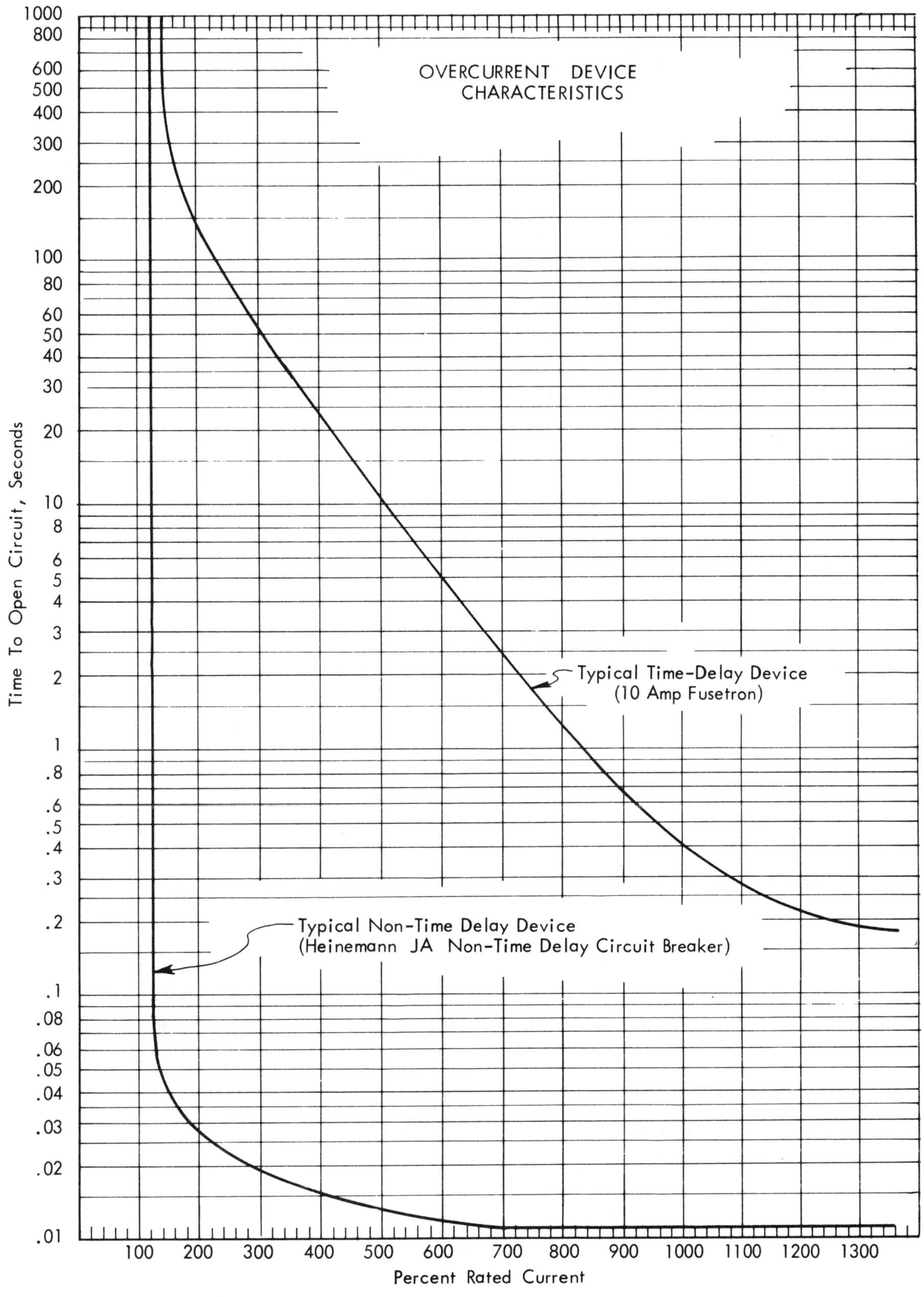

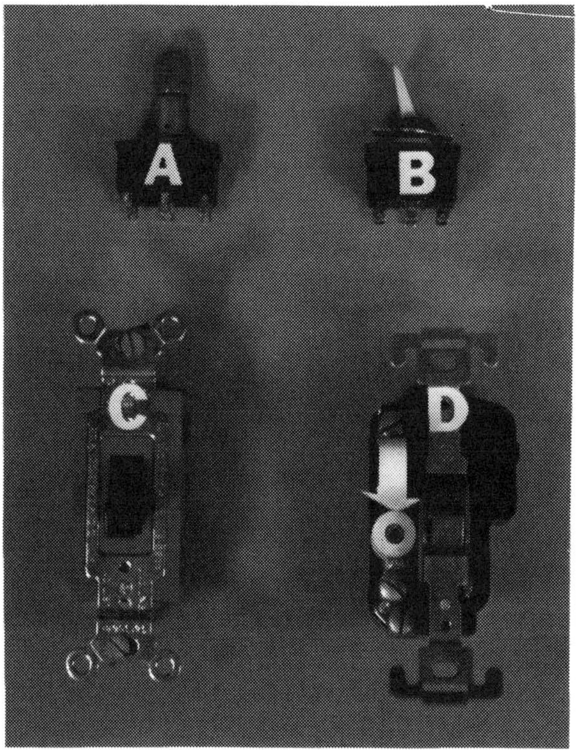

Fig. 10-2. Pushbutton and Toggle Switches

The pushbutton switch (a) toggle switch (b) and snap-action switch (c), in Figure 10-2, do not provide overload protection for the motor. The snap-action switch can have an overload device which will help protect fractional hp single-phase motors, Figure 10-2, (d). These switches are single pole, single throw with labels indicating the on and off position. They are always located on the "hot" conductor side of the service line because the ground or neutral side should never be switched or fused for safety reasons, as required by NEC. Schematic symbols for the SPST and DPDT switches are illustrated in Figure 10-3. The schematic symbol for the SPST (a) switch is used for 120 V. On a 240 V circuit which has two "hot" conductors, a double pole single throw switch, DPST, must be used for safety reasons, Figure 10-3, (b).

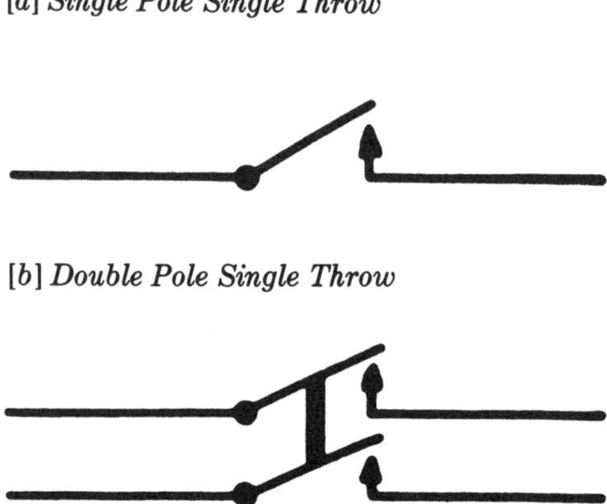

Figure 10-3. Schematics of a SPST and DPST Switches

The schematic for SPST switches, without (a) and with (b) overload protection is sketched in Figure 10-4. The switch nameplate lists the A and V ratings for the switches as detailed in Figure 10-5. The switch without overload protection (a) is used commonly on small fractional hp motors with low starting ampacities, (e.g. small fans). The switch with the overload protection (b) in Figure 10-5, is used on fractional hp motors having a starting ampere draw which is 2 to 7 times the running amperes.

[a] Without Overload Protection

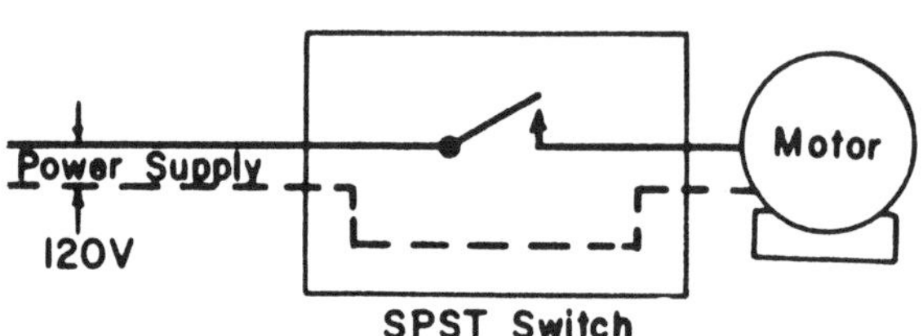

[b] With Overload Protections

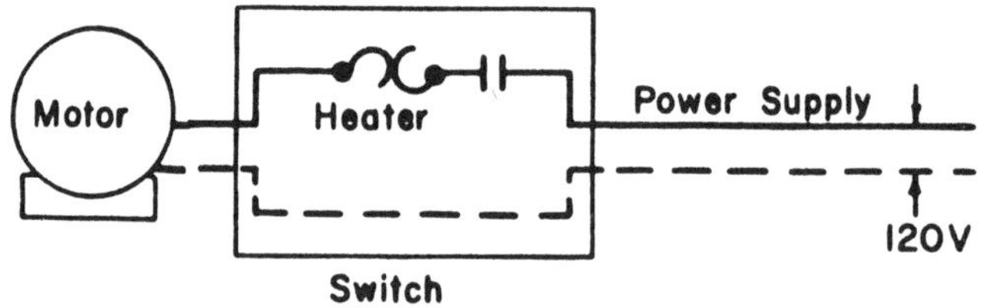

Figure 10-4. SPST Without and With Overload Protection

SWITCHES WITH OVERLOAD PROTECTION

A typical ampere and voltage rating for a SPST switch is shown in Figure 10-5, (a), and a switch with overload protection is illustrated in (b). The snap action switch, (b), has a thermal overload device. The switch cannot be held closed under a sustained motor overload. To reset the overload mechanism, the switch lever is moved to the "OFF" position and then the motor can be restarted by pushing the switch lever to the "ON" position. This type of a switch provides overload protection for small single-phase motors on heaters, stokers, refrigeration compressors, pumps and similar applications. Switches of this type can have a pilot light on the front. If the switch is used to control the motor directly the light indicates the motor is running. If, however, the motor is being controlled by a sensing device (thermostat, humidistat, pressure switch, etc.) the light indicates whether the electrical service is "ON" or "OFF". Switches are designed for special purposes and according to NEMA specifications; (a) Type 1 for general purpose enclosures and surface mount; (b) Type 4 for watertight enclosures; (c) Type 7 and 9 enclosures for hazardous locations; and (d) Type 1 for general purpose enclosures and flush mountings.

[a] *SPST Switch, 15 A and 120 V*

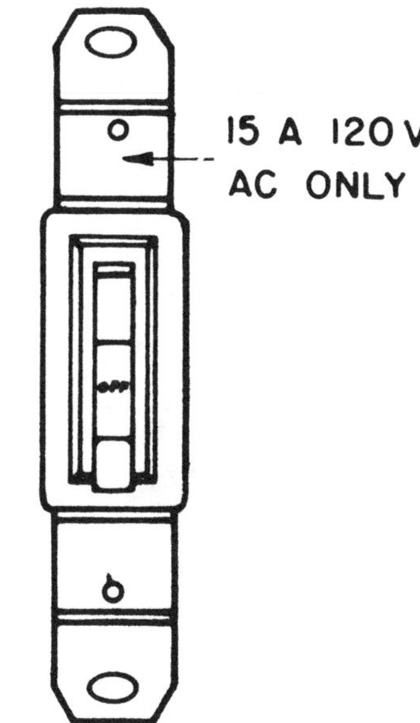

[b] *SPST Switch, 1 hp and 115 V*

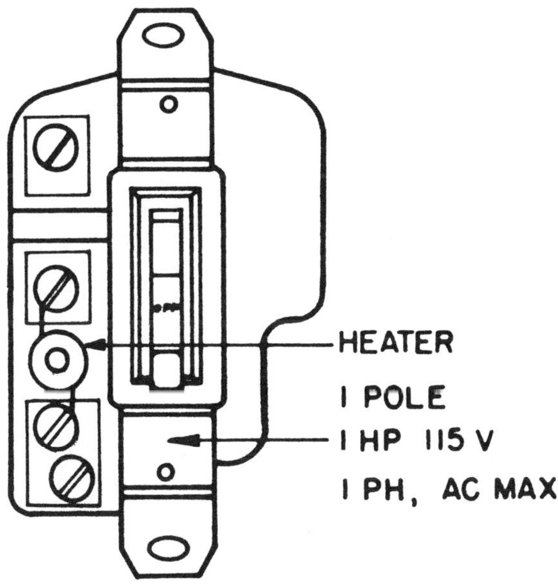

Fig. 10-5 Nameplate Data on Switches

Selecting overload protection for a motor depends on several factors, such as; motor characteristics, type of starter and the starter enclosure. The heater element for the switch must be matched to the motor nameplate full load current and for the other set of conditions. Thus, a specific heater element can be recommended for different full load current ratings in a different condition. Electric motors can be operated in an environment with the same temperature at the controller as at the motor, at a higher temperature at the controller than at the motor, and at a lower temperature at the controller than at the motor. Refer to Table 10-5, for heater element selection requirements for motors and Table 10-6 for heater element type number selection.

Apply the regulations from Table 10-5, for a motor with a service factor of 1.25 and an amperage rating of 5.75 amperes, and determine the correct heater type number in Table 10-6. If there is the same temperature at the motor as at the controller the P 30 heater element would be selected and provide protection between 110 and 120 percent. If there is a higher temperature at the controller than the motor the P 30 heater element would also be selected and provide protection between 115 and 125 percent. If there is a lower temperature at the controller than the motor the P 29 heater element would be selected and provide protection between 105 and 115 percent. If however, the motor would have been rated for continuous duty with a service factor of 1.0 the correct heater element would be a P 28. These heater elements will protect the electric motor for its operating condition and the same heater element can satisfy two different conditions. The overcurrent protection for the branch circuit will not protect the motor unless the motor has the same ampere rating required as the branch circuit conductors which would be an unusual situation. Always match the heater element to provide the motor overcurrent protection and the circuit fuse or circuit breaker protects the current carrying conductors. The devices so far have been called switches. The next devices that can be placed in the circuit to control motors are called controllers.

Table 10-5. Heater Element Selection Requirements for Motors

MOTORS RATED FOR CONTINUOUS DUTY:

MOTORS WITH MARKED SERVICE FACTOR OF NOT LESS THAN 1.15, OR MOTORS WITH A MARKED TEMPERATURE RISE NOT OVER 40°C.

1. **The Same Temperature at the Controller and the Motor** — Select the "Heater Type No." with the listed "Full Load Amps." nearest the full load value shown on the motor nameplate. This will provide integral horsepower motors with protection between 110 and 120% (between 115 and 125% for Type WL elements) of the nameplate full load currents.

2. **Higher Temperature at the Controller than at the Motor** ◼ — If the full load current value shown on the motor nameplate is between the listed "Full Load Amps.", select the "Heater Type No." with the higher value. This will provide integral horsepower motors with protection between 115 and 125% (between 110 and 120% for Type WL elements) of the nameplate full load currents.

3. **Lower Temperature at the Controller than at the Motor** ◼ — If the full load current value shown on the motor nameplate is between the listed "Full Load Amps.", select the "Heater Type No." with the lower value. This will provide integral horsepower motors with protection between 105 and 115% (between 120 and 130% for Type WL elements) of the nameplate full load currents.

ALL OTHER MOTORS RATED FOR CONTINUOUS DUTY (INCLUDES MOTORS WITH MARKED SERVICE FACTOR OF 1.0):

Select the "Heater Type No." one rating smaller than determined by the rules in paragraphs 1, 2 and 3. This will provide protection at current levels 10% lower than indicated above.

"Courtesy of Allen-Bradley Company—form supplied with Bulletin 600, Manual Switches, size 1 Hp in NEMA Type 1 enclosure."

Table 10-6. Heater Element Type Numbers

Heater Type No.	Full Load Amps.	Heater Type No.	Full Load Amps.
P1	0.17	P21	2.58
P2	0.21	P22	2.92
P3	0.25	P23	3.09
P4	0.32	P24	3.32
P5	0.39	P25	3.77
P6	0.46	P26	4.16
P7	0.57	P27	4.51
P8	0.71	P28	4.93
P9	0.79	P29	5.43
P10	0.87	P30	6.03
P11	0.98	P31	6.83
P12	1.08	P32	7.72
P13	1.19	P33	8.24
P14	1.30	P34	8.90
P15	1.43	P35	9.60
P16	1.58	P36	10.80
P17	1.75	P37	12.00
P18	1.88	P38	13.50
P19	2.13	P39	15.20
P20	2.40		

"Courtesy of Allen-Bradley Company—form supplied with Bulletin 600 Manual Switches, size P 1 - P 39 heater elements."

Fig. 10-6. Manual Motor Starter

MOTOR CONTROLLERS

Motor controllers are classified in two general types:

(1) Manual-Snap action drive switching used for up to 10 Hp motor sizes, and

(2) Magnetic-Electromagnetic drive switching used on all hp motors.

Some manual motor starters are switches which provide only an "on or off" control of single- or three-phase AC motors when overload protection is not required or when protection is provided separately, see Figure 10-6.

Other manual motor starters are called integral hp manual starters and these differ from the manual because they have overload relays capable of standing overcurrents and can unlatch the contact to "trip" and "break" current carrying conductors. On the integral hp manual starter when the start button, Figure 10-7, is pushed, this thrust causes the heavy duty snap-action toggle to join current carrying contacts. This controller, unlike the manual type, can "trip" and will stay "tripped" until the reset button is pushed (on some units the stop button must be pushed) to reset for further motor operation. Both the manual starter and integral jp manual starter are designed by most manufacturers to handle ampacities for single-phase motors up to 5 hp and three-phase motors up to 10 hp. These starters are also excellent switching and/or protectors for large resistant heating and lighting circuits of appropriate ampacities. Magnetic motor starters, may be designed with larger contacts and can handle any size of motor, as long as the ampacity rating of the switching device meets the ampacity characteristic of the motor being controlled. Obviously larger motors require large contactors in the starter and smaller motors may have smaller contactors. Generally a motor controller or starter must have a hp rating at or more than the motor it controls. There are exceptions to this rule and the following sections of the National Electrical Code should be consulted; 430-81 430-83, and 430-84. The controller need not open all the

Figure 10-7. Integral Horsepower Manual Starter

ungrounded conductors to a motor unless it is also used as the disconnecting means. This rule permits a single-pole pressure switch to start and stop a 230 V pump when a double pole disconnect is used in a circuit. The motor and its driven machine must be in sight from the controller unless; (a) the controller or its disconnect can be locked in the open position or (b) a manually operated switch to prevent starting is in sight from the motor. A distance of over 50 foot is considered out of sight by the NEC.

The operating principle which distinguished a magnetic controller from a manual controller is the use of an electromagnet, thus the name, magnetic starter. The electromagnet consists of a coil of insulated wire placed around an iron core, Note Figure 10-8 (b). When current flows through this coil, the iron of the entire magnet (the 3 prong configuration) becomes magnetized and can attract the flat iron bar immediately under it. This man-made magnet functions the same way as a permanent horseshoe magnet, Figure 10-8 (a). However, the permanent magnet would have to be physically pulled away from the iron bar. The electromagnet can have its electromagnetic effectiveness reduced to near zero power by de-energizing the circuit. In the motor starters, a spring is used to help provide contact pressure in the pickup mode. This spring also provides an initial push at de-energization to help pull the flat iron bar away and break the switching contacts. It overcomes any residual magnetism held by the electro-d switch used to engage the typical automotive or tractor cranking motor (starter) operates on exactly the same principle. Each magnetic starter (load circuit) is rated in volts, amperes and hp. Depending upon the application, the control circuit may be the same voltage as the load circuit or it may be a lower voltage as is often the case when the service is 120/240 volts. The control circuit's magnetic coil must be voltage rated to the voltage feeding it.

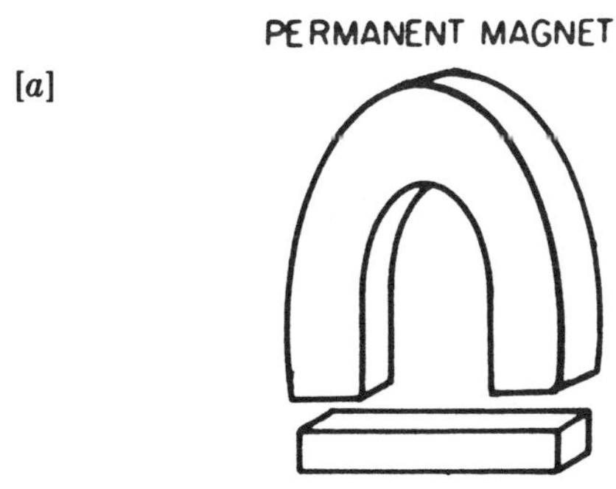

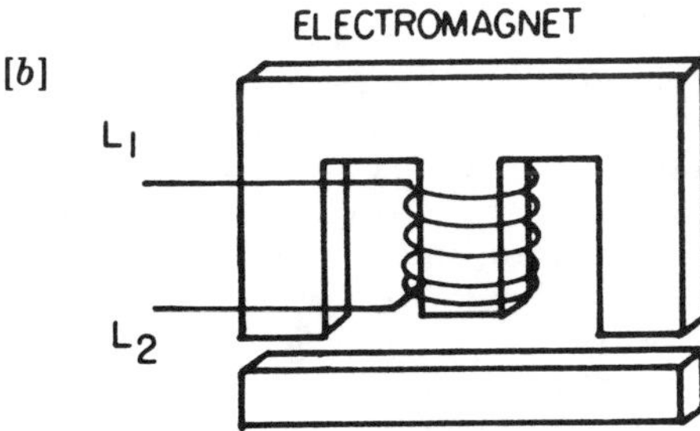

Figure 10-8. Principles of Operation of the Electromagnetic Motor Starter

For many motors, expecially the larger hp, magnetic starters are advantageous for these reasons:

(1) They have a rapid "make" and "break" because the electromagnet is capable of high speed movement of the current carrying contacts being moved together. The "make", reduces arcing problems caused by high starting amperes. They have a spring to rapidly move the contacts apart, the "break", thus minimizing arcing often caused by normal running amperes. "Make" and "break" ampacity ratings of magnetic starting should always be considered as selection criteria, and are just as important as general hp ratings of the controller itself. Manufacturer's engineering design specifications give "make" and "break" values.

(2) They lend themselves to remote control. For many applications of motor use, especially in industry and on farms, controlling from more than one location is important for convenience and safety reasons. An example of convenience might be where four start-stop push button stations are conveniently placed in a facility so workers can start or stop a motor without having to walk to the motor. A safety example might be turning on a motorized fan unit automatically by remote sensing to exhaust harmful gases that have reached a dangerous level. Numerous automatic sensing devices (thermostats, humidistats, micro-limit switches, time clocks, photo-electric cells and other sensors) can be connected to control a motor as easily as traditional start-stop button stations. These applications will be discussed later in this unit.

(3) Manufacturers provide the opportunity for motor users to obtain precision overload and overcurrent protecion by matching rated ampacities of protection of the motor with replaceable heater elements. When motors are changed, and if the controller is large enough for the new motor, the heater element is changed accordingly with ease and at a very low cost. Often heater elements provide an opportunity to protect motors at one-tenth of one ampere precision. When heater elements are used properly, a large amount of money can be saved annually.

Magnetic motor starters and controllers, although primarily used for motor applications, are also commonly used for non-motor electrical loads. One example might be a large bank of lights in a factory or resistance heating on a farm when over 25 A (lowest rating of most motor starters) is needed to remotely control from one or several locations.

The schematic of a magnetic motor controller with two start-stop push buttons is found in Figure 10-9.

The relationship of the actual magnetic motor controller components to the schematic symbols may be confusing. Therefore, the schematic will be broken down to the basics and a new component will be added in sequence. Inside the magnetic starter are three separate poles, these are called contactors and they are normally open (N/O) switches, Figure 10-10, because "at rest" or when the motor is not being operated the contactors are apart.

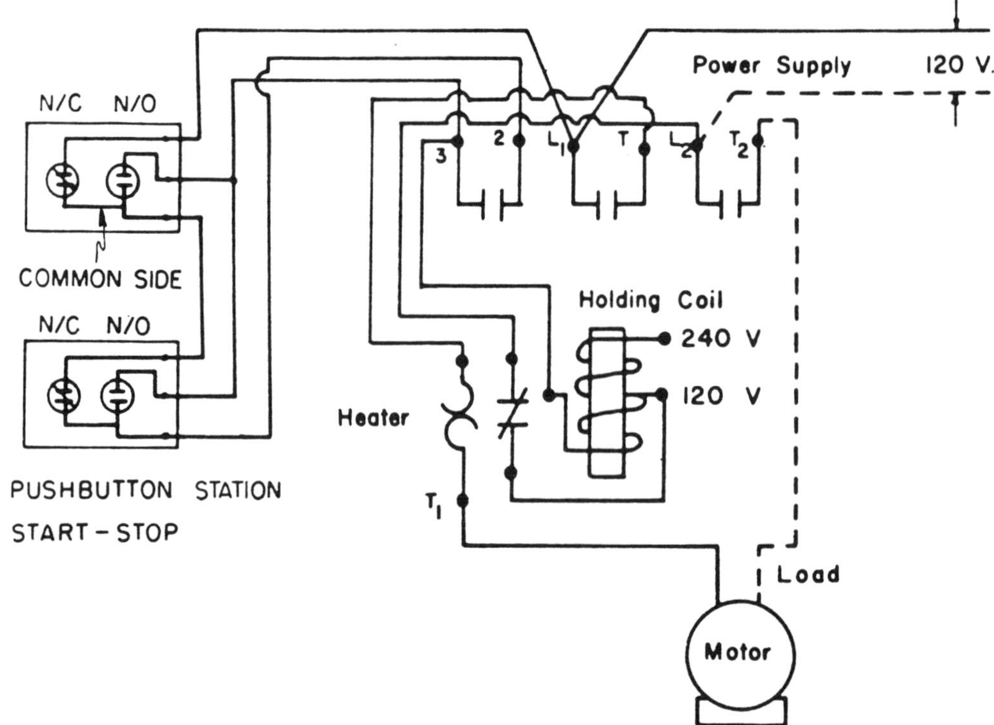

Fig. 10-9. Magnetic Motor Starter with Two-Push Button Stations

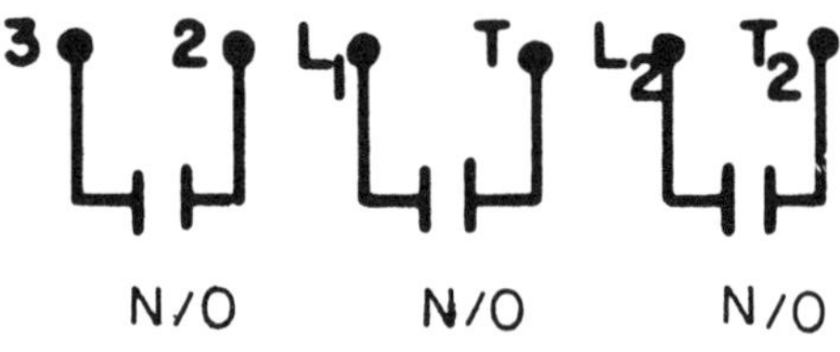

Figure 10-10. Three-Pole Contactors, N/O

The electromagnet, often called a magnetic coil, is used to close the contactors, 3 and 2, L-1 and T and L-2 and T-2, refer to Figure 10-11. All three contactors will close at the same time. Likewise if the coil is de-energized all three will "break" or move apart at the same time. As stated earlier, the magnetic controller provides overload protection, O/L, to the circuit. To perform this function, a heater element is added to the hot conductor, note Figure 10-12. The heater schematic looks like two fish hooks and is selected to fit the size of load, note Table 10-7. Heater Element Selection for Motor Controllers.

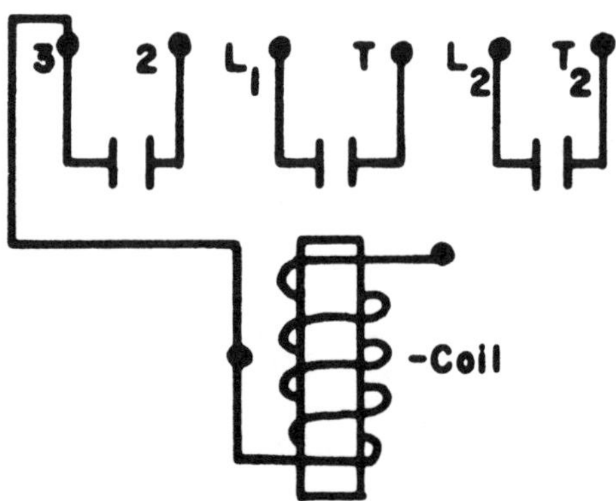

Figure 10-11. Electromagnet To Close the Three-Pole Switch

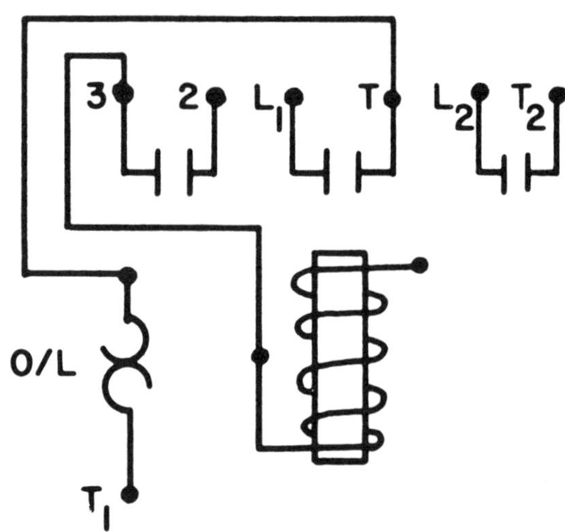

Fig. 10-12. Motor Controller Overload Protection

The heater element (O/L) is connected to the power supply L-1 through "T", when L-1 and T closes, and through terminal T-1 is extended to the motor, note Figure 10-13. Therefore, the O/L protector is in series with the motor. A small normally closed (N/C) switch, is also added within the overload protector, note Figure 10-13. If an overload occurs, this N/C switch opens. This switch can be of a small ampacity because it must carry only the current load of the energized coil which is usually less than three amperes. If there has been an overload on the motor, the N/C switch opens, current cannot flow through the electromagnet and the three-pole switch opens, stopping all current in the circuit. The N/C switch is in series with the electromagnet and is on the neutral side of the circuit in the system being explained. The switch can not become engaged until the eutectic alloy type O/L unit cools and the manual reset button has been pushed engaging the N/C switch. This is

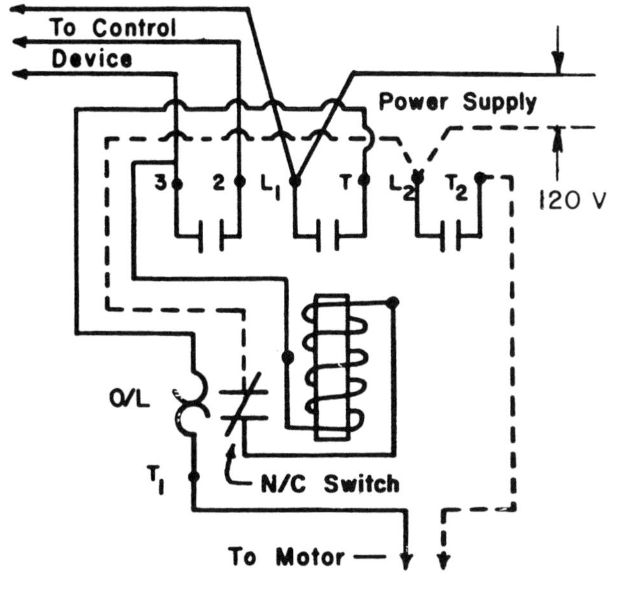

Fig. 10-13. N/C Switch in the Overload Protection Device

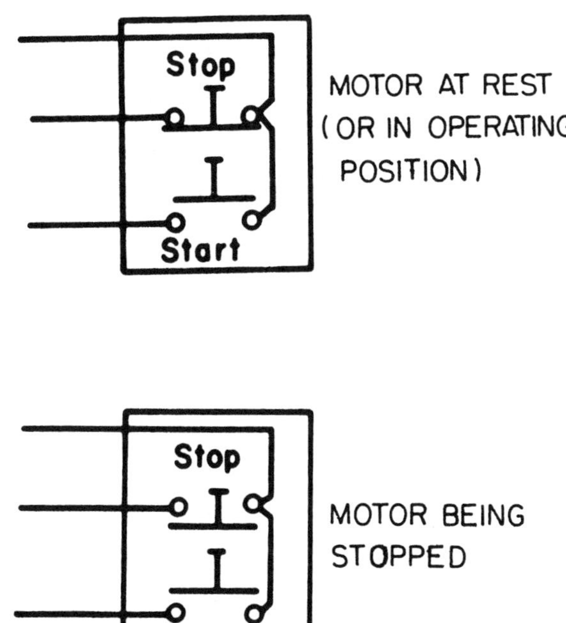

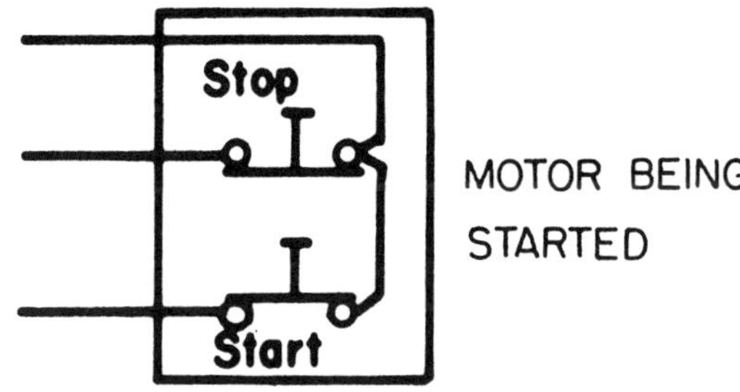

Fig. 10-14. Push-Button Stations With Buttons in Commonly Operated Positions

the important safety feature of the magnetic motor starter. An automatic reset O/L unit could be the bi-metallic type which would start the motor automatically as soon as the heater element has cooled which is undesirable and unsafe. The magnetic controller components and functions have been explained and the push-button switches for the control circuit must be added and schematic explained to operate the system illustrated in Figure 10-14. A start-stop push-button station will be added as a control device to control the contactors of the magnetic starter. The push-button station, Figure 10-14; (a) is in the normal position with the stop button in the N/C position and the start button in the N/O position, (b) the stop button has been depressed and it opens and in (c) the start button is depressed and it closes. This push-button control cannot be used as a switch by itself but requires the magnetic starter to perform its control function. The pushbuttons have a current rating that is low, 1-3 ampere, because the electromagnet in the control circuit requires a small current flow and the load current for the motor, which is higher, does not flow through the push-button contacts. The use of smaller conductors, usually AWG number 14, in the control circuit is good economically because frequently more than one push-button station will be used and these can be located some distance removed from the motor.

The placement of one push-button station in the control circuit is illustrated in Figure 10-15. The stop button side of the stop button (N/C) is connected to L-1, which leads to the "hot" conductor of the power supply. The other side of the stop button feeds to the common side of both buttons and on to terminal 2. When the start button is depressed current flows to the electromagnet through terminal 3 which closes the three-pole contactors including 2 and 3. The electromagnet is now energized by the in-series stop button which provides the 2 to 3 path. Thus, when the start button is no longer depressed and when opened the motor will continue to operate. This circuit which keeps the motor operating is called the "holding circuit" and keeps the magnetic coil energized even though the start button has been released. The terminals 2 and 3 are the so-called "holding circuit" contactors. The motor will operate until (a) the stop button is depressed or opened, (b) the motor is overloaded and the overload heater causes the N/C switch to open, which is designed to be very voltage sensitive, and/or the coil senses low voltage from the power supplier allowing the coil to not hold against the spring tension of the contactors. In any case, the electromagnet is de-energized and the three-pole contacts are opened. Note, when the three-pole contacts are closed, the circuit path for the motor is from contact L-1 to T, T to T-1 and T-1 to the motor. The neutral circuit returns from the motor to T-2, T-2 to L-2 and L-2 to the power source.

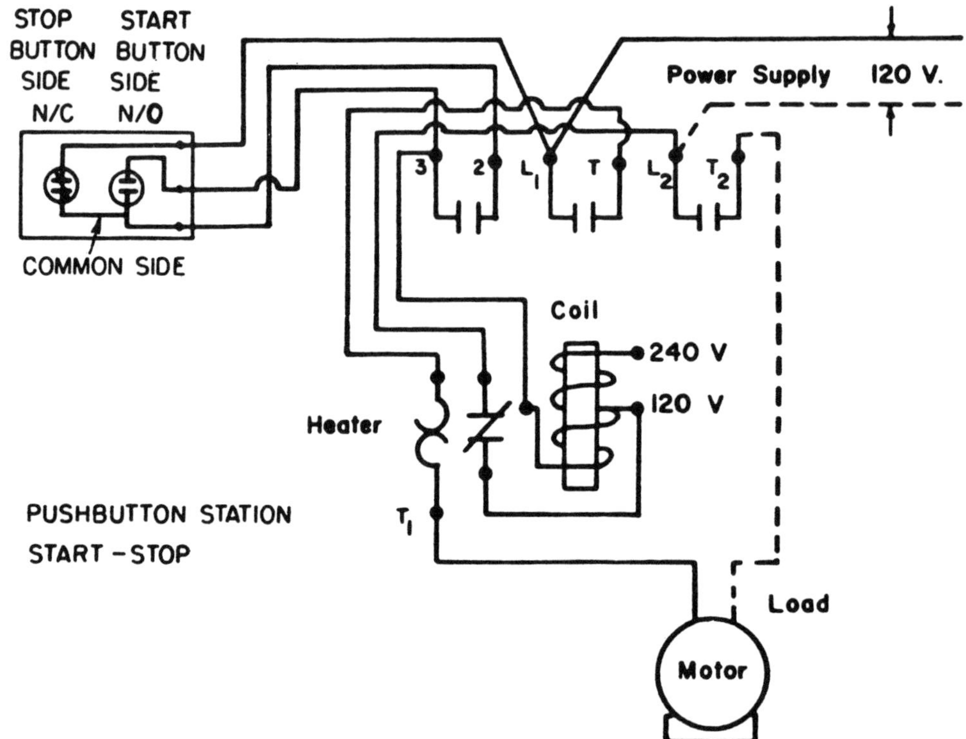

Fig. 10-15. One Push-Button Station in Control Circuit

The second push button has been added to the control circuit in Figure 10-16. The N/C stop buttons in the two stations are placed in series so that current flows continuously through them. The start buttons are placed in parallel which permits current flow to the magnetic starter if either of the buttons are depressed. If more push-button stations were added to the control circuit, they would also have the stop button in series and the start buttons in parallel. In your mind depress the start button on the lower pushbutton station of Figure 10-16 and follow the electrical paths that operate the motor.

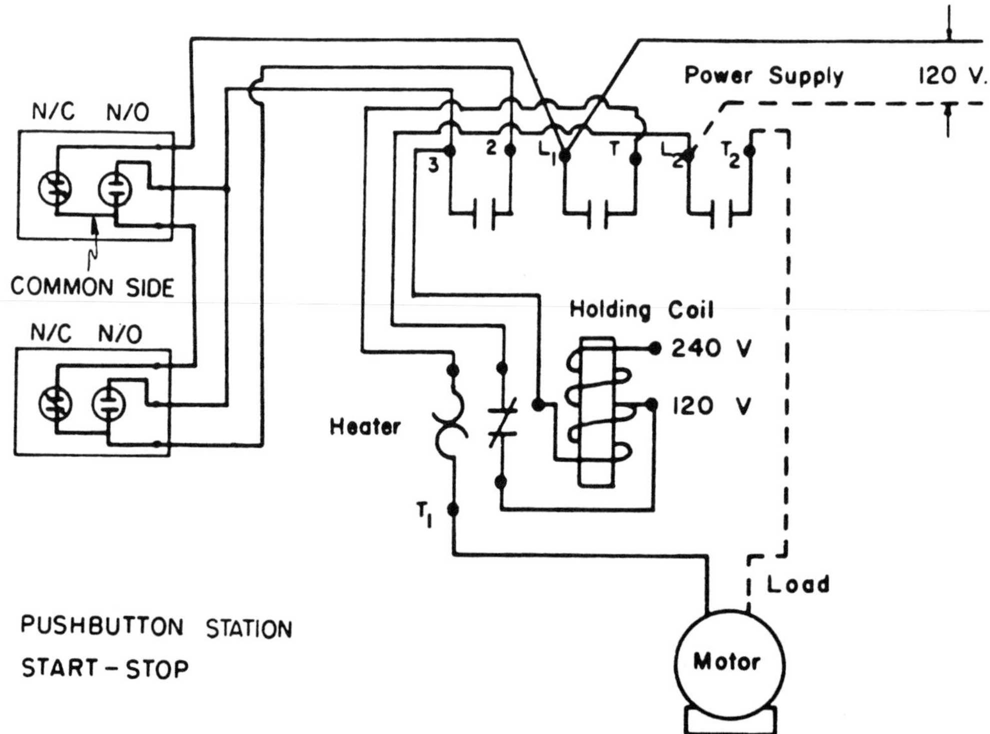

Fig. 10-16. Two-Push Button Station in Control Circuit

Each motor and motor controller must have the correct overload protection. Inside the motor controller cover is a data sheet for temperature conditions for the controller and motor and the listing of heater type numbers, note Table 10-7. Following the rules in Table 10-7 an electric motor with a 100 ampere rating and the same temperature at the controller as the motor would use a W 32 heater type, if the temperature is higher at the controller the heater type would be a W 33 and if the temperature is lower at the controller the heater type would be a W 32. Remember, when selecting a heater element there are three considerations; motor nameplate full load current, starter characteristics and type of starter enclosure. The same number of heater element can serve a variety of ampere loads. In checking the Allen-Bradley Company manual reset relay heater element table an N-10 heater type can be rated for 1.40, 1.32, 1.33, 1.48 or, 1.37 ampere depending upon the table being read. Be sure that all variables are checked when selecting a heater element and if in doubt, consult an electrician or electrical contractor. Select the motor branch circuit overload protection in accordance with the National Electric Code (NEC). The magnetic starter in a circuit must have a manual switch ahead of the magnetic switch. This is a requirement in case of any failure of the electromagnet unit to open where a need is required. The manual switch can then disconnect the supply of power to the motor and its control.

Table 10-7. Heater Element Selection For Motor Controllers

Warning: To provide continued protection against fire or shock hazard, the complete overload relay must be replaced if burnout of any heater element occurs.

Important: When ordering heater elements for this controller always specify the desired "Heater Type No.".

Motors rated for continuous duty:

Motors with marked service factor of not less than 1.15, or motors with a marked temperature rise not over 40°C.

1. The same temperature at the controller and the motor — Select the "Heater Type No." with the listed "Full Load Amps." nearest the full load current value shown on the motor nameplate. This will provide integral horsepower motors with protection between 110 and 120% of the nameplate full load currents.
2. **A** Higher temperature at the controller than at the motor — If the full load current value shown on the motor nameplate is between the listed "Full Load Amps.", select the "Heater Type No." with the higher value. This will provide integral horsepower motors with protection between 115 and 125% of the nameplate full load currents.
3. **A** Lower temperature at the controller than at the motor — If the full load current value shown on the motor nameplate is between the listed "Full Load Amps.", select the "Heater Type No." with the lower value. This will provide integral horsepower motors with protection between 105 and 115% of the nameplate full load currents.

A **Rules 2 & 3 apply when the temperature difference does not exceed 10°C (18°F).** Consult local Allen-Bradley office when the temperature difference is greater.

All other motors rated for continuous duty (includes motors with marked service factor of 1.0)

Select the "Heater Type No." one rating smaller than determined by the rules in paragraphs 1, 2, and 3. This will provide protection at current levels 10% lower than indicated above.

Motors rated for intermittent duty:

Consult local Allen-Bradley office.

Select the motor branch circuit overcurrent protection in accordance with the National Electrical Code. In no case shall the rating of the Class H (non-time delay) or the Class K5 (time delay) fuses exceed 400 amperes.

TABLE 147

Heater Type No.	Full Load Amps.	Heater Type No.	Full Load Amps.	Heater Type No.	Full Load Amp.
W29	74	W35	126	W41	215
W30	81	W36	138	W42	235
W31	88	W37	151	W43	256
W32	97	W38	165	W44	281
W33	106	W39	180		
W34	115	W40	197		

The rating of the relay at 40°C is 115% of the "Full Load Amps." listed for the "Heater Type No."

"Courtesy of Allen-Bradley Company—form supplied with Bulletin 509 motor starter, size 5 in NEMA Type 1 enclosure."

The motors discussed have been controlled nonautomatically by either switches or motor controllers. Many times it is more desirable to turn on the motor with an automatic control device such as limit switches, thermostats, humidistats, time clocks, repeat cycle timers, photo electric cells and time-delay relays.

AUTOMATIC CONTROL DEVICES FOR CONTROLLING MOTORS

An automatic sensing device is an electrical component capable of sensing an environmental change such as humidity, light, time, temperature and pressure. After sensing the change the component will mechanically actuate a switch or contactors. The sensing device can be designed to operate on line voltage or at a lower voltage, 24 volts being common on home heating furnaces, for example. The ability to operate at the lower voltage is a great advantage for safety, size of conductor needed in the control circuit and design of the sensing device to carry a lower current. Check the sensing device's nameplate for its voltage and ampere rating. For circuits to utilize a lower voltage for the control circuit than the load circuit, it is necessary to use a relay. Relays are available with a variety of switching combinations such as SPDT, SPST, DPDT, DPST and others. The SPST magnetic relay is shown in Figure 10-17. The control switch is open in Figure 10-17 (a) and the load circuit contactors are open. The control switch is closed in (b), the electromagnet is energized, attracts the lever arm with the contact and the load circuit contactors are closed.

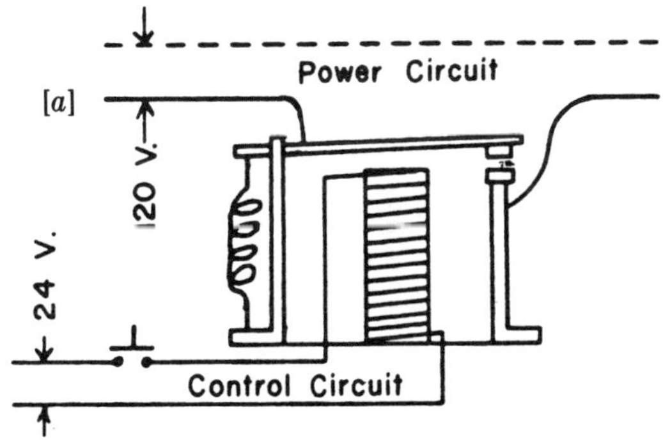

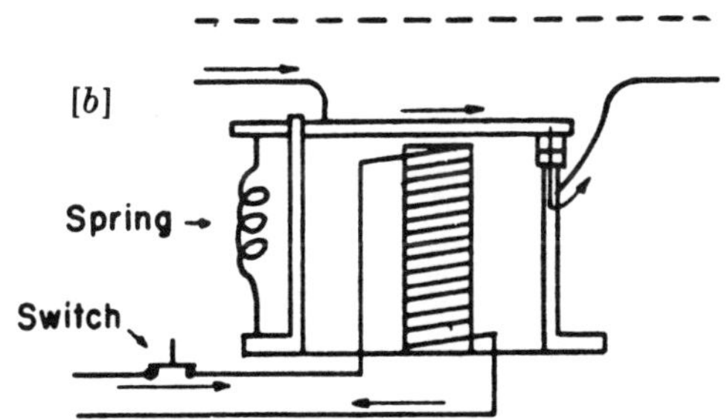

Fig. 10-17. Single-Pole Single Throw Magnetic Relay

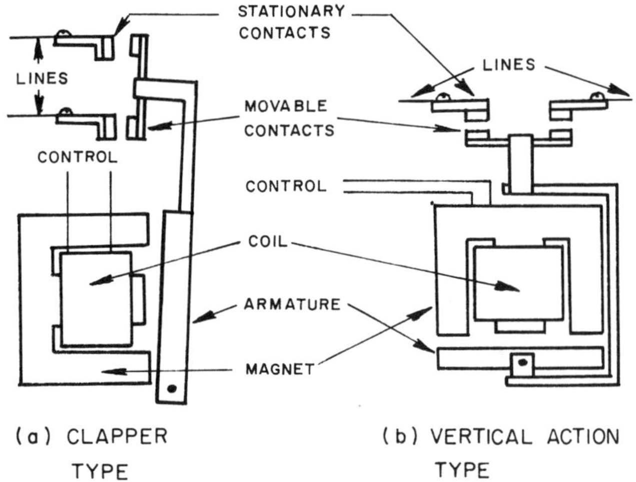

Fig. 10-18. Styles of Electromagnets and Contacts

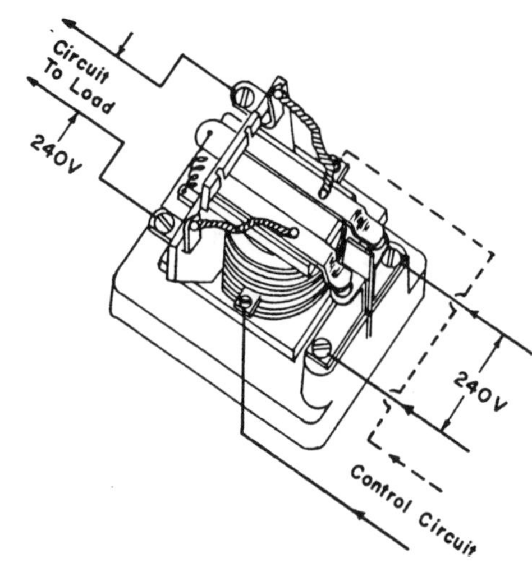

Fig. 10-20. Double-Pole Single Throw Magnetic Relay

Illustrated in Figure 10-18 are two styles of electromagnetic units which can be found in magnetic motor starters. The types are clapper and vertical action. The electromagnet, when energized, attracts the armature which is attached to the movable contacts. The movable contacts are then brought into alignment and contact with the stationary component of the unit. Another electromagnetic relay, SPDT is illustrated in Figure 10-19 and a DPST relay is shown in Figure 10-20. The DPST relay, Figure 10-20, is illustrated for 240 volt service but if necessary, it can be used as a 120 volt SPST relay by utilizing only one side.

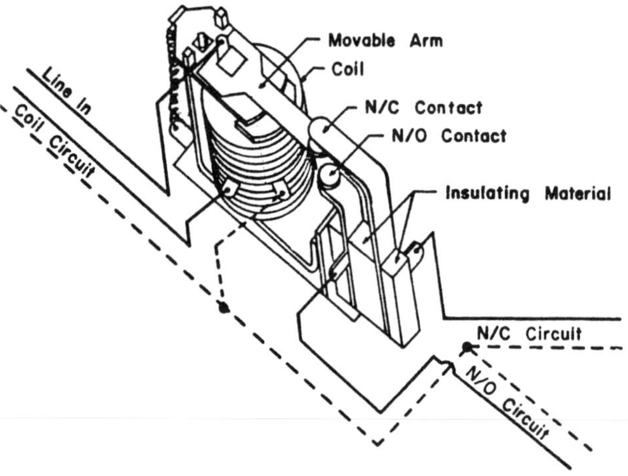

Fig. 10-19. Single-Pole Double Throw Magnetic Relay

The N/O and N/C limit switch is an example of the pressure sensing device. This style of switch is sometimes called a "micro-switch" because its movement for activating the contacts can be a very small distance. The roller or contact point, Figure 10-21, could be attached to an air vane, float in a liquid or movable mechanical platform or conveyor belt. Limit switches can be connected so that the circuit is N/O (a), N/C (b), and with both N/O and N/C (c) connected to two circuit loads so as to operate either one device or the other.

[a] *N/O Circuit*

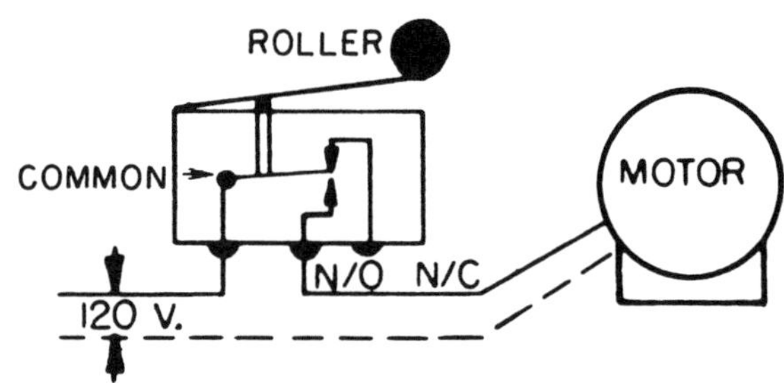

[b] *N/C Circuit*

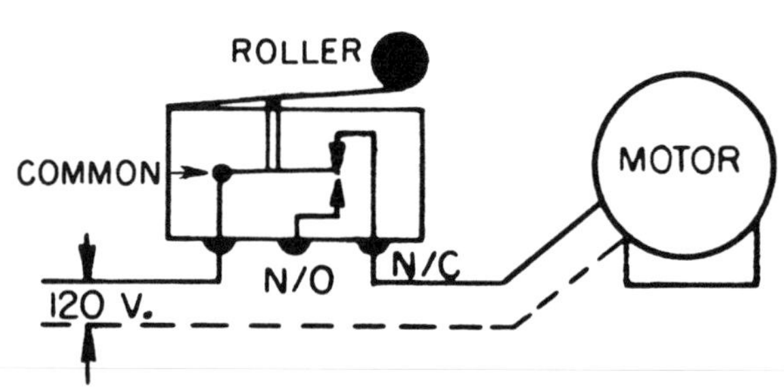

[c] *N/O and N/C Circuits*

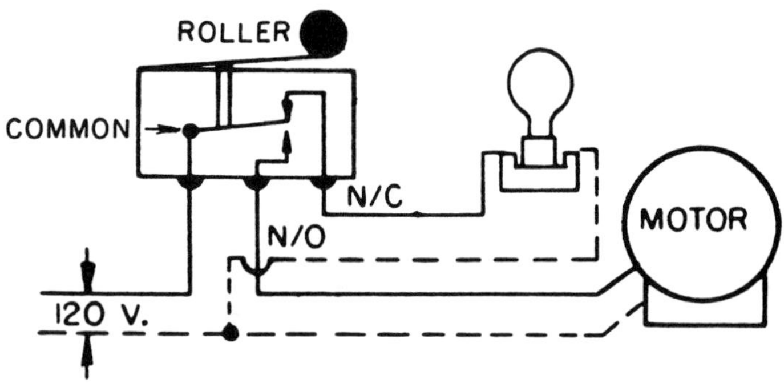

Fig. 10-21. N/O and N/C Limit Switch Circuits

Light can also serve to actuate a sensing device. Note, the circuit in Figure 10-22. The photoelectric switch can be used to control lights, open and shut doors, stop conveyor belts and augers. All sensing devices must be designed to operate at the same ampere value of the motor or load or a relay must be used in the circuit to prevent full load current from traveling through the sensing device. The temperature sensing device is a thermostat and the unit that responds to the change in temperature can be bimetallic (metal strips), liquid in a wafer or bellows (liquid-gas), or a liquid filled tube (hydraulic). The thermostat circuit in Figure 10-23, has both the N/O and N/C circuit connections depending upon temperature rise and temperature fall.

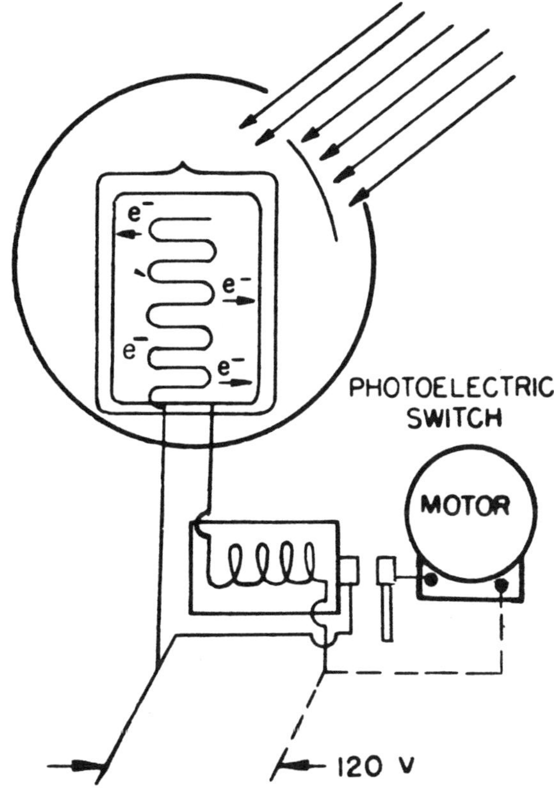

Fig. 10-22. Photoelectric Circuit

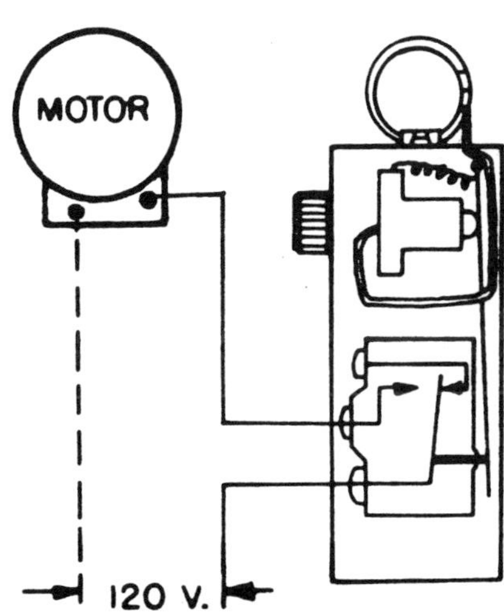

Fig. 10-23. Thermostat Circuit

The sensing device which detects changes in moisture is a humidistat. The unit detecting this change can be hair or made of a synthetic product which responds to humidity change. The humidistat and its circuit are illustrated in Figure 10-24. Sensing devices detecting time can be the 24-hour time clock, Figure 10-25, or a repeat cycle timer, Figure 10-26. The control circuit for both of these devices is the synchronous motor. Time clocks can be provided with two or more trippers which can cause several changes in events. The time clock can control lights, heaters, waterflow valves and motor driven machines. The repeat cycle timer is for a shorter duration of time, for example, ten minutes, and can be set to schedule the two events as designed by the time setting selected. An application of the repeat cycle timer is used in a confinement livestock building, or for a greenhouse water misting of bench plants. For example, a thermostat can control the temperature level with a furnace and fan. The humidity can be controlled with a humidistat and exhaust fan if the moisture content is too high. The operation of either of these sensing devices will cause air movement to replace the oxygen level. However, suppose that neither of these were activated and all the oxygen in a confinement building is consumed and the poultry or livestock suffocate. The use of a repeat cycle timer on a separate fan circuit could have prevented this by supplying a quantity of fresh air each ten minutes regardless of whether the thermostat was activating the furnace or the humidistat the exhaust fan.

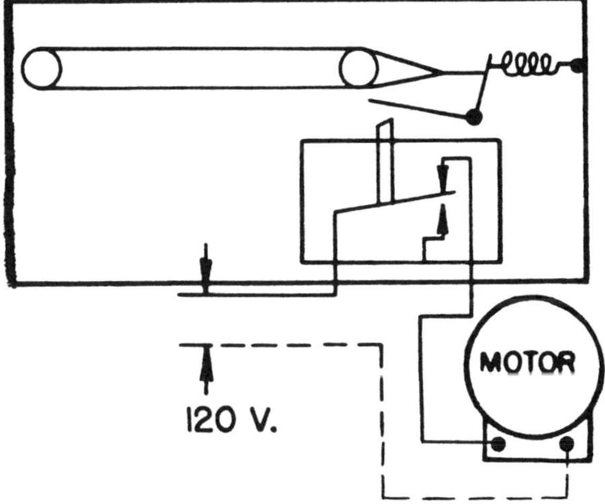

Fig. 10-24. Humidistat Circuit

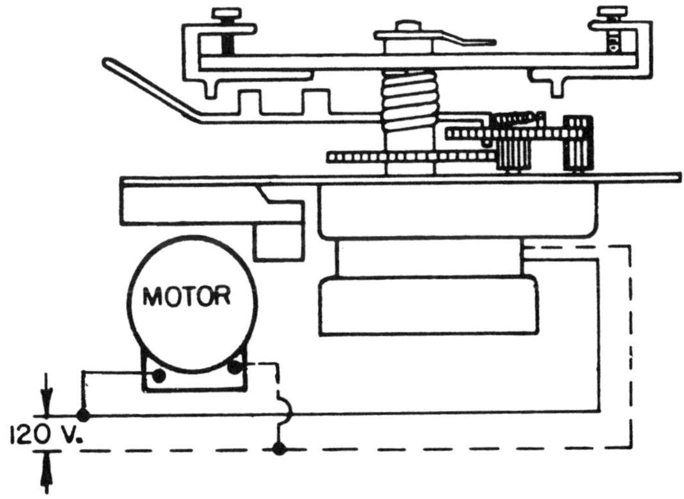

Fig. 10-25. 24 Hour Time Clock Circuit

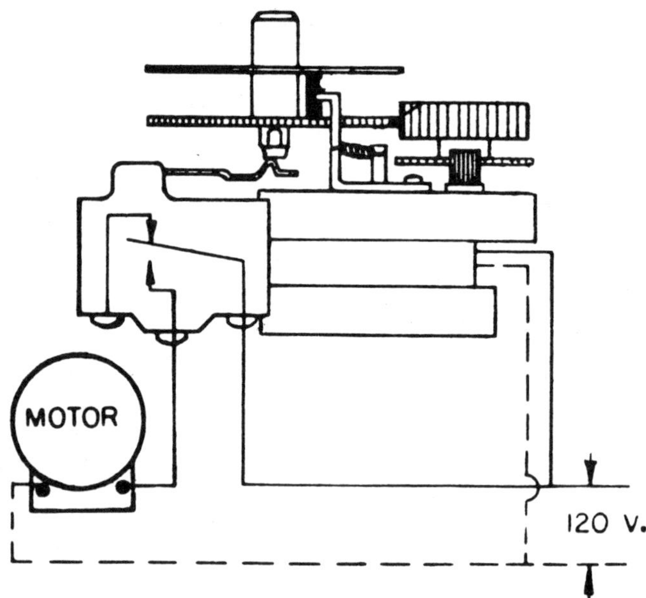

Fig. 10-26. Repeat Cycle Timer Circuit

A time-delay relay device is illustrated in Figure 10-27. It has bimetallic strips that move together, bringing the contacts together. This occurs because the control circuit has a heater coil. Each time-delay relay will be identified as either N/O or N/C. A time-delay relay could be identified as 115/NO/10. The 115 is the recommended line voltage, NO indicates normally open and the 10 represents the seconds before the relay performs its function. Not all time-delay relay devices are the vacuum tube type, some are solid state and others have a clock to sequence the circuits to be energized. The purpose of the time-delay relay device is to sequence when motors or other loads come on the line which can reduce the size of transformers needed or prevent the overloading of the recommended transformer. An example might be the use of three electric motors in an agricultural livestock feeding system. One motor turns on the bunk conveyer, the next motor activates the elevator which carries the feed to the conveyer an the third motor operates the silo unloader. Flipping all three switches on at the same time could overload the distribution transformer. Therefore, the time-delay relay devices perform a necessary function. The proper sequencing of these units as mentioned places the units in motion before a load of feed is applied. An industrial appplication might be the starting of several motors in sequence to start a series flow of materials about ten seconds apart. Time-delay relays can perform these functions too.

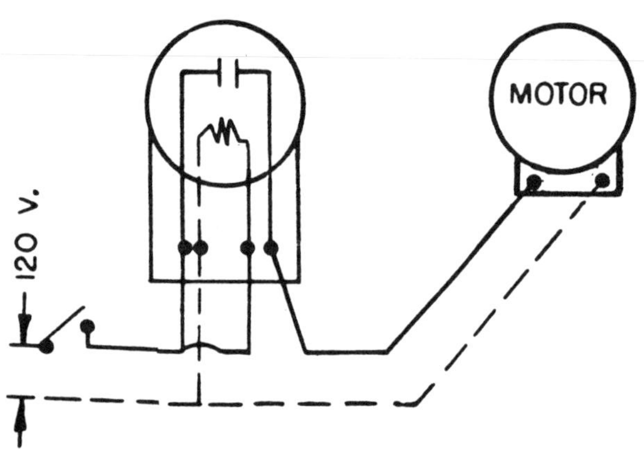

Fig. 10-27. Time Delay Relay Circuit

Motors in industrial installations frequently must be stopped more quickly than if the power were only disconnected. Electric brakes have been in use for about 100 years and they have been called magnetic, friction and mechanical brakes. These brakes have two friction surfaces, shoes or disks, which press against a wheel on the motor shaft. An electric system initiates the release or application of the pressure against the wheel or disk and is held in place by springs and the braking is achieved by friction. Dynamic braking is achieved by disconnecting the motor from the power source and then reconnecting it so that it acts as a generator. A supplemental system does the final stopping. Another braking system is electronic (electric) braking which is a control that can be added to the motor starter. When a motor is to be stopped, direct current is applied to one or all of three phases of the motor after the AC voltage is removed. There is now a stationary DC field in the stator instead of an AC rotating field. As a result, the motor is braked quickly and smoothly to a standstill. An electric braking controller provides a smoother positive stop because the braking torque decreases rapidly as the speed approaches zero. Tapped resistors are generally used to adjust the braking torque. Three-phase electric braking is more efficient and reduces the heat buildup in the motor as compared to single-phase braking.

The electric motor must have protection. Factors to consider are the service factor of the motor, ambient temperature at both the motor and the starter and size of the enclosure. These factors are considered when selecting the heater element for the controller plus the nameplate full load current, starter characteristics and type of starter enclosure. The branch circuit must also have overload current protection from the motor and the machine it drives. The motor will be switched either manually or automatically. The next unit will cover the proper installation of the motor on its load.

DEFINITION OF TERMS

AMPACITY: The safe carrying capacity of an electrical conductor or component in amperes.

CIRCUIT: A path through which electrical current flows.

CIRCUIT, BRANCH: The conductors for carrying electricity to and from a load and the load matches conductor size and overload protection device.

CIRCUIT, BREAKER: A switch-type overload current protector that opens automatically when excessive current flows and can be manually reset. Special forms of circuit breakers will reset automatically.

CIRCUIT, PARALLEL: A hookup of two noninterrupting conductors to which two or more resistors (loads) are individually placed across the circuit.

CIRCUIT, SERIES: An electrical circuit in which two or more resistors are placed end to end or in line with two conductors.

CIRCUIT, SHORT: A connection between two lines or energized parts of such low resistance that excessive current flows, as in the case when hot and neutral conductors make contact.

CONDUCTOR: Materials such as copper and aluminum which allow for easy flow of electricity. Neutral conductors are white, hot conductors are normally black or red except in conduit circuits when the colors could be yellow, orange, brown and purple. The grounding conductor is green or does not have a covering and is bare.

CONTACTOR: Power switching device without overload protection.

CONTROLLER: Refers to either a contactor without overload protection or a starter with overload protection.

CURRENT OVERLOAD PROTECTION: Devices, usually called fuses or circuit breakers, which are designed to open the circuit when a predetermined current has been exceeded.

ELECTROMAGNET: Magnet produced by winding conductors around a metal core and then energizing the circuit.

FUSE: A device designed to open the circuit when a specific current has been exceeded. Non-time delay fuses open quicker than do time-delay types, which can also withstand normal overload currents above their ratings for several seconds before opening the circuit.

FUSE, CARTRIDGE: A single-element fuse that is cylindrically shaped. Those rated more than 60 amperes have a knife-blade terminal attached to each end.

FUSE, DUAL ELEMENT: A time-delay, Edison base, plug type fuse.

FUSE, PLUG: A fuse with an Edison base, similar to the screw portion of an ordinary 120 volt light bulb.

FUSE, TIME DELAY: Any fuse having capability of carrying about twice its rated ampacity for several seconds, but will blow or trip at its rated ampacity under continuous overload current conditions.

FUSETRON: A brand name for an Edison base time-delay fuse.

FUSTAT: A trade name for a time-delay fuse which as a special base that screws into an Edison base. Each fuse fits its special base and are not interchangeable.

GROUND: Any part of the electrical system that has continuity or bonding to earth.

GROUNDED: A conductor, box or any other device that is in continuity with the earth.

GROUND FAULT CIRCUIT INTERRUPTER (GFCI): Safety device that can sense small amounts of current flowing to ground and quickly break or disconnect the hot conductor source. Most styles are overload current protection devices having a specific ampacity rating can protect equipment, and people against electrical shock hazards.

NATIONAL ELECTRICAL CODE (NEC): Publication by the National Fire Protection Association (NFPA) which establishes minimum standards for wiring materials, wiring applications, electrical equipment, and communications systems.

NO LOAD: A situation when an electric motor is running but is not doing useful work.

NORMALLY CLOSED (N/C): Contacts on a switching device that are together and will remain in that position unless moved manually or by same sensing device controlled by electricity, light, temperature, humidity, pressure, liquid or a mechanical device.

NORMALLY OPEN (N/O): Contacts on a switching device that are separated and will remain in that position unless moved manually or by some sensing devices controlled by electricity, light, temperature, humidity, pressure, liquid or a mechanical device.

OVERCURRENT: A current greater than the design of a device or circuit.

OVERLOAD: A power dissipation greater than that for which the circuit or parts of a circuit are designed.

SCHEMATIC: A sketch, plan or graph that helps to explain or illustrate something by outlining its points in the form of a drawing or visual configuration.

SENSING DEVICE: Electrical component that senses heat, humidity, time, light or pressure and acts as a control for an electric motor or other components.

STARTER, MOTOR: Power switching device with overload protection.

CLASSROOM EXERCISE X-A

Motor Overcurrent Protection

1. A new fan requires a 1.2 hp motor. What would be the 125% full-load amps for this single-phase motor will be connected to 115 volts? at size conductor would be required for this circuit?
   ____

2. For the 1/2 hp fan motor, it was discovered that 3-phase current was available and 208 volts. What would be the ampere rating at 125 % of full load? ______________________________

3. A 230 volt single-phase circuit has three motors: 1/4, 3/4, and 1/3 hp. Determine the total current requirement for these motors.

	Motor	% Load	Amperes
a.	1/3	______	______
b.	1/4	______	______
c.	3/4	______	______
		Total Ampere	______

   d. What size conductor would be required for this circuit? ______________

4. What affect does an increase in voltage drop (2% -> 3% -> 4%) have on copper conductor size for a circuit with 45A? ______________________________

   ____

   ____

5. Running current protection provides protection for the ______________ and circuit overcurrent protection provides protection for the circuit ______________________________.

6. What size overcurrent devices would be recommended for these electric motors with a 40°C rise on 230 volts with single-phase service.

	Size of Motor	Rated %	Full Load Current	Overload Device
Over 1 hp	3	______	______	______
Less than 1 hp	1/4	______	______	______

7. What is the advantage of a time-delay fuse over a regular blow? ______________________________

   ____

   ____

8. Which fuse style and type would you recommend for these applications?

   (Choices -- Regular or Time Delay)

Application	Type of Fuse
a. Lighting circuit	______________
b. Motor on attic fan	______________
c. Water pump	______________
d. Feed auger	______________
e. Clock motor	______________
f. Garbage disposal motor, with thermal overload	______________

   (Choices: Cartridge, plug circuit breaker, or circuit breaker with GFCI)

g. House lighting circuit	______________
h. Furnace fan motor	______________
i. Duplex near swimming pool	______________
j. Stove circuit	______________

9. A twenty ampere circuit is loaded to 40 ampere and the circuit has a non-time delay device:

   a. How soon would fuse trip? ______________________________

   b. If a time delay fuse had been used, the fuse would not have tripped until __________ seconds.

10. A short circuit might be considered to load a circuit at ______________ times its rated ampacity.

CLASSROOM EXERCISE X-B

Switch Identification

1. ______________________________

5. ______________________________

1 A B A B 1

2. ______________________________

A A A B B B 1 2 3

6. ______________________________

3. ______________________________

"A"
Contact Points
Closed

7. ______________________________

A B A A B A B 1 2

4. ______________________________

Movable
Contact
"A"
Stationary
Contact
"A"

8. ______________________________

CLASSROOM EXERCISES X-C

Switches and Controllers

1. What is the advantage of overload protection on a SPST switch which controls a motor in a circuit?

____

____

2. What size heater cone would you select for a switch for a general purpose motor with a 1.15 service factor that has 12.00 full load ampere? _______________

3. Motor controllers are normally used on motors of _______________ horsepower and larger.

4. List the advantages of a magnetic type switch:

   a. ____

   b. ____

   c. ____

5. The contactors in magnetic controllers are normally open or normally closed? _______________

6. How does the electromagnetic close the contactors?_______________

____

____

7. Stop buttons are normally _______________ and start buttons are normally _______________ and the stop buttons are connected in _______________ and the start buttons are connected in _______________.

8. What heater type number for a motor controller would you select for a general purpose motor with a 1.15 service factor that has a 8.5 full load ampere?_______________

9. Why are magnetic relays added to circuits? a._______________

____

   b. ____

____

10. What type of sensing device would control these motors?

    a. Motor on a furnace _______________
    b. Motor on a dehumidifier _______________
    c. Motor on a water pump _______________
    d. Motor controlling a door _______________
    e. Motor on a confident housing building fan _______________
    f. Motors which must be sequenced as they come on the line._______________

CLASSROOM EXERCISES X-D

Electric Component Identificaiton

SYMBOL	ITEM	APPLICATION
1.		
2.		
3.		
4.		
5.		
6. M		
7.		
8.		
9.		
10.		
11.		
12.		
13.		
14.		
15. L_1 120 L_2		

CLASSROOM EXERCISE X-E

Sensing Devices

1. Complete the following table on sensing devices.

Environmental Change	Name of Device	Application

2. Most sensing devices have two circuits and their names are ______________ and ______________.

3. Which circuit from question 2 might have a lower voltage than the regular service?

   ____

4. Sensing devices will have switches to control the load circuit and are often identified by letters. Write the name for each set of initials.

   a. SPST ______________________________ d. DPDT ______________________________

   b. SPST ______________________________ e. N/O ______________________________

   c. DPST ______________________________ f. N/C ______________________________

5. List the three essential considerations when selecting a heater element for a switch or magnetic motor controller.

   a. ____

   b. ____

   c. ____

6. When using a magnetic motor starter with two push-button stations, the start-button symbol is ________________ and stop-button symbol is ________________. The start-buttons are wired in ____________________________ and the stop-buttons in ____________________________.

7. Explain how the electromagnet closes the contacts in the magnetic motor starter and how they are opened in case of an overload or the pushing of a stop switch.

LABORATORY EXERCISE X-A
Wiring and Operating Circuits

Wiring Activity	Date Completed	Evaluation
Motor Citcuit With:		
1. SPST switch		
2. SPST switch with overload protection		
3. Magnetic starter with one push button		
4. Magnetic starter with two push buttons		
5. Limit switch, N/O		
6. Limit switch, N/C		
7. Limit switch with both N/O and N/C		
8. Photoelectric and SPST relay*		
9. Thermostat and SPST relay		
10. Humidistat and SPST relay		
11. Twenty-four hour time clock and SPST relay		
12. Repeat cycle timer and SPST relay		
13. Time delay relay		
Additional Developed Circuits:		
14.		
15.		
16.		
17.		
18.		
19.		
20.		
21.		

* If the motor ampere rating exceeds the ampere rating of the sensing element, then a relay must be used to prevent full motor ampere flow from going through the sensing element. A DPST relay can be used as a substitute for a SPST. Other circuits can be developed using the SPDT relay and one of the two loads will be operating at all times.

MOTOR INSTALLATION

There are environmental factors affecting the life of a motor. The major factors are dust, water or moisture, stray oil and air. Each of these factors will be discussed separately to explain why they are often called the "enemies of electric motors".

ENVIRONMENTAL FACTORS

DUST

Dust acts as a heat barrier or insulation when it is blanketed on the exterior of a motor. Likewise, dust drawn into the windings of a motor will reduce the cooling efficiency of the windings and other internal motor parts. Therefore, dust deposited either on the exterior or interior of a motor will promote improper cooling, overheating of the motor beyond the practical limits, and will ultimately result in a shortened life for the motor.

There are several ways that dust can be controlled in the environment. First, in extremely dusty environments, the drip-proof and/or splash-proof enclosure motors should not be used, but rather the totally enclosed motors as discussed in UNIT II, EXTERNAL FEATURES OF ELECTRIC MOTORS. Totally enclosed motors are essentially air tight and with the recirculation of entrapped air dust cannot enter. However, the totally enclosed motors must radiate the heat from the exterior walls to keep the heat at a reasonable level. Therefore, this type of motor must be cleaned externally on a regular maintenance schedule. High pressure washing with detergents and clear water rinses can be used on totally enclosed motors. Brushes and detergent washing solutions can also be used.

Motors of the open enclosure style are more difficult to keep clean and free of dust both externally and internally. The only practical way of cleaning the motor internally is by disassembly, clean-up and reassembly, as discussed in UNIT XIII, ELECTRIC MOTOR MAINTENANCE. Open enclosure motors can be wiped off externally using water and detergents on a regular basis to keep them reasonably free of blanketed dust build up. Do not wash with high pressure washers as water could be accidently blasted through the end bell slots and into windings causing damage to windings or making electrical shorts in the electrical contact mechanisms.

WATER OR MOISTURE

Moisture can result from rain, condensation or high moisture environmental conditions and flooding. When moisture enters an electric motor it creates corrosion, rusting, electrical arcing or shorting and deterioration of the insulation on the windings. Where moisture is a problem, it is advisable to use a totally enclosed motor. This type of electric motor is used on water pumps, on machines where frequent washing of equipment is necessary, and in other high moisture situations. Submersible water pump motors are totally enclosed motors which have water tight seals and packing so water cannot enter beyond the bearing areas. If water becomes a problem in an electric motor, it is possible to remove the moisture with heat from a portable heater or to use an open oven and force air through the motor with a fan. Compressed air can also be used to help dry out motors. When possible, use heat plus air for best results. In reconnecting a motor that has been flooded, it should be disassembled, cleaned and reassembled. It should also be started with the use of a GFCI (Ground Fault Circuit Interruptor) to guard against electrical shock. In areas where water is less of a problem, open enclosed motors can be used if they are protected from the weather by a well ventilated enclosure.

STRAY OIL

Stray oil or excessive oiling can cause deterioration of wire insulation, windings, and bearings. It also adds to dust build up because oil holds the dust upon contact. Mixing dust and oil forms a gum which is abrasive to both bearings and shafts. Gum can shorten the life of the motor. Because it is impossible to lubricate a motor perfectly, it is important to disassemble and clean excess oil, dust and dirt periodically. See UNIT XII, ELECTRIC MOTOR MAINTENANCE, for proper lubrication and cleaning procedures for an electric motor.

AIR

The natural air surrounding the motor can be a serious problem to the environmental health of a motor. Previously, dust and moisture were discussed as enemies of the motor. However, it is the surrounding air and the air flow into and/or around the motor that can carry high amounts of dust, contaminates and moisture.

Air can be filtered and it is often done in certain manufacturing facilities. Industries manufacturing products that must be kept clean will use "clean rooms" or "white rooms", and these facilities are environmentally controlled by filtering or electrostatic equipment to take dust and other precipitant materials out of the air. These industries do not need to use totally enclosed motors, unless the conditions are explosive. In many industries including metal fabrication, agricultural, chemical manufacturing and mining, air quality around motors is a serious problem. In a rock quarry or a farm feed mill use totally enclosed motors and clean the motor's exterior on a regular maintenance schedule. Open enclosure motors are frequently used in industries that have motors under cover and where no explosive atmosphere exists. An example of this might be in light manufacturing or in the farm service center. With these conditions, the surrounding air can be kept relatively clean by good housekeeping, such as sweeping floors with compounds, and by brushing or vacuuming walls and ceiling when the conditions dictate. Chemically contaminated air environments nearly always mandate the use of totally enclosed motors, especially if air contaminants are toxic to motor internal parts and/or if a potential explosive situation exists. Washing of motors on a regular basis is a must in these conditions.

ALIGNMENT

Alignment refers to the general straightness of the drive system. A good example for explanation purposes is two pulley sheaves, one on a motor acting as a driver and the other on a machine acting as a driven pulley. In a belt drive system, Figure 11-1 (a) shows a misalignment and (b) a proper alignment.

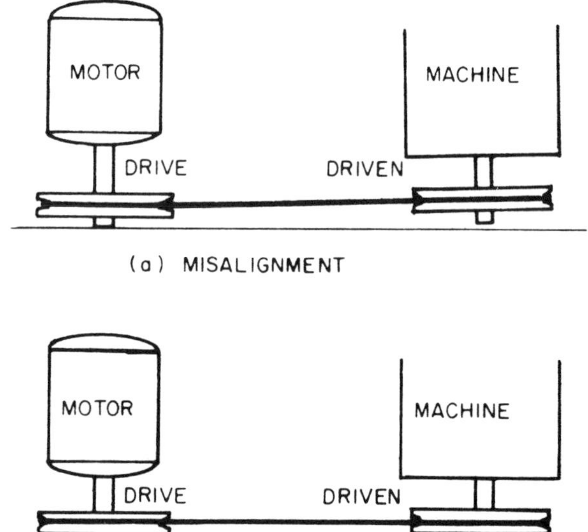

Fig. 11-1. Misalignment and Proper Alignment of Driving Motor to Driven Machine

A misaligned V-belt placed between the driver and driven pulleys would place excessive wear on the sides of the V-belt, friction would be created on the insides of the pulley sheaves and belt and pulley life would be reduced. Another serious problem would be the forces due to misalignment causing the PTO shaft to be forced in one direction or the other which upsets the critical end-play tolerances. Typical sleeve bearing types of motors have shims or thrust washers to position the rotor inside of the end bells and bearings. A sketch in Figure 11-2 illustrates the position of the thrust washer (s).

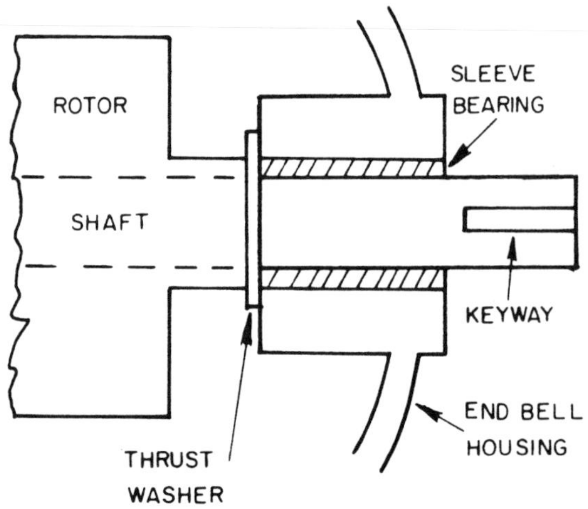

Fig. 11-2. Typical Sleeve Bearing Motor

When forces are too great, due to misaligned driving systems, the thrust washer can be destroyed and recommended end-play changed. The end-play of a motor's shaft is critical for two reasons. First, if too great, the mechanisms used to make electrical contact when the motor is at rest will not make the necessary contact to help start the motor. Also, when the motor is running; it may have its rotor and shaft pulled so far in one direction that the starting winding will become energized and become overheated. Secondly, the rotor is designed by the manufacturer, to run in a certain relative position to the stator field or the result will be overheating of the motor. Damaged or destroyed thrust washers can cause any of the problems; therefore, proper alignment is the best prevention. Anti-friction bearings are less affected by misalignment. However, ball bearings, as illustrated in Unit II, Figure 2-15, can overheat when misalignment occurs. Regardless of bearing type, misalignment causes a reduction in size of the lubrication oil film on bearings and/or thrust washers. When the oil film is destroyed, heat is generated quickly and bearings start to seize, overloading occurs and excessive heat is generated in the windings. These steps lead to motor burn out.

Alignment of the motor to the driven machine is best checked by use of a straight edge, such as a mechanic's metal rule. Figure 11-3 illustrates a situation showing misalignment. The motor and machine should be repositioned so the straight edge will fall along the outside of both the motor's driver pulley and the machine's driven pulley. If there are idler pulleys, they too should be checked for alignment in the drive system.

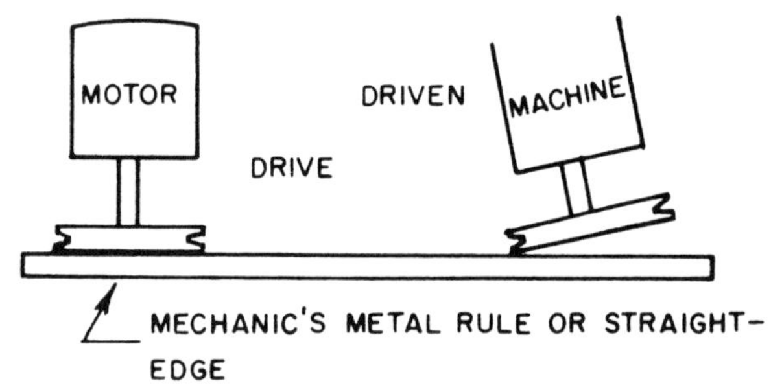

Fig. 11-3. Proper Motor and Machine Alignment

VIBRATION

Vibration literally shakes everything apart. Motors, because of their high revolutions per minute can cause serious vibration problems. Potential causes of motor vibration are listed in Table 11-1. The motor may be an integral part of the machine. The list provides an operator with ideas on how to trouble shoot and correct vibration problems.

Table 11-1. Possible Causes of Vibration in Motors

CAUSE	HOW TO CORRECT
Loose Bearings	Change Bearings
Dry Bearings	Lubricate Properly
Bent Shaft	Replace Shaft and Rotor
Driving Pulley Out-of-Balance	Replace or Balance Pulley
Drive System Chains or Couplers Out-of-Balance or Out-of-Alignment	Replace Belt or Chain
Machine Is Vibrating	1. Mount Motor on Resilient Mounts or Cushions 2. Mount Motor Away from Frame of Machine if Possible

Vibration not only physically damages motor and machine parts, but it can place excessive pressure on the oil film on bearings overheating them, and ultimately leading to a burned out motor.

DRIVE SYSTEMS

Drive systems include any transfer of power from the motor doing the driving or powering to the machine which receives the power. The most common of the driving systems used with electric motors is the belt drive. One reason for this is because of the high feet per minute capability of the belt drive compared to other common drive systems of gears and chains. However, the direct drive system discussed later in this unit is also capable of transmitting high rpm , but the feet per minute is not applicable here. Feet per minute is in reference to the surface movement of a belt, chain or gear. For example, if a five-inch pulley is attached to a motor's power shaft turning 1725 rpm, the feet per minute of the belt is 2256.9, because:

$$\frac{\text{feet}}{\text{min}} = \frac{1\text{ ft}}{12"} \times \frac{15.7^*}{\text{rev}} \times \frac{1725}{\text{min}} = 2256.9$$

*15.7 inches of circumference of 5" D x Pi

Belts run quietly, are relatively inexpensive to replace and have very good shock absorbing characteristics. Among the types of belts available, the V-belt is the most common. The flat-belt drive has lost it popularity to the V-belt. Both systems are shown in Figure ll-4.

Flat belts can have two styles of pulleys, the slightly oval, Figure 11-4 (b), or the "shouldered", (c). The V-belt (a) does an outstanding job of transmitting power because when the belt is tensioned, it tends to squeeze downward into the "Vee" of the pulley making considerable surface contact to the sides. When the belt wears out, it will ride deeper in the groove and make contact only on the bottom, and not the sides, therefore will "slip". Always be suspicious of a V-belt riding too deep into its groove because it is probably slipping.

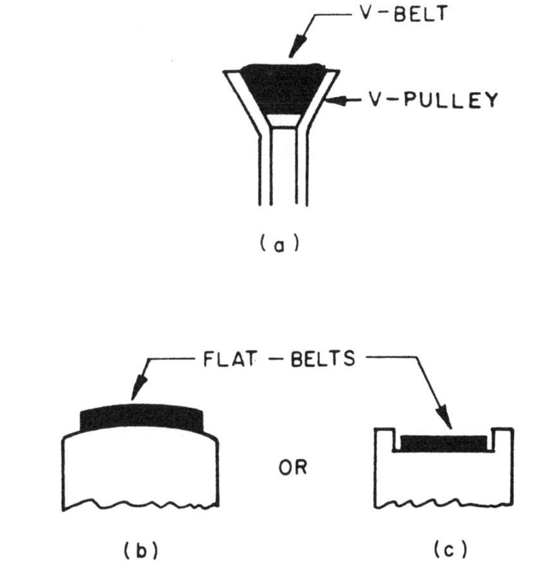

Fig. 11-4. Common Belts and Pulleys Used in Driving Systems

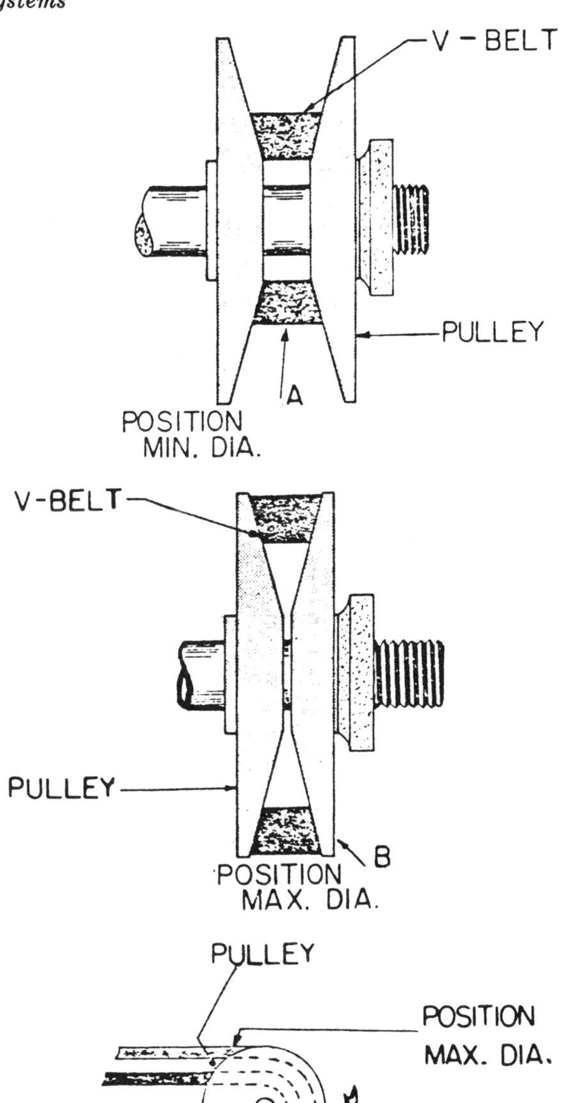

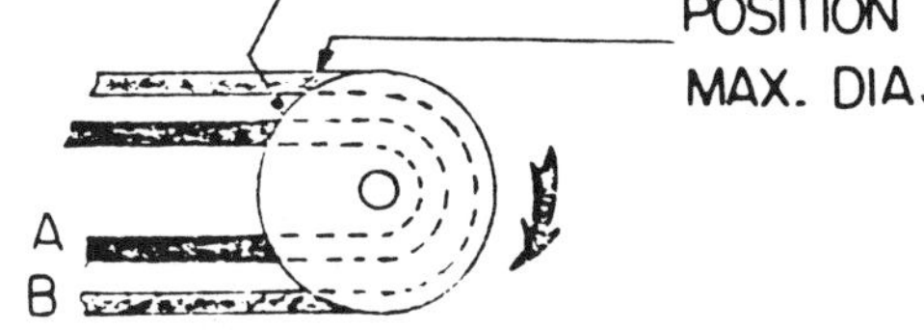

Fig. 11-5. Adjustable or Variable Sheave Pulley

The adjustable or variable sheave pulley, illustrated in Figure 11-5, is used to change the driven speed of a machine. This type of pulley may be mounted to the motor's shaft, the machine's input shaft, or both. The purpose is to cause the V-belt to ride lower or higher within the sheave and, therefore, to change the effective diameter of the pulley easily by sheave adjustment. Assuming the pulley is a motor driver, as in Figure 11-5, if the pulley is adjusted for position A (riding lower within the sheave), the pulley would act as a small diameter pulley and drive the machine at a slower speed. If the sheaves are closed, position B, the belt would ride higher, acting as a larger diameter pulley and would, therefore, drive the machine faster. One common application of this arrangement is with motors driving squirrel cage fans on home heater units and the variable speed drill press. The user can match the load characteristics of the motor and the load without changing pulleys on either the motor or the driven load.

Some of the common V-belts and pulleys are presented in Figure 11-6. Note, the width of the top part of the belt matches the top of the groove of the pulley. The 3/8 inch belt is a fractional horsepower belt (FHP), and used commonly with motors of less than one horsepower. A 3/8 inch belt is 3/8 inch on its top measurement, and this system holds true in sizing and replacing all various widths of belts. The A-section belt is used with larger fractional horsepower applications, especially one-half and three-quarter hp motors. The B-section belt is generally used with smaller integral hp motors. Larger motors require larger belts and these common sizes continue with 7/8", 1-1/4" and 1-3/4" widths.

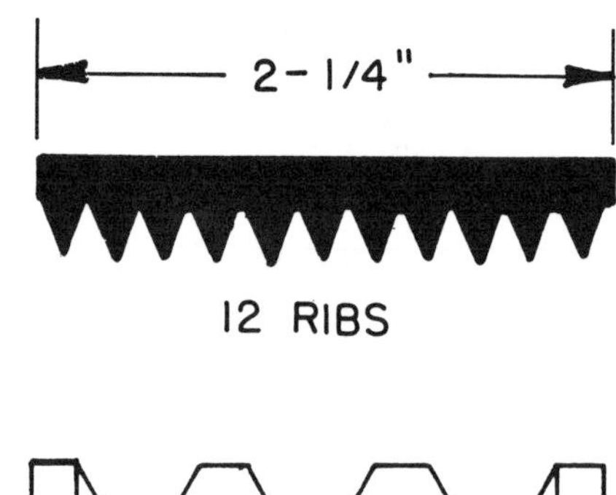

Fig. 11-7. Multi-Groove Belt

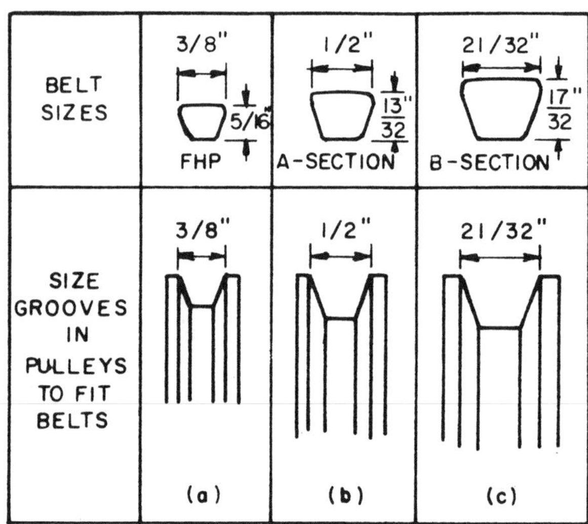

Fig. 11-6. Common V-Belts and Pulleys

The webbed multi-v-belt system is also used for high horsepower applications. For example, three belts may be side-by-side, each running in a multi-groove belt pulley as illustrated in Figure 11-7. When belts wear out on these multiple belt drive systems, it is important that all belts be purchased as a matched webbed set and replaced as a set, otherwise, serious belt slippage may exist. The V-belt stepped pulley is sketched in Figure 11-8. There may be two to five steps on the pulley. Usually, both the driver and driven will have a stepped pulley on their respective shafts, but the steps will be opposite each other. If the belt is on the largest diameter of the driver and the smallest of the driven and you want to change to a smaller step on the driver and a larger step on the driven, the speed change to a lower driven rpm may be made with a minimal re-adjustment for belt length. Many drill presses and wood or metal lathes use this arrangement for quick and easy speed changes.

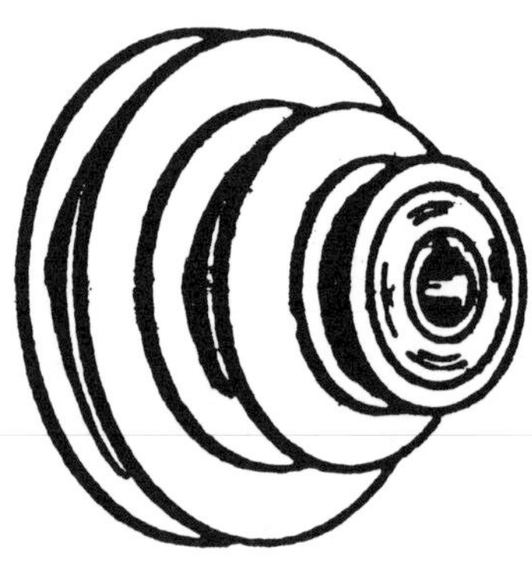

Fig. 11-8. V-Belt Stepped Pulley

In recent years, the ribbed belt has become popular on large hp motor applications. A belt for a 12-ribbed pulley is sketched in Figure 11-7. Although, the one shown is 2-1/4" wide, various widths and differing number of ribs are available. Notice that each rib of the belt is pointed, and an outstanding contact of the belt to the pulley is made when the belt is properly tensioned.

CHAIN AND GEAR DRIVES

Other types of drive systems to transfer power from the motor to a machine are chain drive, gear drive, and direct drive systems. Chains go between sprockets from the driver of the motor to the driven of the machine, Figure 11-9. The standard pitch roller chain is good for linear surface speeds up to 4500 feet per minute. Basically, this means that a driver pulley of four inches in diameter attached to a 3500 rpm motor is approaching the critical speed in feet per minute. Rather, industry uses the lower speed synchronous motors of less than 1800 rpm and smaller diameter drivers in applications of the motor chain drive system. High quality chains must also be used. Chains tend to be noisy, alignment is critical and regular maintenance of the roller chain is a must. The main advantage of the chain drive system is the no slip, because of the positive engagement of the sprocket teeth to the chain. Additionally, the transfer of power from driver to driven is about 98 to 99 percent efficient in a well-maintained drive. V-belts lose about five to ten percent efficiency in this transfer of power, because of the friction and heat loss as the belt is slipping within the groove of the pulley.

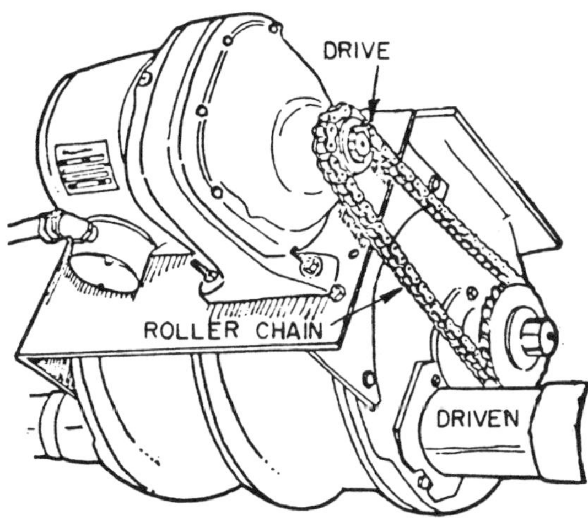

Fig. 11-9. Roller Chain and Sprocket Drive Systems

There are numerous gear drive systems used with electric motors. To discuss the different systems is beyond the scope of this manual. However, the most commonly used arrangements are the worm gear and the spur gear, note Figure 11-10. The worm gear, Figure 11-10 (a) is often used where a large value gear reduction is desired, for example an 8:1 ratio. This means that for every eight revolutions of the driver worm attached to the motor's shaft, the output shaft of the large center piece will turn only once. The motor and gear drive system shown in Figure 11-11 is a typical application of the worm drive gear system.

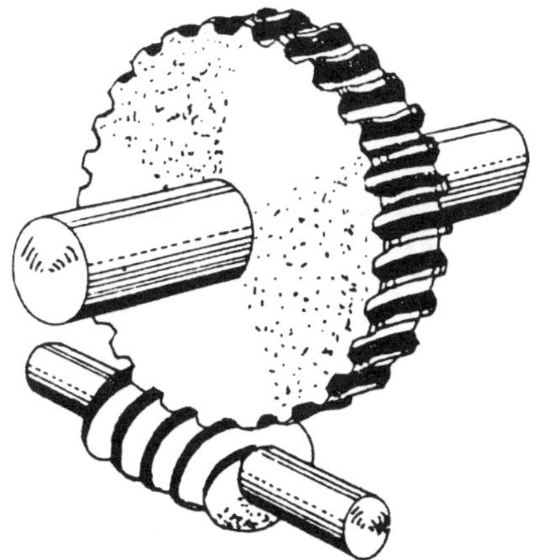

(a) WORM GEAR

(b) SPUR GEAR

Fig. 11-10. Two Commonly Used Types of Gears Driven by Electric Motors

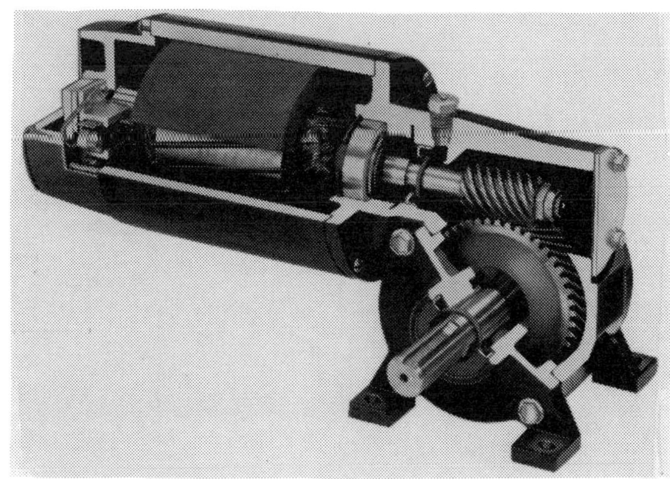

Courtesy of the Bodine Electric Company, Chicago.

Fig. 11-11. Single Reduction Right Angle Gearmotor

The spur gear, Figure 11-10 (b), is used frequently on portable electric power hand tools, such as illustrated by the exploded view of the electric drill drive in Figure 11-12. Note the reference number 8, which is a stepped-spur gear assembly. The universal motor's shaft with a small spur gear attached to it, not shown in the illustration, drives the large gear of this stepped gear assembly. The smaller gear of assembly number 8, drives gear number 10, making a further step-down in speed. The motor of a drill such as this may run at 2000 to 3000 rpm. However, through the step-down spur gears, the drill chuck, number 15 may operate at a speed of 500 rpm. The driving systems discussed previously all have a driver and a driven pulley. The direct-drive system discussed in the latter part of this unit, technically has only a driver. Certain rules and formulas apply to better understanding of directional rotation, change of power flow, rpm or speed and slippage.

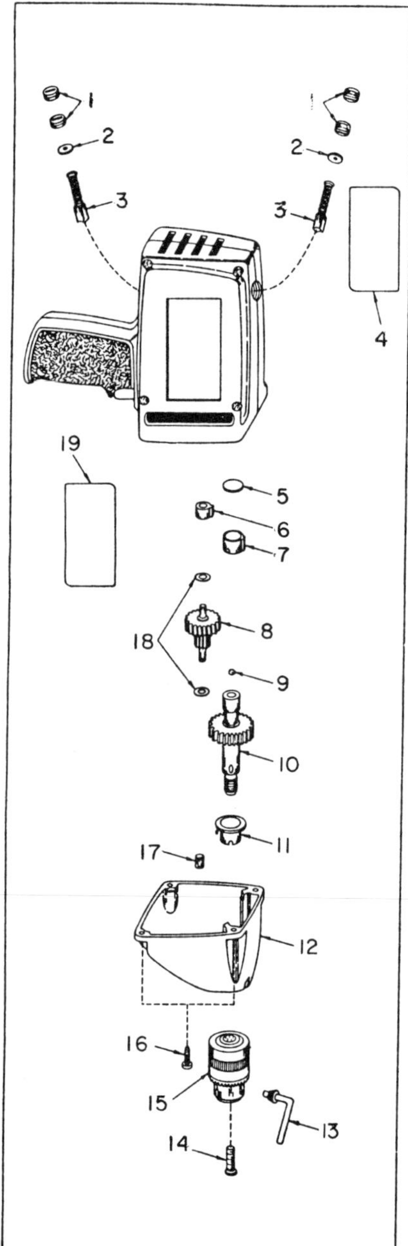

Fig. 11-12. Electric Drill Drive

Table 11-2. Characteristics of Types of Common Motor Drives

TYPE OF DRIVE	SLIPPAGE PERCENT	EFFICIENCY PERCENT	ROTATIONAL CHANGES	DIRECTIONAL CHANGES
Belt	5 to 10	90 to 95	None	Parallel
Worm Gear	None	92 to 95	Either	Right Angle
Spur Gear	None	95 to 97	Reverse	Parallel
Chain	None	98 to 99	None	Parallel

Two formulas which are much alike, may be used for pulleys and V-belts (assuming no slippage) and spur gears and chains, relative to rpm or speed changes for motors driving machines. For belts use:

D(Dr) x RPM(Dr) = D(Dn) x RPM(Dn)

where D equals pulley diameter in inches, RPM equals revolutions per minute, DR equals driver and DN equals driven.

For chains and spur gears use:

T(Dr) x RPM(Dr) = T(Dn) x RPM(DN)

where T equals number of teeth on sprocket or gear, RPM equals revolutions per minute, Dr equals driver and Dn equals driven.

For example, to operate a belt driven fan at 650 rpm, with a 1725 rpm motor and a 3-1/2 inch pulley determine the size of driven pulley that should be used on the fan. To solve this problem use the belt formula and insert the three known factors in the equation and solve for the fourth:

3.5" D(Dr) x 1725 RPM(Dr) =

D(Dn) x 650 RPM(Dn) And

D(DN) = 3.5 x 1725 / 650 = 9.29"

Because a pulley cannot be purchased in a 9.29" size, the user would probably purchase an 8-1/2 or 9 inch pulley. After some normal slippage of the belt, this would approximate the desired 650 rpm.

For another practical example, a motor that operates at 1150 rpm, has a sprocket with twelve teeth and drives another sprocket with eighteen teeth. How fast will the driven sprocket turn? Add the three known quantities to the equation and solve for the fourth. Use the chain or spur gear formula:

12 Teeth T(Dr) x 1150 RPM(Dr) =

18 Teeth T(Dn) x ?? RPM(Dn) and

RPM(Dn) = 12 x 1150 / 18 = 767 RPM

DIRECT DRIVE

Motors are often directly driven to the machine being powered. An example of this is with various pump applications, such as illustrated in Figure 11-13. The pump's impeller is a threaded arbor and the motor's rotor shaft is threaded externally. When threaded together, the direct drive connection is made. A pump seal is used to keep the water or other liquid away from the motor's bearing.

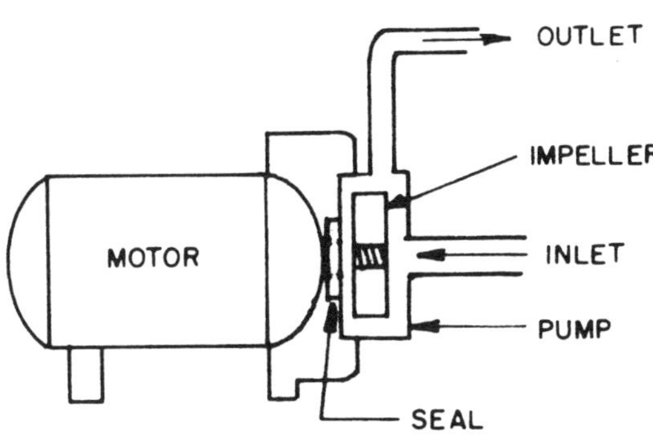

Fig. 11-13. Water Pump Using The Direct Drive System

Other types of direct drives are used with pumps and various powered machines. These systems frequently have some type of a coupling to join the motor's PTO shaft to the input shaft of the machine. Several of these are pictured in Figure 11-14. All of these have one thing in common. They are classified as resilient because fiber, rubber or some other material that will help absorb shock is in the drive line. The resilient drive also helps to overcome imperfect alignment between the motor and the machine. In Figure 11-14, (a) is called a flange coupling; (b) is a braided hose and is held on the shafts with mechanical clamps; (c) is a cushion flange and is made of a high density rubber material; and (d) is a flexible coupling with a floating non-metallic center member, either of a rubber or a fiber material.

There are several other applications of direct drives that are extremely common. They are the portable electric hand saw and the typical wheel grinder. In both cases, the saw blade or the emery wheel is held in place with a nut threaded on the direct drive shaft of the motor. The flexible shaft drive as illustrated in Figure 11-15, is a form of the direct drive system. A cable is housed inside of another cable. The inner cable is rotated by a direct connection to the motor's shaft, while the outer cable is stationary yet flexible to be moved about by the user. Applications of this drive system are mainly used for grinders and rotary wire brush powered units.

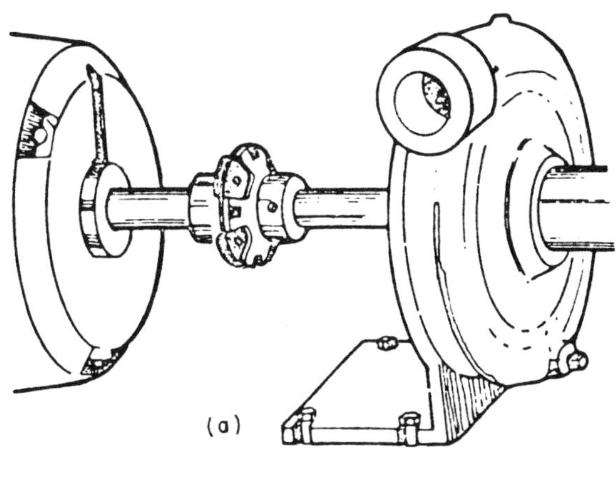

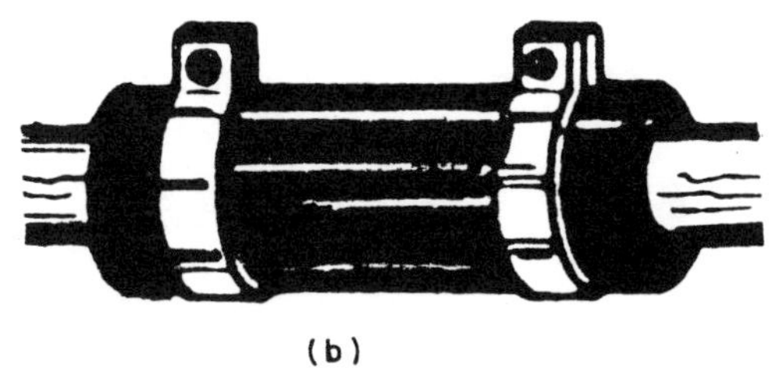

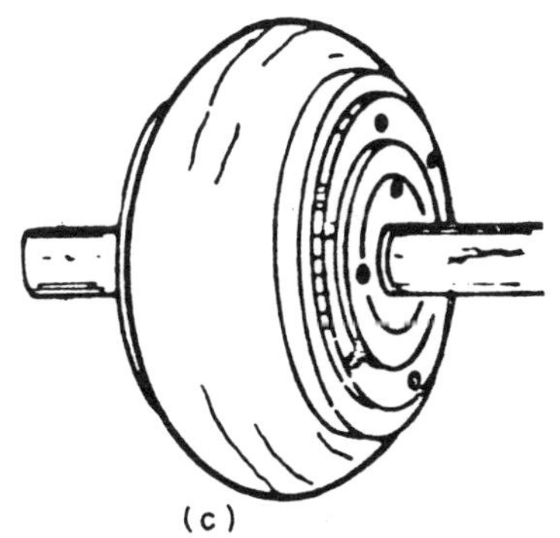

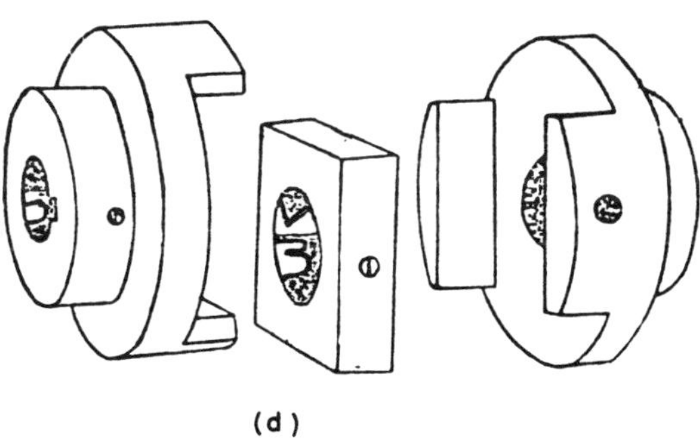

Fig. 11-14. Common Types of Couplers

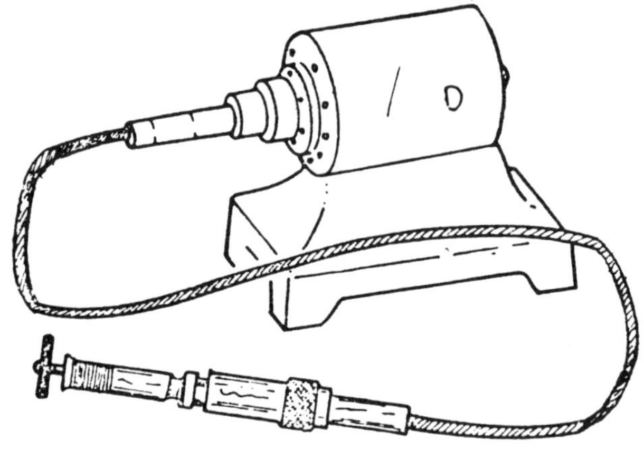

Fig. 11-15. Flexible Shaft Drive System For Grinders and Brushes

Consider all the environmental factors which are "enemies of the electric motor" when making an initial motor installation or when replacing an electric motor which has failed. With all the environmental factor problems solved the motor can still fail if misaligned to the load or the incorrect style of drive system has been selected. Remember the manufacturer can make a near perfect motor and place it on the shelf. You can destroy this motor with improper installation and lack of service and maintenance. Learn about ELECTRIC MOTOR MAINTENANCE in Unit XII and TROUBLE SHOOTING in Unit XIII.

****

DEFINITION OF TERMS

ALIGNMENT: Two or more parts falling or positioned along the same plane or line.

AMBIENT: The existing conditions. For example, if the ambient condition of air is 70 degrees F, the ambient temperature is said to be 70 degree F. Ambient conditions are also commonly expressed for relative humidity and other factors of the air, including unnatural contaminants.

CLEAN ROOMS: Relatively closed rooms or facilities in which the environment is kept clean by filtering and air treatment techniques in an attempt to keep air-borne contaminates at very low level.

DRIVEN: A reference made to a pulley or sprocket attached to a machine being powered by an external force such as a motor.

DRIVER: A reference made to a pulley or sprocket attached to a power unit such as a motor.

ELECTROSTATIC FILTER: A filter which is charged with either a negative or positive charge having the capability of picking up air-borne contaminants.

END-PLAY: The movement in linear distance of one part in relation to a stationary part.

ENVIRONMENT: The air, space, and other conditions in and around the area described. For example, a dirty working environment could be a condition where air is contaminated with dust particles.

SLIP (BELT): The inability of a belt to maintain linear speed equivalent to that provided by the sheaves.

SPROCKETS: A wheel which has teeth or cogs around the circumference and normally drives or is driven by chains or mating teeth of gears.

THRUST WASHERS: Flat washers which take up space between the end of the rotor and bearing area of an electric motor. They are typically used for adjusting end-play and reducing wear of parts due to axial forces or thrust.

VIBRATION: Back and forth movements that tend to shake motors and equipment whose rotary moving parts are not perfectly balanced.

WHITE ROOMS: Same as CLEAN ROOMS. However, in white rooms, workers usually wear white work clothing.

NOTES

CLASSROOM EXERCISE XI-A

Motor Installation

1. The environmental factors affecting the life of a motor are____________________,

   ____________________, ____________________,

   and ____.

2. Match the best description on the right with an environmental problem on the left.

Dust ____________________	a. Becomes gum when mixed with dirt
Moisture ____________________	b. Possible shorting
Stray oil ____________________	c. Motors operating in a contaminated area
Air ____________________	d. Reduces motor cooling

3. What is meant by a "white room"? ______________________________

   ____

   ____

4. Which type of motor can be high pressure washed to clean its exterior? Open enclosure or totally enclosed______________________________________

5. Define alignment. ____________________________________

   ____

   ____

   ____

6. What harmful effect results from a misaligned motor? Give two

   a. ____

   b. ____

7. If a motor is misaligned and no end-play is present, the ____________________ inside the motor may be destroyed.

8. One of the following motor bearing types is least affected by misalignment. Sleeve type or anti-friction bearings. Why?

____

9. What condition literally shakes everything apart? ______________________

10. What is the purpose of a drive system? ______________________

____

____

11. Compute the linear surface movement of a four-inch diameter pulley in feet per minute if:

    a. The motor operates at 1725 rpm ______________________

    b. The motor operates at 3450 rpm ______________________

12. List two advantages that drive belts have over other drive systems.

    a. ____

    b. ____

13. What is the measured height of a 21/32" V-belt? ______________________

____

14. What is the name of the belt which rides in several grooves and is used for high horsepower application?______________________

15. List two shop tool applications where a V-belt stepped pulley is commonly used.

    a. ____

    b. ____

16. The standard pitch roller chain is good for linear speeds up to ______________ feet per minute . Could it be used on a 1750 rpm motor driving a three-inch sprocket? Yes or No.

Show calculations above

17. Which is noisier: Chain or belt drives? ______________________________ WHY?

____

18. Two commonly used gear type driven electric motors are ______________________

and ____

19. Compute the rpm of a driven shaft with a two-inch pulley when the motor turns at 1730 and has a six-inch driver.

Show calculations below. Answer:______________________________

20 A spur gear drive motor operates at 1725 and has 15 teeth directly attached to the motor's shaft. The output shaft of the gear box is running at 575. How many teeth are on the output shaft's gear?

Show calculations below. Answer:______________________________

21. In the transfer of power from the driver to the driven and when considering belts, chains or gears:

a. Which two systems are most efficient?

______________________and______________________

b. Which system has the most slippage? ______________________

c. Which system has the least maintenance? ______________________

d. Which system is the most expensive? ______________________

e. How can you reverse the direction of the driven pulley as compared to the drive pulley?

____

LABORATORY EXERCISE XI-A

Motor Installation

1. Go to the laboratory, home, farm, or local store and study the drive systems of at least four different tools that are belt, chain, or gear driven and powered by electric motors. Complete the chart below.

POWER TOOL	TYPE OF DRIVE SYSTEM	DRIVER DIAMETER (No. of Teeth)	MOTOR DRIVER RPM	DRIVEN DIAMETER (No. of Teeth)	TOOL DRIVEN RPM

2. List three direct drive applications of motors to tools. Give full load rpm and wattage of each. Assume power factor of 0.65.

a. ____________________ RPM ____________________ Wattage ____________________

b. ____________________ RPM ____________________ Wattage ____________________

c. ____________________ RPM ____________________ Wattage ____________________

3. Check two or more motors in the school shop or the home shop and note if they are being affected by dust, moisture, stray oil, or contaminated air.

MOTOR APPLICATION	ENVIRONMENTAL FACTORS AFFECTING THE MOTOR
a.	
b.	
c.	

ELECTRIC MOTOR MAINTENANCE

ANALYZING THE SITUATION

An electric motor, note Figure 12-1, needs service and maintenance as does other equipment. Motors have moving parts and these parts must have regular and routine maintenance, otherwise break-downs can be expected. There has been a problem with a machine driven by an electric motor. The problem has been identified as motor malfunction, therefore, make the following quick checks: (1) the bearings are in good condition and operating properly; (2) there is no mechanical obstruction to prevent rotation; (3) all bolts and nuts are tightened securely; and (4) proper connection to drive machine or load has been made. Motors generate heat and if heat is not properly and continually removed, the motor overheats, often causing serious damage to internal windings and other parts. The causes of motor heating are listed in Table 12-1. Frequent starting and overloading is caused by the motor application. Ambient temperature problems are caused by the environment and ventilation. The loss of ventilation can be corrected by proper motor maintenance. However, regular maintenance, discussed in this unit, should help in reducing the amount of "down-time", expensive repair costs and the expense of motor replacement.

Fig. 12-1. Motor in Need of Servicing

Table 12-1. Causes of Motor Heating

---

1. Frequent Starting

2. Overloading

3. High Ambient Temperature

4. Loss of Ventilation

In UNIT XI, MOTOR INSTALLATION, the numerous enemies of electric motors were discussed: (1) guarding against the environmental conditions where a motor will be installed; (2) problems associated with alignment, vibration and drive system; and, (3) providing satisfactory protection. When proper installation is accomplished it is indeed the first important step of motor maintenance. However, motor maintenance does not end here because things can and will go wrong. For example, consider a dusty environment where a motor with an air flow-through cooling system is being used. After normal use, perhaps for only a few months, dust particles drawn through the motor's internal parts begin to build-up and adhere to windings and work into non-sealed bearings. If the dust is not removed from the motor the air-flow through the motor will be reduced and the motor will overheat causing damage to insulation of the windings and ultimately burned-out windings. Likewise, dust contaminated bearings will reduce the free movement and relatively friction-free action normally expected of bearings; and, the result will be bearings with an excessive amount of friction. Additional friction is wasted energy and the hp needed to run the driven machine would be inadequate. The motor continues to try and do its job but it overheats and leads to the sequence of events ending in motor winding burn-out. The few minutes spent on motor cleaning and maintenance would have avoided the problem. Motor maintenance safety rules are listed in Table 12-2.

Table 12-2. Motor Maintenance Safety Rules

---

1. Do not perform any maintenance or service on any motor before disconnecting the power source.

2. Discharge all capacitors before servicing a motor.

3. Always keep hands and clothing away from moving parts.

4. Electrical repairs should be performed by trained and qualified personnel only.

5. Failure to follow instructions and safe electrical procedures could result in serious injury.

6. If safety guards are required, be sure the guards are in use.

Earlier it was mentioned that proper installation is the first step in good motor maintenance. The second step is checking for electrical troubles, be sure that: (1) the line voltage and frequency correspond to the voltage and frequency stamped on the nameplate of the motor; (2) the voltage is actually available at the motor terminals; (3) the fuses and other protective devices are in the proper condition; and (4) all connections and contacts are properly made in the circuits between the control apparatus and motor. Next it will be essential to know the amperage draw of the motor. Most problems leading to serious motor malfunctions can be detected early with an ammeter. The clamp-on style ammeter is easy to use, relatively inexpensive (as compared to motors) and provides the maintenance person an easy check of the amperage draw during normal motor usage. A typical ammeter is shown in Figure 12-2 being used to measure amperage draw by a motor in the 120 volt circuit.

Fig. 12-2. Using the Clamp-On Style Ammeter to Check Amperage Draw By a Motor Driving a Machine

The motor nameplate information lists the amperage at 100 percent design rated load. If the motor is drawing more than 100 percent of the rated load, expect problems and look for reasons that may be causing the overload. The following unit, TROUBLE SHOOTING, presents logical methods in a SYMPTOM—-POSSIBLE CAUSE—-REMEDY format that should help the maintenance person arrive at what to do about an overloaded motor that is probably overheating. If motors are installed properly, the overload circuit protectors, whether an integral part of the motor's wiring or placed in the feeder line, should trip and break the electrical service to the motor at about 125 percent of the rated load. However, overload protectors are man made and are not infallible, especially as they get old and start to wear. Routine and regular checking with the ammeter will not cause the maintenance man to depend only on the overload protectors. The knowledge of what is happening on amperage draw is extremely important in providing high quality motor maintenance. Medium and large sized factories and industries do an outstanding job of measuring amperage draw of motors during maintenance because they realize the cost of "down-time" due to motor malfunction. Perhaps, one of the poorest occupational groups in checking the amperage of motors is the agricutural worker. The authors believe that agriculturists, who depend heavily upon motors, should own and regularly use clamp-on ammeters for checking motors.

If the motor problem had not been electrical it could have been the bearings. Continue to learn how to analyze and handle bearing problems.

BEARING MAINTENANCE

Bearings, whether the sleeve or roller type, need regular attention. The reader should reacquaint himself with information found in Fig. 2-12 through Fig. 2-16 for the purpose of part identification for the sleeve and the anti-friction types commonly used in motors.

The worst condition that can happen is for the bearing to go "dry". On a sleeve bearing, the motor's shaft would be sliding in a metal to metal contact surface if no oil film were present. There would be excessive heat generated from the moving metal parts and the parts would groove into each other and binding could occur. Ball bearing must roll freely. Without oil they will heat, the balls will expand, but are limited to the confines of the inner and outer race. Therefore, dry ball or roller bearings lacking lubricants will also bind. The binding of moving parts would result in extra heavy loads, beyond those normally needed for the driven machine. This situation leads to the consequences of overload and usually results in burned out motor windings. The best way to maintain motor bearings is to follow the directions, service literature and/or operator's manual supplied with the motor. If not readily available, manufacturers are cooperative in supplying service literature upon request. Sleeve bearings are mostly used on fractional horsepower motors, and as discussed in Unit II, are capable of continual outstanding service when installed and maintained properly. Generally, sleeve bearings can withstand no more than about 20 pounds of force with a belt pulling to one side. The oil film that separates the outside of the shaft and the inside of the sleeve bearing, lined with babbitt material, is an outstanding bearing when lubricated as recommended. It would be improper to recommend any type, grade or viscosity of oil for lubricating sleeve type bearings in this manual. The recommended service literature can only be obtained from the manufacturer. However, over the years manufacturers have often, but not always, recommended relatively low viscosity (5W, 10W, or 20W), non-detergent, mineral oils. Some manufacturers have recommended oils with lithium, lead, copper and graphite as oil additives. It is common for motor manufacturers to dictate a certain oil by brand name, grade and viscosity. In emergencies lubricate smaller hp motors with 5W or a 10W and larger hp motors with 20 W non-detergent oil, until the proper type can be determined. After all, a lubricated bearing, even with the wrong oil, is better than to have the "dry" bearing situation develop and lead to an eventual motor burn out. Do believe and follow the service recommendations from the manufacturer.

Ball and roller bearing require a small amount of lubricant. It has been calculated that 1/1000 drop of oil will lubricate all surfaces of a 10mm bearing. However, the lubricant must be constantly replenished. While the intervals between addition or replacement of lubricant in bearings can be long, lubricants must always be present. Otherwise, bearing life may be affected by wear which might easily have been avoided.

It is not possible to precisely predetermine when a bearing lubricant must be renewed since the lubricant in a bearing does not suddenly lose its lubricating ability. The lubricants effectiveness is gradually lost during motor operation. The length of time a bearing can be run without the addition or replacement of lubricant depends on lubricant properties and the bearings operating condition as speed, load, and temperature. Most equipment manufacturers determine lubrication intervals by merging the recommendations of the bearing manufacturer with their own test experience. The results are passed on as recommendations to the ultimate users of the equipment in the maintenance instructions accompanying the equipment. A ball bearing's lubricant, whether grease or oil, is usually fortified with additives to improve its properties. Storage capabilities are also taken into account in predicting total life because bearings or equipment may be stocked for long periods before use. Greases in grease lubricated motors may tend to harden and even separate when stored for prolonged periods. Temperature is the greatest enemy of lubricants, whether generated with the equipment containing the bearings or transmitted from the surrounding environment. Generally, the most important single property of an oil, whether used alone or mixed with a "soap" to make grease, is viscosity and its relationship to operating temperature. An increase in temperature can cause a rapid thinning of the oil, resulting in rapid loss from the bearing through leakage or evaporation. For this reason, oils with very high viscosity index (VI) numbers are specified for ball bearing lubrication. High VI number oils are desirable because they thin out relatively less with increasing temperature than low VI number oils. "Natural" base oils do not generally have very high VI numbers (though some, such as paraffinic oils, have higher VI's than napthenic oils). Chemical additives increase the viscosity index, fortify oxidation resistance, increase film strength, provide detergent properties, furnish corrosion resistance, provide extreme pressure properties, and lower the pour point.

For ball bearing lubrication in electric motors, grease is generally preferred over oil to overcome the problem of providing an adequate supply of lubricant for long maintenance free service. This preference is due to rapid progress in the development of better ball bearing greases, simplification of bearing housing design, and elimination of the "human maintenance factor" which is frequently responsible for too much lubrication, not enough lubrication, or the wrong kind of lubricant. Prelubrication of the bearings with the correct amount of the correct lubricant and the elimination of grease fittings provides for long, maintenance free service of ball bearings which sometimes last for years.

When used in ball bearings, several characteristics of lubricating greases are important. They are:

1. Consistency-The grease must have enough body to stay in the bearing and not run out, but it must not be so stiff that it will impede the movement of the bearing components.
2. Dropping Point-The temperature at which the first drop of fluid flows from the grease.
3. Shear Stability-The ability of a grease to resist structural breakdown resulting from the shearing forces present when a grease is "worked" in a ball bearing.
4. Oil and Soap Separation-A grease should be relatively resistant to oil-soap separation.
5. Channeling-While a certain amount of channeling of a grease is desirable, there must be enough lubricant flow to keep the bearing surfaces lubricated.

The major functions of a ball bearing lubricant are; (a) to dissipate heat caused by friction or bearing members under load, and (b) to protect bearing members from rust or corrosion.

"Premature" bearing failures are caused by one or more of the following conditions: (a) Foreign matter in the bearing from dirty grease or ineffective seals; (b) Deterioration of grease due to excessive temperature or contamination; and (c) Overheated bearings as a result of too much grease or an overload. Some danger signals are: (a) A sudden increase in the temperature differential between the motor and bearing temperatures is an indication of malfunction of the bearing or lubricant; (b) A temperature higher than that recommended for the lubricant warns of a reduction in bearing life (rule of thumb is that grease life is cut in half for each 25 degrees F increase in operating temperature); and (c) An increase in bearing noise, accompanied by a bearing temperature rise, is an indication of a serious malfunction of the bearing.

In summary, successful ball bearing lubrication depends on the proper selection and application of the lubricant. The best lubricant available will fail if improperly matched to operating conditions. Correst lubrication and maintenance practice will determine to a great extent, not only the bearing life, but also the life of the motor. Because of the many factors involved, and variations found in lubricants follow the equipment manufacturer's recommendations for best results.

CLEANING THE MOTOR

The free flow of air through open enclosure motors, whether drip proof or splash proof, is a must for motors to maintain or stay below their proper designed temperature rise rating. Therefore, most motors will from time to time need to be disassembled, cleaned, lubricated and properly reassembled unless they are operating in very ideal environmental conditions. Contaminants, such as dust particles being drawn through a motor's interior, will eventually adhere to motor parts clogging or retarding the free flow of air currents around critical windings which must be cooled or motor burn out will result. Motors can be difficult to disassemble. Part sequencing during removal is very important or motor parts can be damaged. Often motor removal from the machine, especially when the motor is an integral part of a piece of equipment, as a portable sander or a water pump, can be difficult. Unless the service literature or exploded views are available, it would be best for the layman to leave removal and repair of motors that are an integral part of the machine to the professional repair person. When motors have been operating in less than ideal environmental conditions, such as high moisture, fastening parts and through bolts may be rusted causing problems in disassembly. Motors operating around agricultural fertilizer and similar chemical conditions experience corrosion and serious difficulties exist when being "taken apart" for clean-up purposes. The layman or student who receives adequate training in electric motor work and has the common and necessary specialized tools is no longer a layman, but could be called a trainee or apprentice. Furthermore, when he gets good enough, he will call himself a professional.

The Laboratory Exercises, A, B, and C at the end of this unit are excellent for fractional hp sleeve type bearing motor disassembly, cleaning, lubricating and reassembly. It is beyond the scope of this manual to cover the repair activities of the hundreds of different motor configurations and machines the small appliance repairman would encounter. It is not the intention of this manual to provide information on motor work for large hp motors, whose parts during disassemble must be hoisted and windings rewound by persons trained specifically in the technical field. It was the intent to give students an opportunity to work on motors which are relatively easy to trouble shoot, disassemble, detect what lack of maintenance does to a motor, ability to cleanup internal and external parts, lubricate bearings, reassemble and test run.

In a suggested sequence, a typical single-phase fractional hp motor (split phase-start or capacitor-start) will be disassembled, cleaned, minor repair activities performed, reassembled and operated.

****

CONSTRUCTING EXPERIMENTAL MOTORS

All information for the experimental electric motors: From Loper, Orla; Ahr, Arthur; and Glendenning, Lee. Introduction to Electricity and Electronics (c) 1979 by Delmar Publications, Inc. Used with permission of Delmar Publishers, Inc.

NOTES

UNIT XII

Electric Motor Service Repair

STEPS	ACTIVITY	SUGGESTIONS
1	Remove motor from machine and electrical service.	Note and sketch the electrical lines, power cords, fusing devices and work SAFELY around all electrical service.
2	Remove any belt pulleys. File and emery cloth PTO shaft(s) and check with micrometer to make sure the shafts will slide out the bearing journal.	See Unit II for pulley removal.
3	Remove power cord from terminal block of motor and use temporary cord so that the amperes of the motor when running can be learned by using the induction clamp-type ammeter.	Note and sketch power cord lines, the terminal block and leads by colors and/or by labeling. Record the amperage draw. See Unit XIII, TROUBLE SHOOTING, for possible problems, causes, and cures.
4	Wipe dirty material from exterior of motor.	Use soft shop towels or paper toweling.
5	Remove motor base cradle if a resilient mount system is used.	Note and sketch or mark with metal scribe the relationship of parts.
6	Determine end-play of motor shaft. Most motors should be about 1/64 to 1/16 of an inch on sleeve bearing motors and near nothing on bearing motors.	Note and record the end-play.
7	If end-play is excessive, operate the motor and determine where end thrust washers need to go. Rotors will line up properly with the magnetic field when running.	Note and record how many, the thickness, diameter, and where thrust washers need to go.
8	Move the PTO shaft(s) up and down and left and right to note any slack in the bearing to shaft relationship.	Normal or good bearings should have no noticeable or felt slack. If there is, the bearings are bad and repair of the bearings and/or rotor shaft is needed.
9	Mark disassembly orientation marks with center punch on end bells and stator frame to assist for alignment during reassembly. See Laboratory Exercise XII-A at the end of this unit.	Use two sets of marks on the electric end and one set of marks opposite the electric end. The electric end is where the power cord or line voltage attaches to the terminal block.
10	Slightly loosen all through bolts, one at a time, in a staggered sequence. Loosen fully and remove.	Note and sketch the direction through bolts pass through the motor as to bolt end and nut end. Record any odd length through bolt.

STEPS	ACTIVITY	SUGGESTIONS
11	Usually remove the end bell with electrical connections first, sometimes the PTO end will be removed first.	Use a hardwood dowel (3/8 or 1/2 inch diameter) as a punch to lessen damage to motor end bells of stator frame.
12	Pull the rotor from the stator. Drag the rotor over the field winding surfaces as little as possible in this procedure. One ball bearing will stay with the rotor shaft.	Some motors will have a clip on the opposite end making this manuever impossible until the clip is removed. This is frequently true of ball bearing types of motors.
13	Inspect the rotor and stator field winding surfaces for possible rotor drag.	If found, new bearings are needed.
14	Blow out the excessive "dirt" with air pressure of less than 35 psi. Dirt may also be wiped off with soft toweling. Approved motor solvents only are to be used when cleaning	DO NOT USE gasoline, diesel fuel or any other solvent that may damage and melt enamel on the windings.
15	Inspect the shaft areas where they ride on sleeve bearing after removing the thrust washers. Measure the shaft bearing area with micrometer. There should be no more than 0.003 inch less than at the adjacent shaft diameter area.	Sketch and label all thrust washers and their relative position to each end of the rotor shaft.
16	Check the centrifugal flyweight starting mechanism attached to the rotor to note if it is free to swing outward. Check the small return springs for wear, rusting, and stretch.	If springs are bad or rusted, replace them. Reattach any spring ends that may be loose. Lightly lubricate these parts.
17	Check the electrical starting switch contact points in the electric end of the end bell. If contacts show poor continuity, they should be filed with an emery board or ignition point file. Check the closing contact spring tension and replace if weak.	Place a continuity tester (or ohmmeter) across the leads going to the contact points. The continuity should show full scale on an ohmmeter or a bright light on a continuity tester.
18	A. Checking the quality of the windings: All windings must be isolated from each other for this check. A winding will have two leads (one going to the winding and one leaving the winding.) Use the HOBAR BE-61 electrical continuity tester for performing this test.	Removing leads from the terminal block may be necessary. Some leads are soldered and may be unsoldered for test purposes and resoldered later. Make a detailed drawing or sketch and identify lead positions by numbering, noting existing colors, (continued)

STEPS	ACTIVITY	SUGGESTIONS
18 (CONT'D)	The starting winding should show a very dim light. The running winding(s) should show a fairly bright light. No light means an "open" winding, and the motor is normally not repairable unless the problem is in an exterior lead. Make sure the capacitor (if there is one) is by-passed when checking continuity of the starting winding. The capacitor may also be disconnected and its leads joined. Still another method is to test on the back side of the capacitor on the lead going directly to the starting winding. A DC powered continuity tester will not flow a current through capacitor.	or by tagging to assist in terminal reattachment. The starting switch is normally closed at rest and should permit the flow of a DC test current. If not, go back to Step #17. A "shorted" or bad capacitor can be detected with the continuity tester, because it will conduct the DC current through it. Microfads values and leaks can only be determined with special test equipment. Always wear approved eyewear when completing electric motor service and maintenance activities.

See the following sketches for methods of checking the continuity of the windings.

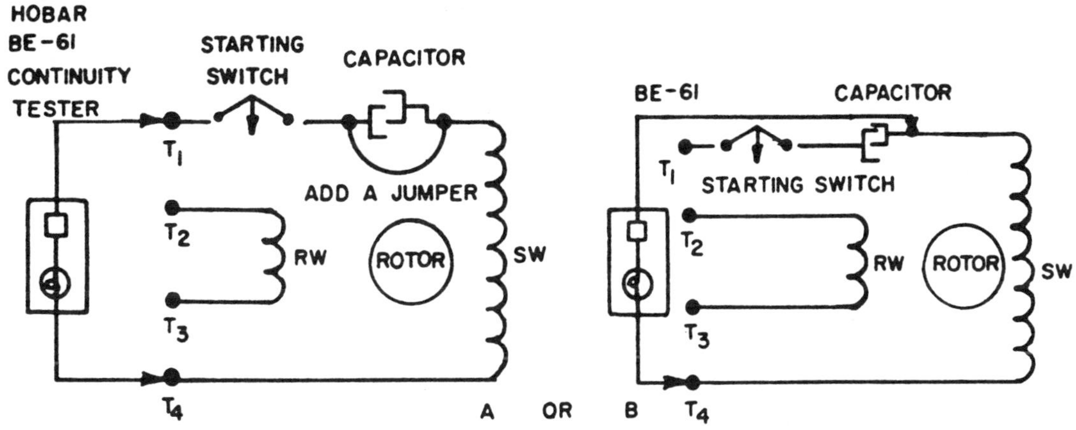

CHECKING THE CONTINUITY OF THE STARTING WINDING

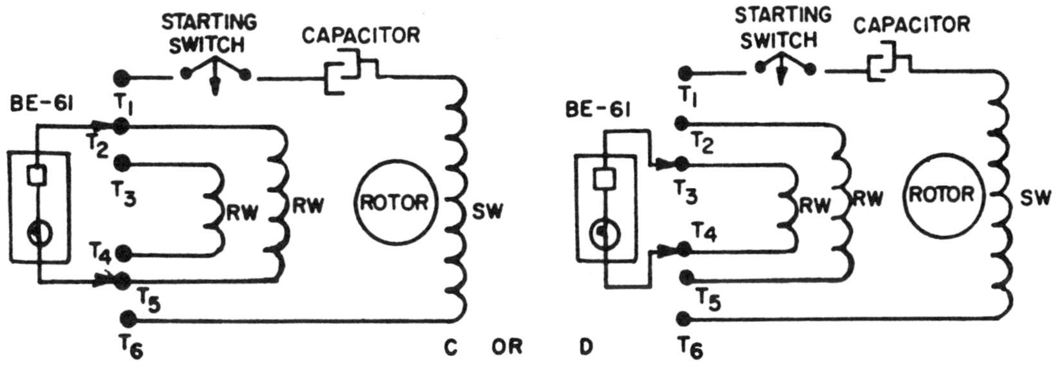

CHECKING THE CONTINUITY OF THE RUNNING WINDINGS

STEPS	ACTIVITY	SUGGESTIONS
18 (CONT'D)	B. Look for discoloration of the windings. Overheated motors with burned out or shorted windings will be blackened and the insulation will slow heat damage.	If either the starting windings and/or running winding(s) are blackened, the motor is unrepairable.
19	Make a detailed drawing or sketched of what combinations of continuity for winding lead sets as determined in Step #18. On a <u>single voltage</u> motor, use T and T for starting winding leads and T and T for the running winding leads. On a <u>dual voltage</u> motor, use T and T for the startingwinding leads and T and T for one set of running windings and T and T for the other set.	Identify the leads by existing colors, temporarily tagging with masking tape or by use of I.D. tag strips designed for this purpose.
20	Motors having thermo-overload devices built into them, can be a problem in performing the previous steps. Unsoldering of the leads from the attachment points of the overload device may have to be done. The thermo-overload device, once isolated can also be checked with the continuity tester, because a full continuity should be shown between all terminals.	Bad or suspect thermo-overload devices can be removed and discarded. A new wiring schematic, such as any of the ones presented will be used. REMEMBER, that another suitable overload protector must be used when the motor is placed in practical use again.
21	Leads must frequently be serviced on used motors. Basically, leads are soldered to terminals or crimp-type terminals are used for joining purposes. Terminals may be bolts and nuts fastened to the terminal block. Spade and friction fasteners are sometimes used. Never repair a terminal lead without soldering. Soldering (or commonly called tinning) is used to make stranded conductors within a larger conductor as one. Loose strands of a motor lead makes for shorts and electrical shock accidents. Additionally, stranded power cord leads going to the electric terminal block should always be tinned for safety.	Soldering (tinning) should be done with a soldering gun or pencil-type of soldering iron. <u>Do not</u> use soldering paste flux*. Use resin core solder and/or liquid types of fluxes designed for joining dissimilar metals to copper alloys. Perform the tinning steps as follows for making general lead and terminal fastening repairs. *Flux must always be removed after soldering and sometimes the paste is difficult to remove in a motor. There are liquid fluxes recommended for copper. Always check the flux label to see if it matches the metals being joined.

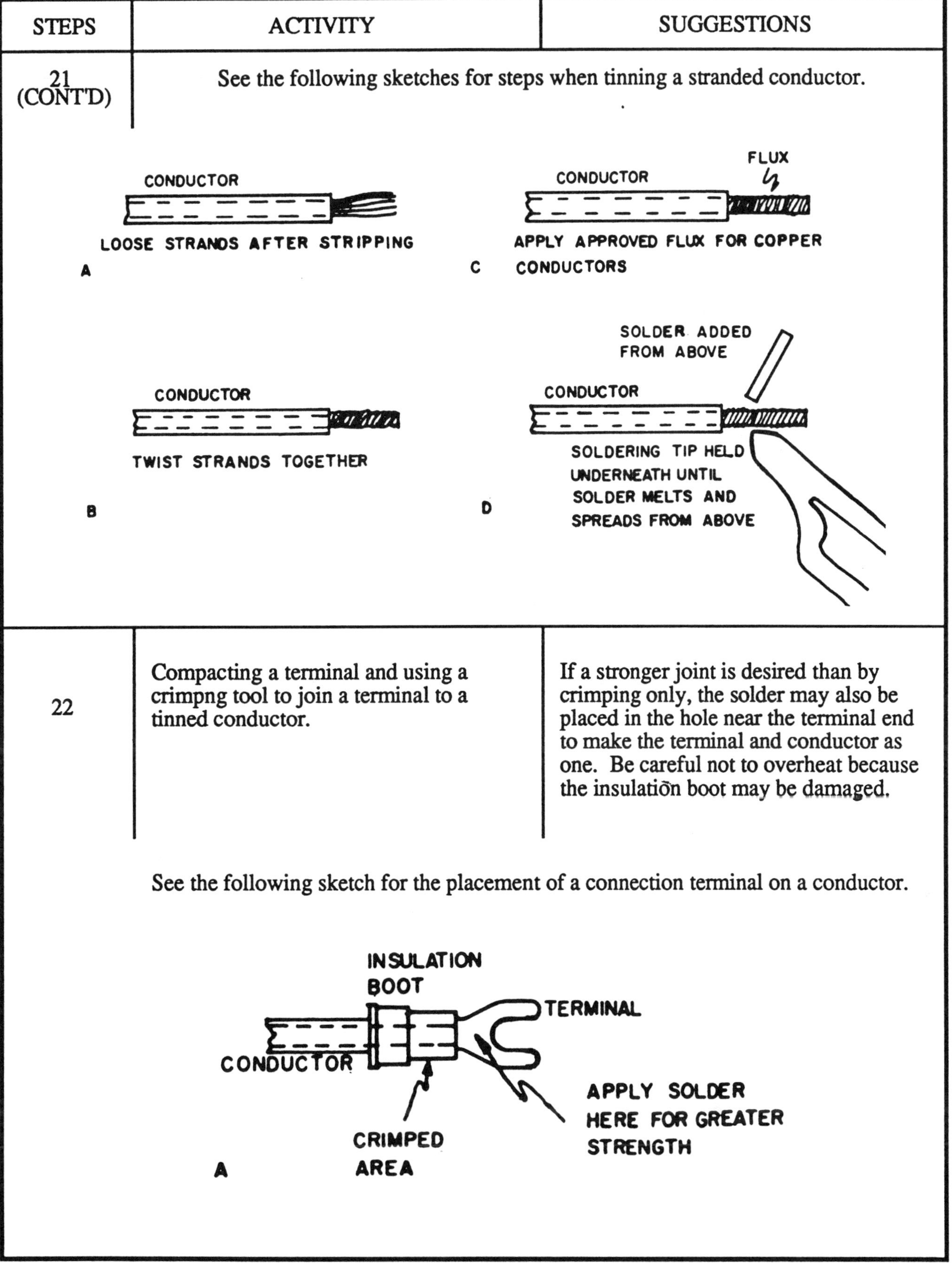

STEPS	ACTIVITY	SUGGESTIONS
21 (CONT'D)	See the following sketches for steps when tinning a stranded conductor.	
22	Compacting a terminal and using a crimpng tool to join a terminal to a tinned conductor.	If a stronger joint is desired than by crimping only, the solder may also be placed in the hole near the terminal end to make the terminal and conductor as one. Be careful not to overheat because the insulation boot may be damaged.

See the following sketch for the placement of a connection terminal on a conductor.

STEPS	ACTIVITY	SUGGESTIONS
23	Repairing and extending the length of a typical lead. Select lead from stock that is rated for 105° C or better and having an AWG as large or larger than lead being repaired. See the following steps when adding to the conductor length . NEVER use black plastic tape inside of a motor. Use sleeving of the proper rating*.	Sometimes the entire lead back to winding must be repaired. The winding material is covered with an enamel insulation and this enamel must be removed by scraping before new stock is soldered to it. CAUTION: Use heat sparingly so enamel will not be damaged on eindings in the area of the repair vicinity.

See the following sketches for the steps for splicing a conductor.

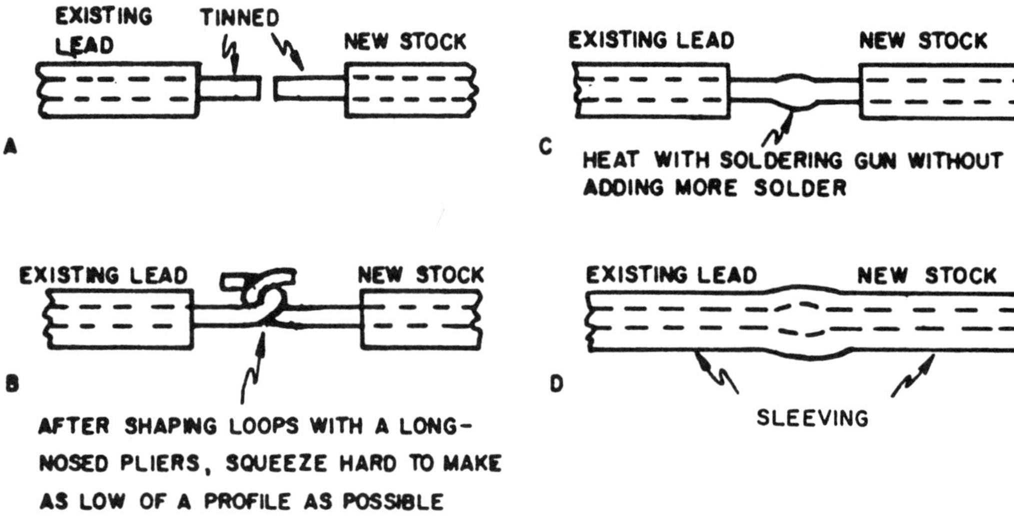

*Slip SLEEVING (D) over thenew stock end and use a piece that will cover about 1-1/2 inches centered over the joint. Fiberglass tape is used to fasted and hold the sleeving in place. The fiberglass tape meets the specifications for Class H (180°C) insulation and has excess protection for Class F (155°C), B (130°C), and A (105°C). It is used for all applications by most repair shops because the cost is not prohibitive. Cotton string can be used to hold widings together for the Class A insulation motors, but it is not satisfactory for Class B, F, and H. Select the rated material which exceeds the rating. The insulated leads are expensive and cost dictates that material meet only the requirements. For soldering, the tin/lead alloy is used for Class A, B, and F and the silver alloy is required for Class H. Some repair shops will use silver alloy for both Class H and F.

STEPS	ACTIVITY	SUGGESTIONS
24	When leads are removed from terminal block bolt heads, removal is done by the use of a soldering gun. To rejoin, use the sketch recommendations:	Heat with gun until the tinned conductor and the solder on the head of the terminal bolt melts and joins as one. Hold the pressure of the screwdriver as steady as possible after removing the heat. A bright shiny surface is a good job. Whereas, a dull finish is a "cold" solder joint and should be redone.
	See the following sketch for the soldering of leads to the terminal block. 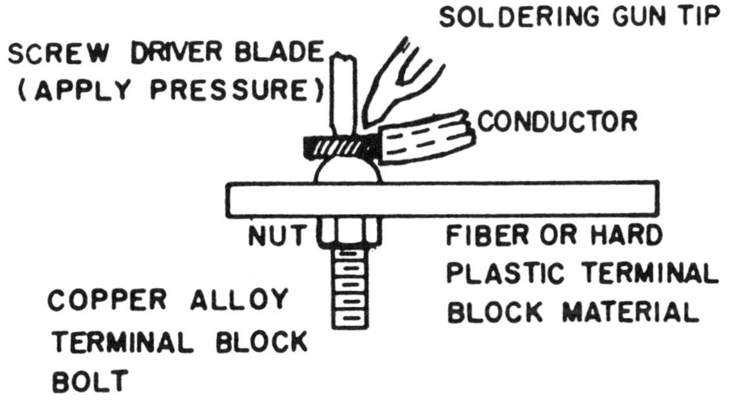 SOLDERING GUN TIP SCREW DRIVER BLADE (APPLY PRESSURE) CONDUCTOR NUT FIBER OR HARD PLASTIC TERMINAL BLOCK MATERIAL COPPER ALLOY TERMINAL BLOCK BOLT A	
25	<u>Reassembly</u> of the motor is generally done in the reverse order of disassembly. <u>First</u>, be sure to lubricate bearings and journals lightly with oil recommended by the motor manufacturer, or with 5W, 10W, or 20W nondetergent machine or engine lubricating oil.	If wick oilers can be removed, and if noticeably dirty, clean in parts washing solvent. Dry in paper toweling by squeezing, wash in warm water and detergents, rinse in warn water, squeeze in paper toweling, and blow dry with air pressure. Re-oil before replacement. Saw dust oil reservoirs may need to be repacked with fresh fine saw duse and re-oiled.

STEPS	ACTIVITY	SUGGESTIONS
26	On ball bearing type motors, the bearings need to be checked for smoothness of operation. Sealed bearings cannot be reoiled. Replace any bad or reluctant bearings. Make sure new bearings are replaced in exactly the same location as the old bearings. Make and note precise measurements from strategic orientation points. This must be done to assure proper placement.	Rotate the bearings with the hand and feel for smoothness. Special pullers are needed for removal of bearings. A mechanical or hydraulic press is needed for pressing the bearings over the rotor shaft.
27	Place the end bell opposite the electric end on the stator housing. Watch for the center punch marks or aligning guides made earlier. Check this alignment by temporary placement of several through bolts. Tap the end bell into a firm holding position with a soft-faced hammer.	Slight readjustment of the parts may be needed if through bolts will not freely enter. If so, remove the end bell and try this step again.
28	Carefully place the rotor into the stator housing, causing the shaft to enter the correct end bell. Make sure all thrust washers are accounted for and in their proper positions.	Avoid dragging the rotor surfaces on the field stator surfaces as much as possible.
29	Position the electrical leads so they will not interfere with the centrifugal switch mechanisms, the cooling fan, and through bolts when placing the PTO end bell in position. Watch for center punch marks previously made for alignment purposes.	Slight readjustment of the parts, including leads, may need to be made if through bolts will not freely enter and pass through the motor encasement. NEVER force through bolts, because if reluctant, they are probably binding on leads which can easily be damaged and cause electrical shorts.
30	Tap the PTO end bell into a firm position with a soft-faced hammer. Try to rotate the shaft.	If a shaft is reluctant, try tapping with soft-faced hammer in different areas of the end bell perimeter to attempt to free rotor so is will spin freely.
31	Place all through bolts in position and finger tighten.	Try rotating the shaft. It should continue to rotate easily.

STEPS	ACTIVITY	SUGGESTIONS
32	In staggered sequence and using normal torque stages, tighten the through bolts with a torque wrench. Rotate the shaft between each torquing sequence go-around. Any reluctance to rotation is a possible problem which must be analyzed and corrected. Binding and incorrect end play can be caused by the addition of too many or the wrong size of thrust washer shims.	A sequence of the typical 4-bolt pattern is shown below. A suggested torque value on many of the 3/16-inch diameter through bolts is as follows: (a) finger tight, (b) 15 inch-pounds, (c) 25 inch-pounds, (d) 30 inch-pounds, and (e) 40 inch-pounds respectively on each go-around of the sequence. 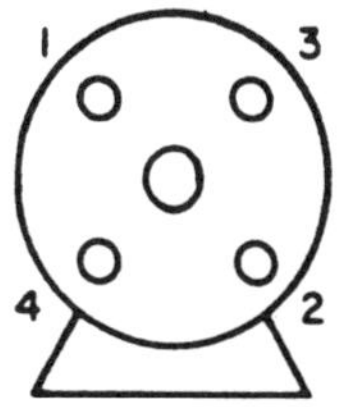
33	Connect power line or power cord to motor. Replace electric cover plate.	Remember, tinning of the power cord stranded wires is the best repair method.
34	Using the Ground Fault Circuit Interruptor (GFCI), try to run the motor. If the GFCI does not "trip," go to Step #36.	If the GFCI "trips," place the motor on an insulated (wooden, for example) table and test for encasement shorts by using a temporary 120 V, 2 wire power lead without the grounding wire as shown below. Either a neon light or an AC voltmeter may be used to test for shorts.
	See the following sketch for the testing of a motor which might have a "short" to the case. 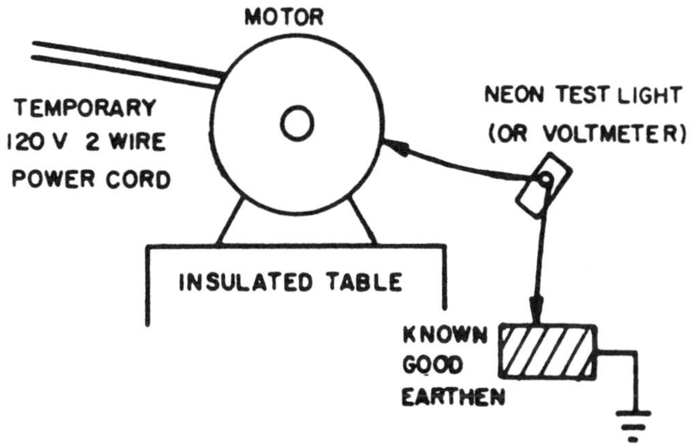	

STEPS	ACTIVITY	SUGGESTIONS
35	If the neon test light or the voltmeter indicates any voltage, try to find the short(s) by disassembly and reassembly.	If the short(s) cannot be found and corrected, throw the motor away. A decent burial for a typical fractional motor is a sacrifice of no more than $100, whereas human funeral services are much higher. Remember, one out of five electrical accidents is fatal.
36	Connect the motor to a 3-wire system of a black, white, and green (or bare) for 120 volts; or black, red, and green (or bare) for 240 volts.	An OSHA approved dead front plug cap is recommended for power cord connections. Use the continuity tester to check the third prong to equipment grounding of the motor encasement.
37	Use the induction clamp-style ammeter to check the amperes for the motor.	Generally, a typical split-phase start or capacitor-start motor should draw about 3/4 to 7/8 of its rated load at a no load condition. Do not run the motor without load longer than five minutes.
38	Replace the motor on the equipment from which it was removed and check amperage again. A Prony brake load test as suggested in Unit IX, Electric Motor Selection and Performance Testing, Laboratory Exercises, could be done to test the motor.	If the motor does not start, run and test out satisfactorily, dispose of it in an approved style and/or recycle the copper conductor and metal parts.
The next Unit XIII will lead you through the correct trouble shooting procedure.		

DEFINITION OF TERMS

ADDITIVES: Chemicals and/or minerals intentionally added to a product, such as lubricating oil, to improve its performance.

ALIGNMENT: The proper positioning of belts, drive shafts, etc. of two or more parts falling or positioned along the same plane or line.

AMBIENT TEMPERATURE: The temperature of the surrounding cooling medium, such as gas or liquid, which comes into contact with the heated parts of the motor. The cooling medium is usually the air surrounding the motor. The standard NEMA rating for ambient temperature is to not exceed 40 degrees Celsius.

AMMETER, CLAMP-On: Meter which can be clamped around one conductor and measures current flow by induction.

CAPACITOR: Electrical component that stores electrical charges. If it is charged, and you touch the leads you might serve as the ground and get shocked.

CONTAMINATED: In reference to air, gas, liquids or solids which have undesirable materials not normally found in them.

CONTINUITY: A low resistance path where electrons are free to flow. When there is a flow of electrons through this path there is continuity. For example, the lead ends of a short copper wire would have a high continuity; whereas, an insulated product would have little or no continuity.

CONTINUITY TESTER: A tester with a low voltage power source, such as a dry cell battery, that will light a lamp, sound a buzzer or provide another sensual detection when continuity is made. See CONTINUITY.

DOWN-TIME: The time a machine cannot be operated productively because repairs are being performed.

GUARDS: A part usually of metal, plastic or wood, that protects people from accidents by restricting the opportunity to be caught by moving dangerous parts, such as shafts, gears and chains.

LUBRICANT: Petroleum based hydrocarbons refined to give a free sliding action of metallic parts and hold an oil-film between the parts.

MAINTENANCE: To practice skills and tasks that are expected to keep machinery and equipment working trouble free.

MALFUNCTION: To not function, operate, work improperly or break-down.

OPERATOR'S MANUAL: Literature which gives instructions to users on how to properly operate, maintain, adjust and make minor repairs to equipment.

SHORT: An accidental connection between two lines or energized parts of such low resistance that excessive current flows, as in the case when a hot and neutral conductor make contact.

SLEEVING (MOTOR LEAD): Tube-like fiberglass material placed around a repaired conductor joint to provide an insulated covering.

SOLDERING: The joining of two like or unlike metals with an alloy whose melting point is lower than the metals being joined. An appropriate heat source must be selected for the size of materials being joined.

TINNING: The chemical cleaning and placement of a thin layer of a solder alloy on a metal's surface. See SOLDERING.

TORQUE SEQUENCE: The suggested staggering from one fastener to the next in an orderly sequence while tightening each to the same torque value. Sometimes called a "go-a-round".

VIBRATION: Frequencies or quick back and forth movements that tend to shake motors and equipment whose rotary moving parts are not perfectly balanced.

VISCOSITY: The ability of a liquid to flow at a given temperature. The higher the viscosity index number the harder it is for the liquid to flow.

NOTES

LABORATORY EXERCISE XII-A

Cleaning and Servicing an Electric Motor

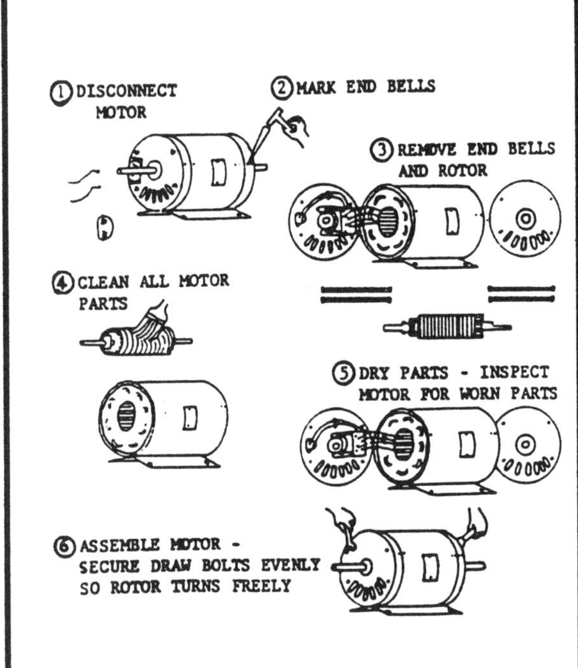

Materials:

1. An electric motor in need of cleaning and servicing.
2. Tools for disassembling an electric motor.
3. A parts cleaner brush, cleaning solvent and parts washer or pan.

Operation Teaches: Ability to ...

1. Disconnect a motor from an electrical source.
2. Correctly mark motor frame and end bell.
3. Explain the importance of carefully removing the end bells so as to not damage leads or motor windings.
4. Remove rotor from motor without damaging the centrifugal switch or motor brushes.
5. Identify the correct type of solvent to use in cleaning an electric motor.
6. Thoroughly clean and dry all motor parts.
7. Identify and inspect parts for wear or damage.
8. Reassemble motor parts in reverse order.
9. Lubricate motor using the correct type and amount of lubricant.
10. Connect motor to power source.

Operational Procedure:

1. Disconnect motor leads at terminal cover plate. Note position of leads on terminals.
2. Remove motor from mount or machine.
3. Wipe all dirt from outside of motor.
4. Mark end bells with center punch-one mark on end bell and stator frame on terminal connection end. Use 2 marks on other end of motor.
5. Remove end bell bolts.
6. Using a wood block of soft-faced hammer, remove end bell containing starting switch or brush ring.
7. Remove rotor and other end bell. Make sketch of thrust washers.
8. Clean all motor parts with an approved solvent. Do not use gasoline. Avoid soaking the windings or leads.
9. Dry all parts with a cloth or compressed air.
10. Identify and examine all parts for wear, damage or loose connections and make necessary repairs.
11. Reassemble motor in reverse to disassembly. Tighten end bell bolts evenly checking to see that the rotor turns freely.
12. Lubricate motor bearings with #10 oil or grease depending upon type. Refer to manufacturer's recommendation.
13. Mount motor and connect electrical leads. Test motor by turning on the switch.

Evalutaion Score Sheet

Item	Points Possible	Points Earned
1. Correctly disconnecting motor from electrical source-mark leads.	5	______
2. Correctly marking end bells.	5	______
3. Carefully removing end bells in correct order.		
4. Selecting correct solvent for cleaning motor.	5	______
5. Thorough cleaning and drying of motor parts.	15	______
6. Inspection and identificaiton of parts-made repairs.	20	______
7. Correct reassembling of motor, rotor free to turn.	10	______
8. Proper type and amount of lubrication used on motor.	5	______
9. Connecting motor to power source-motor runs properly.	15	______
10. Work habits and attitude.	10	______
Total	100	______

Name ______________________________

Date ____________ Grade ________________

LABORATORY EXERCISE XII-B

Electric Motor Service and Maintenance

Use the steps as outlined in Unit XII, Electric Motor Service and Repair, for the correct service and maintenance sequence. Respond to each step as you disassemble, clean, repair, and test operate the electric motor. Make a check mark after the number for a step if you are asked to physically or mechanically perform a task. Make sketches, drawings and/or write notes for steps as requested.

2. ________ 9. ________ 15. ________ 22. ________ 28. ________ 34. ________

3. ________ 11. ________ 16. ________ 23. ________ 29. ________ 35. ________

4. ________ 12. ________ 17. ________ 24. ________ 30. ________ 36. ________

6. ________ 13. ________ 18. ________ 25. ________ 31. ________ 37. ________

8. ________ 14. ________ 21. ________ 27. ________ 33. ________ 38. ________

STEP	SKETCH AND NOTES
1.	
5.	
7.	
10.	
19.	

STEP	SKETCH AND NOTES
20.	
26. For bearing type motors	
32. Torque values used on each go-round or stage	

COMPLETE LABORATORY EXERCISE XII-C

LABORATORY EXERCISE XII-C

Electric Motor Repair Report

Select an electric motor for repairing, testing, and operating exercise.

1. Motor nameplate information:

a. Manufacturer	____________________	f. Voltage	____________________
b. Type of motor	____________________	g. Amperage	____________________
c. Horsepower	____________________	h. RPM	____________________
d. Temperature rise	____________________	i. Time rating	____________________
e. Phase	____________________		

2. Does the electric motor operate properly? Yes or No

   a. Describe the operational or non-operational condition of the electric motor.

   ____

   ____

3. Sketch the electrical connections which exist for this motor:

4. Sketch the electrical connections for reversing this motor:

5. List disassembly instructions and special conditions which existed:

6. List repair activities performed:

7. Test operation of the motor: Better __________ Same __________ Worse __________

8. Estimated value and repair costs:

   Beginning Value __________ Repair Cost __________ Labor Cost __________

   Final Value ______________

9. Student Name ________________________________ Date ________________________________

TROUBLE SHOOTING

The following outline is a systematic approach to trouble shooting an electric motor. The previous units, Unit XI, MOTOR INSTALLATION and Unit XII, ELECTRIC MOTOR MAINTENANCE, are closely associated to troubleshooting. If motors are installed correctly and maintained properly, the need for trouble shooting due to malfunctioning should be greatly reduced.

Even the best electric motors, properly maintained, can develop occasional troubles. Sometimes, however, the problem lies outside of the motor, and if not corrected, serious damage will occur to the motor.

Basically, motor problems can be classified into symptoms easily recognized through the normal senses of:

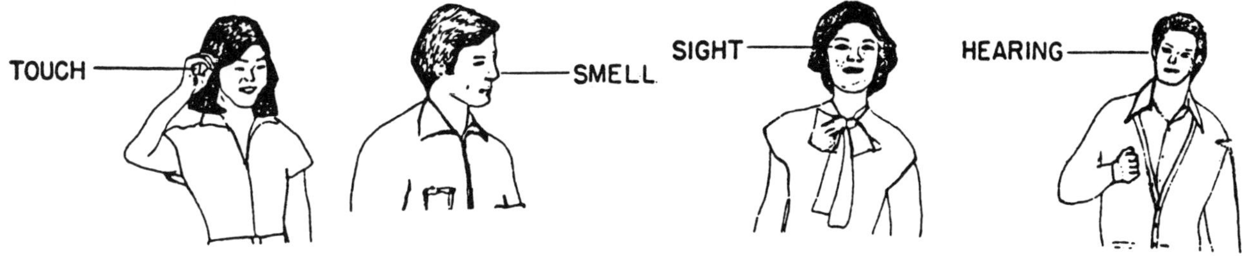

The symptoms normally detected in trouble shooting can be identified with these four senses. Possible causes along with remedies for the problem are presented after each symptom.

SYMPTOMS YOU CAN SEE	POSSIBLE CAUSE	REMEDY
Motor will not start	Usually power trouble -- loose or disconected wire or a blown fuse	Check voltage ar motor. Replace blown fuses. Tighten all wire connections. Check disconnect switches and relay contacts.
	Load too heavy	Disconnect part or all of load to see if it will start. Reduce load or replace with motor of greater horsepower.
	Motor overload due to equipment being plugged	FIRST, for safety, disconnect the power. Remove the plugged condition. Reconnect power. If overload current protector is blown or tripped, replace or reset.

SYMPTOMS YOU CAN SEE (CONT'D)	POSSIBLE CAUSE	REMEDY
Motor will start, but will not obtain normal operating speed and is humming. For split phase or capacitor-start motors only.	Running winding is burned out.	Replace motor, if practical, have winding rewound.
	Loose connection of running winding lead.	Disassemble motor and make necessary repair.
Motor hums and will not start until you spin the shaft in either direction then it runs at rated speed. For split phase or capacitor-start motors only.	Starting winding burned out.	Replace motor, or have winding re-wound, if practical.
	Centrifugal starting switch contacts are bad, not making contact or loose connection of starting winding lead.	Disassemble motor and make necessary repair.
Excessive sparking or flashing at brushes. Blackened commutator. For brush and commutator motors, which includes universal, AC-DC, and repulsion run motors.	Rough commutator.	Sand or turn down, depending on depth of surface roughness.
	Low bar on commutator.	Grind or turn down balance of commutator.
	High bar on commutator.	If extreme, lower with mallet--tightening clamp ring. Grind true.
	High mica.	Undercut.
	Brushes too short.	Replace with harder grade if wearing too rapidly and not by a rough commutator. Sometimes underloaded brushes wear abnormally fast. Ordinarily, set should last for a minimum of 2500 hrs.
	Insufficient brush tensions.	Adjust.
	Weak brush springs.	Replace.
	Brushes sticking in holders.	Free them. Clean brushes and holders.
	Dirt or oil on commutator.	Wipe off. Clean brushed if glazed.
	Water dripping on commutator.	Eliminate source or protect with shield.
	Shorted armature and/or field windings.	Test for short. After removal of metallic contact between commutator bars. Repair armature.

SYMPTOMS YOU CAN HEAR	POSSIBLE CAUSE	REMEDY
Excessive hum.	Motor not properly connected.	Recheck all connections against wiring diagram. Check for proper voltage connections.
	Winding fault or short circuit within motor.	Check for higher than normal amperage draw. If so, replace motor, or have windings re-wound.
	Uneven air gap between rotor and field stators.	Replace bearings before rotor drags on stators.
	Unbalanced rotor or bent PTO shaft.	Balance rotor according to appropriate service for type of motor. Straighten PTO shaft.
Regular Clicking.	Foreign matter in air gap.	Inspect and remove.
	Centrifugal switch problem.	Check end-play, disassemble motor and make necessary repairs and adjustments.
	Cooling fan striking housing or end-bell.	On external fans, realign fan cover. On internal fans, disassemble motor and make necessary repairs and adjustments.
Rapid knocking or rumbling on ball bearing motors, only.	Bearings worn due to lack of or excessive lubrication.	Replace bearings and service with recommended grade and amount of new lubricant.
	Bearings worn due to mechanical overload or excessive temperatures.	Same as above. Correct the overload situation or replace with a larger horse-power motor with bigger and better bearings.
	Foreign matter in lubricant or bearing housing.	Clean bearing housing and replace bearings as listed above.
Rapid knocking on all types of motors.	Misalignment. Probably causing shoulder of shaft to pound periodically against bearing end. Excess end play.	Adjust for proper end-play and correct alignment
Brush chatter.	Loose bearings and extreme vibration.	See vibration under "Symptoms You Can Feel"

SYMPTOMS YOU CAN FEEL	POSSIBLE CAUSE	REMEDY
Vibration.	Misalingment between motor and machine.	Check for alignment.
	Vibration in driven machine. To isolate, disconnect drivers to be sure the motor is not the source of vibration.	Eliminate source of vibration in machine. Some machines are designed to vibrate. Check and service resillient mounts.
Motor overheating.	Overload. Measure the electrical line load and compare with nameplate rating.	Reduce load or replace with motor of greater horespower. Make sure current overload protectors are functioning properly to trip on excessive overload situations.
	Restricted air flow. Check flow of ventilating air in area of motor.	Wipe excessive dirt from surface of motor, clean interior of motor if needed by blowing air through it or by disassembly and cleaning.
	Bearing failure causing excessive overload.	Replace bearings and service with recommended grade and amount of lubricant. Replace end shields if warped or damaged.
	Shorted windings.	Replace motor or have windings rewound by authorized service dealer.
Bearing overheating.	Misalignment of bearings within end-bells to the shaft.	Make proper alignments of bearings and shaft.
	Over lubrication of bearings (ball or roller bearing.)	Relieve excess supply to amount recommended by manufacturer.

SYMPTOMS YOU CAN SMELL

The "smell" symptom is commonly the first warning that something has gone wrong or is going wrong, especially with motors used indoors and relatively hidden because of being an integral part of a large piece of equipment.

The "smell" symptoms need not be written in a Symptom -- Possible Causes -- Remedy structure as presented for the other senses. However, most smell symptoms go hand-in-hand with other symptoms. Therefore, they can be cross referenced to the appropriate symptom.

Common "smells" cross-referenced to other problems which should lead to logical causes and satisfactory remedies are listed:

SMELL	CROSS REFERNCE TO SYMPTOM OF
Hot burning and/or overheating smell.	Feel. Motor overheating and bearing overheating.
Ozone smell. This is an unusual smell, when smelled once is easily identified again. The smell is associated with "arcing" of electricity.	See. (1) Excessive sparking or flashing at brushes. The ozone smell is highly prevalent around brush and commutator types of motors. In fact, some ozone smells are normal, but an excess is not. (2) Motor hums and will not start until spun. Excessive arcing of a centrifugal starting switch contact will give off the ozone smell.
	Hear. Brush chatter.

CLASSROOM EXERCISES XIII-A

Trouble Shooting

1. What four (4) senses can you use to determine motor problems? a. ______________________

   b. ________________ c. ________________ d. ______________________

2. What might be two (2) possible motor problems if a motor does not start and the power supply is satisfactory? a. ____

   b. ____

3. You have discovered the following problems. List the correcting procedures that should be completed.

CAUSE	REMEDY
a. Loose connection of running winding lead	______________________
b. Centrifugal switch problem	______________________
c. Worn bearings	______________________
d. Restricted air flow	______________________
e. Shorted windings	______________________
f. Over-lubricated ball or roller bearings	______________________

4. The ozone smell is associated with ______________________ of electricity.

5. Name two (2) types of motors that have brushes and commutators. (See previous units for answers.)

   a. ____

   b. ____

6. What is the best way to take care of weak brush springs on the repulsion style motors? ____________

   ____

   ____

7. Your friend tells you that the starting winding is burned out on a split-phase motor. You ask him what made him think that was the problem. He should tell you that: ______________________

   ____

   ____

   ____

8. The motor has a knocking sound. What could be the problem? ______________________ and ____

   a. How can it be corrected? ____

   ____

   ____

LABORATORY EXERCISE XIII-A

Trouble Shooting

GO TO THE LABORATORY AND COMPLETE THIS TABLE WITH THE ELECTRIC MOTORS SUPPLIED		
CONDITION	MOTOR A	MOTOR B
1. Type of motor? (See other Units)		
2. Voltage rating(s)		
3. Does the motor start properly?		
4. Does the motor run under the load?		
5. If the motor does not start properly, why not?		
6. If the motor does not run under the load, why not?		
7. Amperage Meter Readings only if motor operates	Starting amps ____________ Running amps ___________	Starting amps ____________ Running amps ___________
8. What corrections were made to make the motor operate properly?		
9. Amperage Meter Reading after repair.	Starting amps ____________ Running amps ___________	Starting amps ____________ Running amps ___________
10. Amperage Meter Reading after 10 minutes of operation	Starting amps ____________ Running amps ___________	Starting amps ____________ Running amps ___________
11. Final motor analysis	Operates according to specifications? Yes ______ No ______	Operates according to specifications? Yes ______ No ______

APPENDIX

REFERENCES

Alerich, Walter N., 1983. Electric Motor Control. Delmar Publishers Inc., Albany, New York.

Andreas, John C., 1982. Energy-Efficient Electric Motors. Marcel Dekker, Inc., New York, New York.

Anderson, Edwin P., 1968. Electric Motors. Theodore Audel and Company, Division of Howard W. Sams and Company Inc., Indianapolis, Indiana.

Baldor, 1978. Industrial Electric Motors Catalog 500. Baldor, Fort Smith, Arkansas.

Bodine, Clay, 1978. Small Motor, Gearmotor and Control Handbook. Bodine Electric Company, Chicago, Illinois.

Butchbaker, A.F. 1977. Electricity and Electronics for Agriculture. Iowa State University Press, Ames, Iowa.

Duff, John R. and Milton Kaufman, 1980. Alternating Current Fundamentals. Delmar Publishers Inc., Albany, New York.

Dunsheath, Percy, 1962. A History of Electrical Power Engineering. The M.I.T. Press, Cambrige, Massachusetts.

Edison. Electrical and Basic Controls Used in Agricultural Production. Electric Institute, New York. Available from Hobar Publications, St. Paul, Minnesota.

Fairbanks, Morse. Catechism of Electrical Machinery -- Bulletin E100F. Fairbanks Whitney Electrical Division, Freeport, Illinois.

Food and Energy Council, 1978. Agricultural Wiring Handbook. American Society of Agricultural Engineers, National Rural Electric Coop Assn., Edison Electric Institute, Columbia, Missouri.

General Electric, 1979. Five-Star Motor Catalog, GEP-500E. General Electric.

Gerrish, Howard H., 1968. Electricity and Electronics. Goodheart-Wilcox Company, Inc., Homewood, Illinois.

Graham, Kennard C., 1961. Understanding and Servicing Fractional Horsepower Motors. American Technical Society, Chicago, Illinois.

Gould Century, 1981. Stock Motor Catalog, Bulletin 943. Gould, Inc., Mendota Heights, MN.

Herman, Stephen L. and Walter N. Alerich, 1985. Industrial Motor Control. Delmar Publishers Inc., Albany, New York.

Hoerner, Harry, 1977. Basic Electricity and Practical Wiring. Hobar Publications, St. Paul, Minnesota.

Jacobs, Clinton O., 1977. Electric Motors. The Department of Agricultural Education, University of Arizona, Tuscon, Arizona.

Lamkey, F.R., 1967. A Refresher Course on Alternating Current and Electric Motors. Century Electric Co., St. Louis, Missouri.

Leader's Guide, 1969. Motors and Motor Controls, Selection, Application, and Maintenance. The Electrification Council, New York, New York.

Lloyd, Tom C., 1969. Electric Motors and Their Application. Wiley -- Interscience, New York, New York.

Loper, Ahr, Clendenning, 1979. Introduction to Electricity and Electronics. Delmar Publishers Inc., Albany, New York.

Machine Design, 1964. Electric Motors, Reference Issue. Penton Publishing Co., Cleveland, Ohio.

Marathon Electric, 1982. Fractional HP Motor Catalog. Marathon Electric, Wausau, Wisconsin.

Marathon Electric, 1982. Integral HP Motor Catalog. Marathon Electric, Wausau, Wisconsin.

Meyer, Herbert W., 1971. A History of Electricity and Magnetism. The M.I.T. Press, Cambridge, Massachusetts.

National Electrical Manufacturers Association, 1978. ANSI/NEMA Standards Publication/No. MGI-1978. National Electrical Manufacturers Association, Washington, D.C.

National Fire Protection Association, 1981. National Electrical Code. Quincy, Massachusetts.

Parady, Harold W. and Howard J. Turner, 1978. Electric Motors, Selection Protection, and Drives. American Association for Vocational Instructional Materials, Athens, Georgia.

Reliance Electric, Employee Training Department, 1975. Part One -- How an AC Motor Works, Parts of an AC Motor. Reliance Electric, Cleveland Ohio.

Reliance Electric, Employee Training Department, 1975. Part Three -- Types of AC Motors. Reliance Electric, Cleveland, Ohio.

Reliance Electric, Employee Training Department, 1975. Part Five -- Frame Sizes, Mounting Methods. Reliance Electric, Cleveland, Ohio.

Reliance Electric, Employee Training Department, 1975. Part Six -- Protective Enclosures. Insulation Systems, Reliance Electric, Cleveland, Ohio.

Reliance Electric, Employee Training Department, 1975. Part Seven -- Motor Performance. Reliance Electric, Cleveland, Ohio.,

Reliance electric, Employee Training Department, 1975. Part Eight -- Matching AC Motors to Load Requirements. Reliance Electric, Cleveland, Ohio.

Rosenberg, Robert, 1970. Electric Motor Repair. Holt, Rinehart, and Winston, New York, New York.

Sodderholm, L.H. and H.B. Puckett, 1974. Selecting and Using Electric Motors. U.S. Department of Agriculture, Farmer's Bulletin No. 2257, U.S. Government Printing Office, Washington, D.C.

Square D. Company, 1978. Motor Control Fundamentals. Square D. Company, Milwaukee, Wisconsin.

Technical Manual, 1972. Electric Motor and Generator Repair. Headquarters, Department of the Army, Washington D.C.

Van Valkenburgh, Nooger and Neville, Inc. Basic Electricity, Volume 1. John F. Rider Publisher, Inc., New York, New York.

Van Valkenburgh, Nooger and Neville, Inc. Basic Electricity, Volume 2. John F. Rider Publisher, Inc., New York, New York.

Van Valkenburgh, Nooger and Neville, Inc. Basic Electricity, Volume 3. John F. Rider Publisher, Inc., New York, New York.

Van Valkenburgh, Nooger and Neville Inc. Basic Electricity, Volume 4. John F. Rider Publisher, Inc., New York, New York.

Van Valkenburgh, Nooger and Neville Inc. Basic Electricity, Volume 5. John F. Rider Publisher, Inc., New York, New York.

Veinott, Cyril G., 1939. Fractional Horsepower Electric Motors. McGraw-Hill, New York, New York.

Westinghouse Electric Corporation, 1981. Westinghouse Stock Motor Catalog 2720. Westinghouse Electric Corporation, Buffalo, New York.

CONSTRUCTING EXPERIMENTAL MOTORS

PROJECTS

DESIGN AND CONSTRUCT A SIMPLE EXPERIMENTAL MOTOR

This motor, figure 13-32, is easily constructed. It can be made as a simple series-wound type or, by arranging separate terminals from the armature and field coil, it can be operated either as a series-wound or shunt-wound motor. If a second field coil is wound over the primary field coil and separate leads brought out to terminals, it can be operated as a compound-wound motor. If the commutator segments are set in the same relationship to the armature as they appear on the drawing, figure 13-33, the motor will be self-starting.

MATERIALS

1 - Pc. soft steel, 3/8" x 2 3/4" (19 mm x 70 mm)
1 - Pc. soft steel, 1/8" x 3/4" x 9 5/8" (3 mm x 19 mm x 244 mm)
1 - Pc. steel rod, 1/8" x 3" (3 mm x 76 mm)
1 - Pc. brass or copper tubing, 3/8" OD x 1/2" (9 mm OD x 12 mm)
1 - Pc. dowel or plastic to fit inside diameter of tubing
2 - Pcs. sheet iron, #20 or #22 gauge x 1" x 2 7/8" (25 mm x 73 mm)
2 - Pcs. spring brass, #30 gauge x 1/4" x 2 1/4" (6 mm x 57 mm)
2 - Wood screws, #6 x 3/4" RH
6 - Wood screws, #4 x 1/2" RH
6 - Binding posts or Fahnestock clips
1 - Wooden base, 3/4" x 3 1/2" x 6" (20 mm x 90 mm x 150 mm)
Magnet wire, #22 AWG Formvar

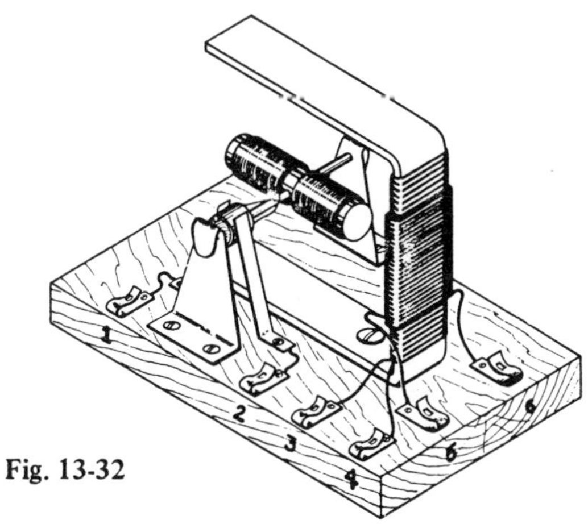

Fig. 13-32

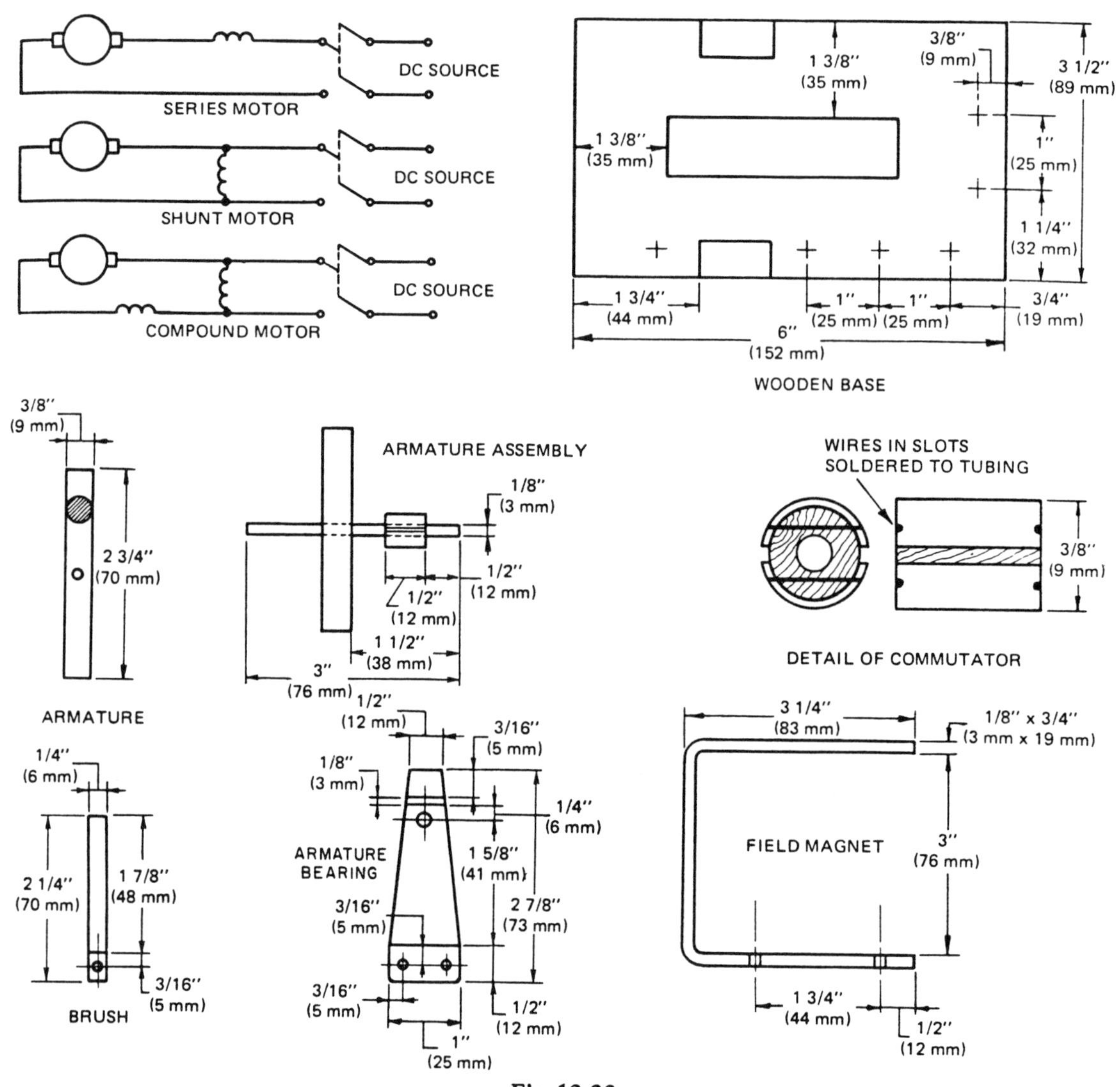

Fig. 13-33

PROCEDURE

Make the field magnet, armature bearings, brushes, and base.

Next, make the armature assembly. The armature must be a tight fit on the shaft. To make the commutator, select a piece of brass or copper tubing. Cut a piece about 17/32" (13.5 mm) long. Square the ends and remove the burrs from the inside. Select a piece of hardwood dowel or plastic larger than the ID of the tubing. Place the material in the chuck on an engine lathe and turn it to fit the inside of the dowel. Drill a hole through the center while it is in the lathe, to make a press fit on the armature shaft. Press the tubing over the dowel, face the end, and cut it off even with the other side of the tubing. Saw slots across each end as indicated on the drawing. Press a piece of copper wire into each slot and solder the ends to the tubing. Remove the excess solder and the ends of the wires with a file.

Saw a slot in the tubing on each side as indicated on the drawing. Carefully remove the burrs.

Wind a layer of plastic tape on the field magnet where the coil will be wound and on each side of the armature.

Wind two layers of #22 magnet wire on each side of the armature. Start in the middle and end in the middle, figure 13-34. Care must be taken to continue winding in the same direction when going from one pole to the other. Secure the ends with cotton tape. *Note:* Leave about three inches of wire at the end of each coil.

Starting at the bottom, wind two layers of #22 gauge magnet wire on the field magnet. The first layer should be 2 3/4″ (70 mm) long. Starting 1/2″ (12 mm) from the bottom and ending 1/2″ (12 mm) from the top, wind two layers of #22 gauge magnet wire over the first coil and in the same direction, figure 13-35. Secure ends of each coil with cotton tape.

Coat the coils on the field magnet and armature with shellac or varnish.

Assemble, test, and evaluate.

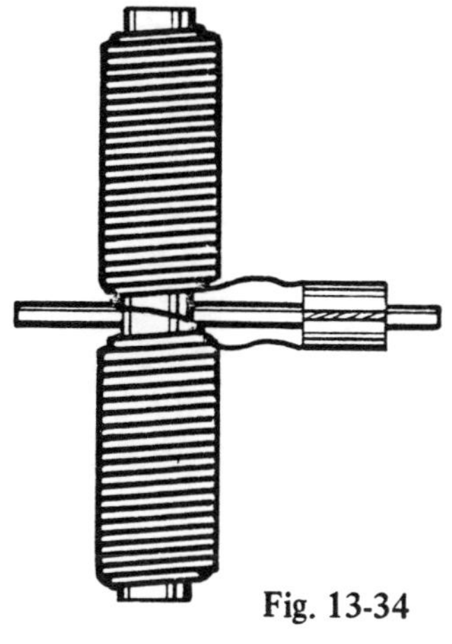

Fig. 13-34

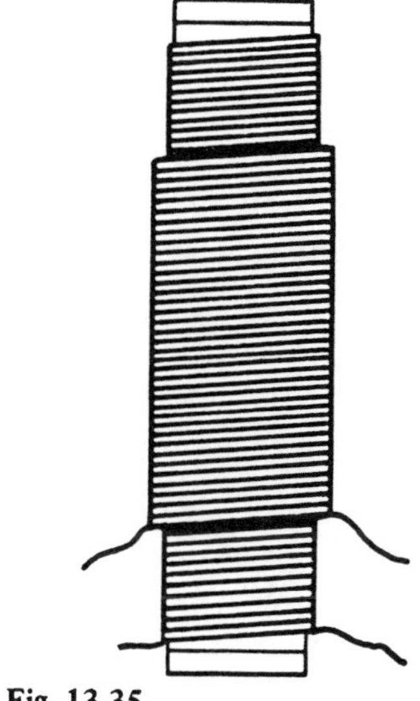

Fig. 13-35

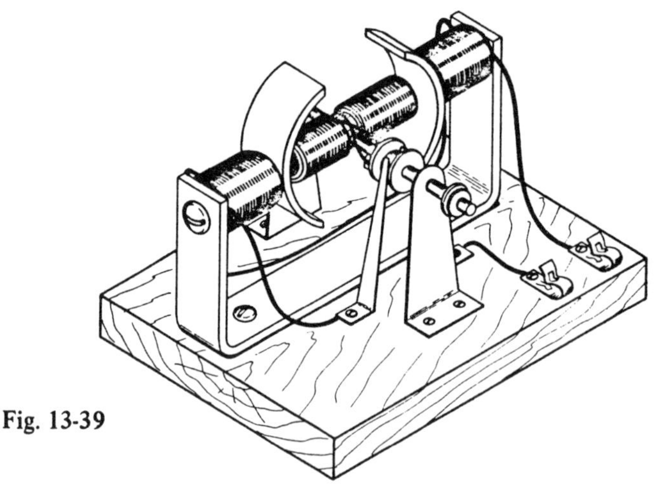

Fig. 13-39

1 - Pc. soft steel, 3/8" dia. or 3/8" x 3/8" x 1 7/8"
1 - Pc. maple dowel, 3/8" x 2" (9 mm x 50 mm)
1 - Pc. brass or copper tubing, 3/8" OD x 9/16" (9 mm OD x 14 mm)
2 - Pcs. sheet iron, #20 or #22 gauge, 1" x 3" (25 mm x 76 mm)
2 - Pcs. spring brass, #30 gauge, 1/4" x 2 1/4" (6 mm x 57 mm)
2 - Pcs. fiber, 1/16" x 3/4" (1.5 mm x 19 mm) square
4 - Pcs. fiber, 1/16" x 1" (1.5 mm x 25 mm) square
1 - Pc. iron pipe, 2" ID x 3/4" (50 mm ID x 20 mm)
1 - Pc. band iron, 3/16" x 3/4" x 9 1/2" (5 mm x 19 mm x 214 mm)
1 - Pc. wood, 3/4" x 3 3/4" x 6" (20 mm x 95 mm x 150 mm)
2 - Machine bolts, 1/4" x 1 1/2" (M6.3 x 40 mm L)
2 - Fahnestock clips
8 - Wood screws, #4 x 1/2" RH
2 - Wood screws, #6 x 3/4" RH
Magnet wire, #22 gauge AWG Formvar

EXPERIMENTAL MOTOR NO. 3

This motor, figure 13-39, is unique because of the type of field coils and the use of sections of iron pipe as part of the field metal. Another feature is the manner in which the commutator segments are held in place with fiber rings.

A three-pole armature similar to the one in motor No. 2 could be substituted for the two-pole armature suggested in this motor; a three-segment tubular-type commutator could be used instead of the disc.

MATERIALS

1 - Pc. soft steel, 3/8" (9 mm) diameter or 3/8" x 3/8" x 1 7/8" (9 mm x 9 mm x 48 mm)
1 - Pc. CRS, 1/8" x 3 1/4" (3 mm x 85 mm)

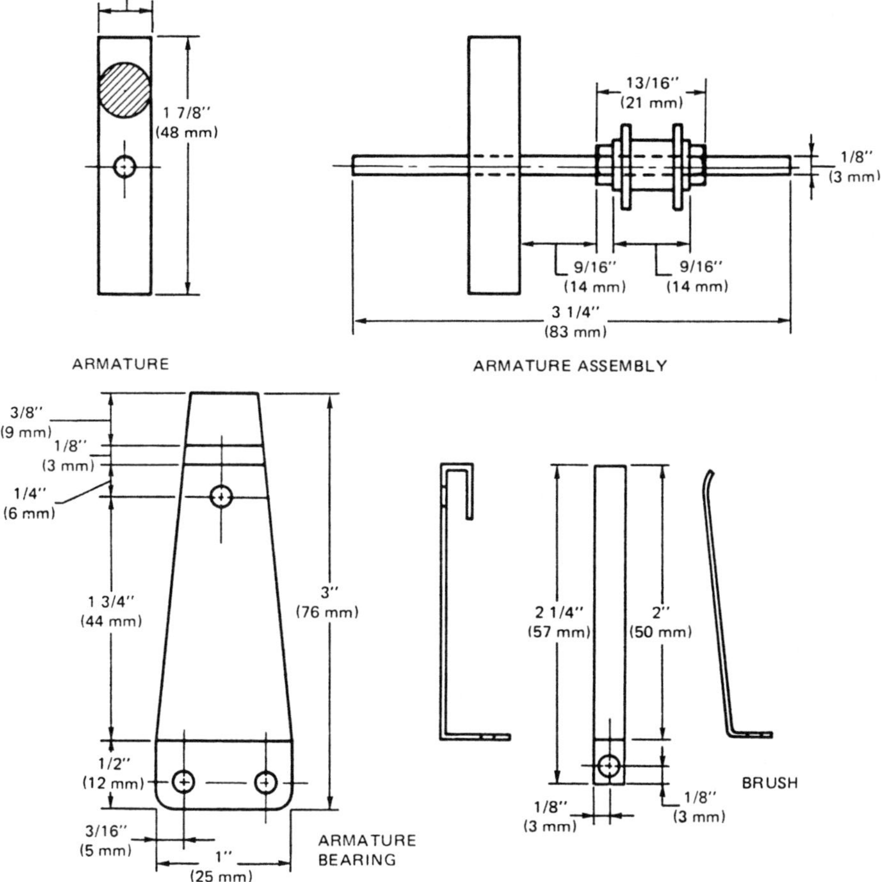

Fig. 13-40

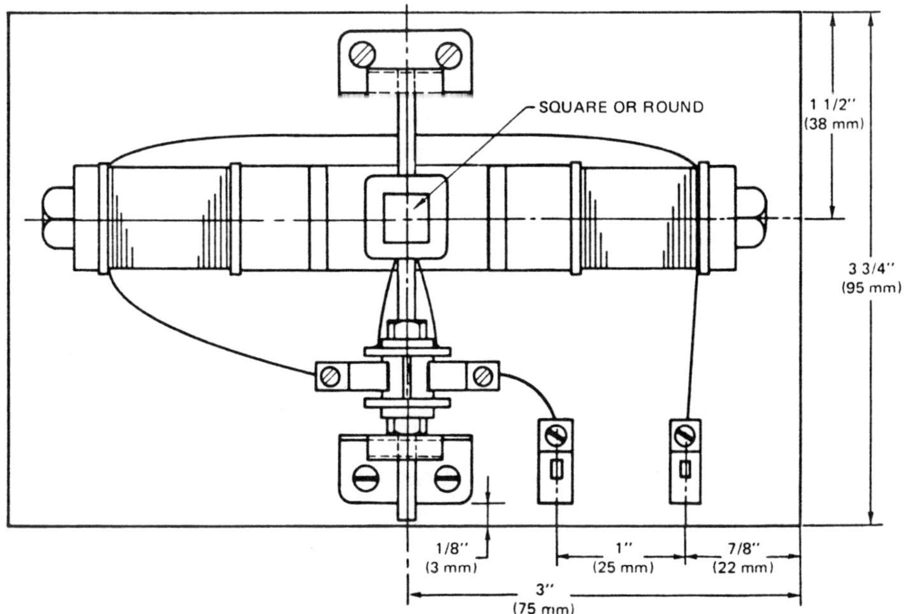

Fig. 13-41

PROCEDURE

Study figures 13-40 through 13-42. Procure the required materials, and make the metal parts and the base.

Next, make the coils. Both coils should be wound in the same direction. For a series motor, the beginning of one coil should be connected to the end of the other.

The commutator is made similar to the one in previous experimental motor, except that fiber rings are used to hold the segments in place instead of wires soldered to the segments. This system has an advantage when more than two segments are needed.

The armature, except for the dimensions, is made exactly the same as in the first experimental motor.

When all the parts are finished, assemble and test. Adjust the tension of the brushes until the rotor spins easily.

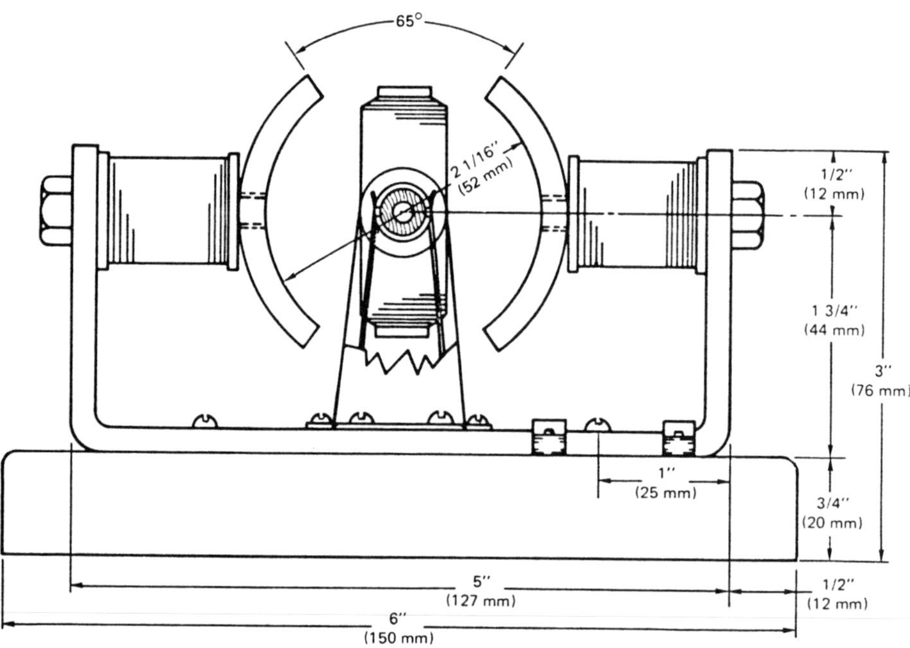

Fig. 13-42

Experiment with the position of the commutator segments in relation to the armature poles until the motor is self-starting and has a good torque. Evaluate.

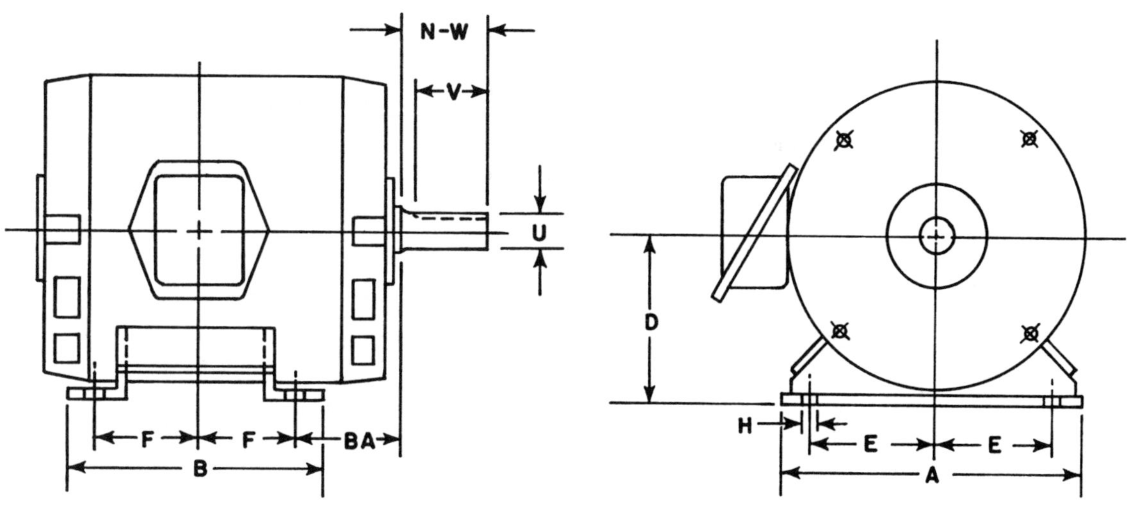

Table 1 —Dimensions for Foot-Mounted A-C Motors

Frame Number	Dimensions (in.) A max	B max	D*	E	F	BA	H	N-W†	U†	V† min	Key Width†	Key Thick-ness†	Key Length†
42	....	....	2⅝	1¾	27/32	2 1/16	9/32 slot	1⅛	⅜	...	..	3/64 flat	...
48	....	....	3	2⅛	1⅜	2½	11/32 slot	1½	½	...	..	3/64 flat	...
48H	....	....	3	2⅛	2⅜	2½	11/32 slot	1½	½	...	..	3/64 flat	...
56	....	....	3½	2 7/16	1½	2¾	11/32 slot	1⅞	⅝	...	3/16	3/16	1⅜§
56H	....	....	3½	2 7/16	2½	2¾	11/32 slot	1⅞	⅝	...	3/16	3/16	1⅜§
66	....	....	4⅛	2 15/16	2½	3⅛	13/32 slot	2¼	¾	...	3/16	3/16	1⅞§
143, T	7	6	3½	2¾	2	2¼	11/32	2, 2¼	¾, ⅞	1¾, 2	3/16, 3/16	3/16, 3/16	1⅜, 1⅜
145, T	7	6	3½	2¾	2½	2¼	11/32	2, 2¼	¾, ⅞	1¾, 2	3/16, 3/16	3/16, 3/16	1⅜, 1⅜
182, T	9	6½	4½	3¾	2¼	2¾	13/32	2¼, 2¾	⅞, 1⅛	2, 2½	3/16, ¼	3/16, ¼	1⅜, 1¾
184, T	9	7½	4½	3¾	2¾	2¾	13/32	2¼, 2¾	⅞, 1⅛	2, 2½	3/16, ¼	3/16, ¼	1⅜, 1¾
213, T	10½	7½	5¼	4¼	2¾	3½	13/32	3, 3⅜	1⅛, 1⅜	2¾, 3⅛	¼, 5/16	¼, 5/16	2, 2⅜
215, T	10½	9	5¼	4¼	3½	3½	13/32	3, 3⅜	1⅛, 1⅜	2¾, 3⅛	¼, 5/16	¼, 5/16	2, 2⅜
254U, T	12½	10¾	6¼	5	4⅛	4¼	17/32	3¾, 4	1⅜, 1⅝	3½, 3¾	5/16, ⅜	5/16, ⅜	2¾, 2⅞
256U, T	12½	12½	6¼	5	5	4¼	17/32	3¾, 4	1⅜, 1⅝	3½, 3¾	5/16, ⅜	5/16, ⅜	2¾, 2⅞
284U, T	14	12½	7	5½	4¾	4¾	17/32	4⅞, 4⅝	1⅝, 1⅞	4⅝, 4⅜	⅜, ½	⅜, ½	3¾, 3¼
284TS	14	12½	7	5½	4¾	4¾	17/32	3¼	1⅝	3	⅜	⅜	1⅞
286U, T	14	14	7	5½	5½	4¾	17/32	4⅞, 4⅝	1⅝, 1⅞	4⅝, 4⅜	⅜, ½	⅜, ½	3¾, 3¼
286TS	14	14	7	5½	5½	4¾	17/32	3¼	1⅝	3	⅜	⅜	1⅞
324U, T	16	14	8	6¼	5¼	5¼	21/32	5⅝, 5¼	1⅞, 2⅛	5⅜, 5	½, ½	½, ½	4¼, 3⅞
324S, TS	16	14	8	6¼	5¼	5¼	21/32	3¼, 3¾	1⅝, 1⅞	3, 3½	⅜, ½	⅜, ½	1⅞, 2
326U, T	16	15½	8	6¼	6	5¼	21/32	5⅝, 5¼	1⅞, 2⅛	5⅜, 5	½, ½	½, ½	4¼, 3⅞
326S, TS	16	15½	8	6¼	6	5¼	21/32	3¼, 3¾	1⅝, 1⅞	3, 3½	⅜, ½	⅜, ½	1⅞, 2
364U, T	18	15¼	9	7	5⅝	5⅞	21/32	6⅜, 5⅞	2⅛, 2⅜	6⅛, 5⅝	½, ⅝	½, ⅝	5, 4¼
364US, TS	18	15¼	9	7	5⅝	5⅞	21/32	3¾, 3¾	1⅞, 1⅞	3½, 3½	½, ½	½, ½	2, 2
365U, T	18	16¼	9	7	6⅛	5⅞	21/32	6⅜, 5⅞	2⅛, 2⅜	6⅛, 5⅝	½, ⅝	½, ⅝	5, 4¼
365US, TS	18	16¼	9	7	6⅛	5⅞	21/32	3¾, 3¾	1⅞, 1⅞	3½, 3½	½, ½	½, ½	2, 2
404U, T	20	16¼	10	8	6⅛	6⅝	13/16	7⅛, 7¼	2⅜, 2⅞	6⅞, 7	⅝, ¾	⅝, ¾	5½, 5⅝
404US, TS	20	16¼	10	8	6⅛	6⅝	13/16	4¼, 4¼	2⅛, 2⅛	4, 4	½, ½	½, ½	2¾, 2¾
405U, T	20	17¾	10	8	6⅞	6⅝	13/16	7⅛, 7¼	2⅜, 2⅞	6⅞, 7	⅝, ¾	⅝, ¾	5½, 5⅝
405US, TS	20	17¾	10	8	6⅞	6⅝	13/16	4¼, 4¼	2⅛, 2⅛	4, 4	½, ½	½, ½	2¾, 2¾
444U, T	22	18½	11	9	7¼	7½	13/16	8⅝, 8½	2⅞, 3⅜	8⅜, 8¼	¾, ⅞	¾, ⅞	7, 6⅞
444US, TS	22	18½	11	9	7¼	7½	13/16	4¼, 4¾	2⅛, 2⅜	4, 4½	½, ⅝	½, ⅝	2¾, 3
445U, T	22	20½	11	9	8¼	7½	13/16	8⅝, 8½	2⅞, 3⅜	8⅜, 8¼	¾, ⅞	¾, ⅞	7, 6⅞
445US, TS	22	20½	11	9	8¼	7½	13/16	4¼, 4¾	2⅛, 2⅜	4, 4½	½, ⅝	½, ⅝	2¾, 3
504U	25	21	12½	10	8	8½	15/16	8⅝	2⅞	8⅜	¾	¾	7¼
504S	25	21	12½	10	8	8½	15/16	4¼	2⅛	4	½	½	2¾
505	25	23	12½	10	9	8½	15/16	8⅝	2⅞	8⅜	¾	¾	7¼
505S	25	23	12½	10	9	8½	15/16	4¼	2⅛	4	½	½	2¾

Adapted from MG 1-11.31 and 11.31a.

*Dimension D will never be greater than the values listed, but it may be less so that shims are usually required for coupled or geared machines. When exact dimension is required, shims up to 1/32 in. may be necessary on frame sizes whose dimension D is 8 in. and less; on larger frames, shims up to 1/16 in. may be necessary.

†Second value, where present, is for rerated T frames. Values for frames 143T through 326TS are final; values for 364T through 445TS are tentative.

§Effective length of keyway.

Table 2

THREE-PHASE SQUIRREL CAGE AC MOTOR LOCKED-ROTOR TORQUE AND BREAKDOWN TORQUE

	locked-rotor torque, % of full load torque										breakdown torque, % of full load torque*									
	designs A and B							design C			design B							design C		
syn 60 cy speeds 50 cy (rpm)	3600 3000	1800 1500	1200 1000	900 750	720 ...	600 ...	514 ...	1800 1500	1200 1000	900 750	3600 3000	1800 1500	1200 1000	900 750	720 ...	600 ...	514 ...	1800 1500	1200 1000	900 750
hp.																				
½	...	...	...	140	140	115	110	...	...	...	...	...	...	225	200	200	200	...	...	...
¾	...	...	175	135	135	115	110	...	...	...	...	...	275	220	200	200	200	...	...	...
1	...	275	170	135	135	115	110	...	...	...	...	300	265	215	200	200	200	...	...	...
1½	175	250	165	130	130	115	110	...	...	...	250	280	250	210	200	200	200	...	...	...
2	170	235	160	130	125	115	110	...	...	...	240	270	240	210	200	200	200	...	...	...
3	160	215	155	130	125	115	110	...	250	225	230	250	230	205	200	200	200	...	225	200
5	150	185	150	130	125	115	110	250	250	225	215	225	215	205	200	200	200	200	200	200
7½	140	175	150	125	120	115	110	250	225	200	200	215	205	200	200	200	200	190	190	190
10	135	165	150	125	120	115	110	250	225	200	200	200	200	200	200	200	200	190	190	190
15	130	160	140	125	120	115	110	225	200	200	200	200	200	200	200	200	200	190	190	190
20	130	150	135	125	120	115	110	200	200	200	200	200	200	200	200	200	200	190	190	190
25	130	150	135	125	120	115	110	200	200	200	200	200	200	200	200	200	200	190	190	190
30	130	150	135	125	120	115	110	200	200	200	200	200	200	200	200	200	200	190	190	190
40	125	140	135	125	120	115	110	200	200	200	200	200	200	200	200	200	200	190	190	190
50	120	140	135	125	120	115	110	200	200	200	200	200	200	200	200	200	200	190	190	190

[MG1-12.35]

*The breakdown torque of general purpose, polyphase, squirrel-cage, fractional-horsepower motors, with rated voltage and frequency applied, shall not be less than 140 percent of the breakdown torque of a single-phase general purpose fractional-horsepower, motor of the same horsepower and speed rating.

Note 1: Motors rated 208 volts may have 10 percent lower torques than 220 volt motors having same horsepower and speed rating.

Note 2: Speed at breakdown torque is ordinarily much lower for fractional-horsepower polyphase motors than for fractional-horsepower, single-phase motors. Higher breakdown torques are required for polyphase motors so that polyphase and single-phase motors will have interchangeable running characteristics, rating for rating, when applied to normal single-phase motor loads.

Table 3

LOCKED-ROTOR TORQUE FOR SINGLE PHASE GENERAL PURPOSE MOTORS.

	minimum locked—rotor torque						
	60-cycle speed, rpm			50-cycle speed, rpm			
hp	3600 3450	1800 1725	1200 1140	3000 2850	1500 1425	1000 950	
1/8	......	24	32	......	29	39	torque in oz. ft.
1/6	15	33	43	18	39	51	
1/4	21	46	59	25	55	70	
1/3	26	57	73	31	69	88	
1/2	37	85	100	44	102	120	
3/4	50	119	8.0	60	143	......	
1	61	9.0	9.5	73	......	......	
1-1/2	4.5	12.5	13.0	......	......	......	torque in lb. ft.
2	5.5	16.0	16.0	......	......	......	
3	7.5	22.0	23.0	......	......	......	
5	11.0	33.0		......	......	......	
7-1/2	16.0	45.0		......	......	......	

[MG1-12.31]

Note 1. Approximate full speeds are for fractional horsepower motor ratings only.

Table 4

THREE-PHASE SQUIRREL CAGE AC MOTOR LOCKED ROTOR CURRENT*1

Hp	Locked-rotor Current, Amperes *2	Design Letters
½	20	B, D
¾	25	B, D
1	30	B, D
1½	40	B, D
2	50	B, D
3	64	B, C, D
5	92	B, C, D
7½	127	B, C, D
10	162	B, C, D
15	232	B, C, D
20	290	B, C, D
25	365	B, C, D
30	435	B, C, D
40	580	B, C, D
50	725	B, C, D

[MG1-12.34]

*1. Single-speed and constant speed design.

*2. *Locked-rotor current of motors designed for voltages other than 220 volts shall be inversely proportional to the voltages, except for motors rated at 230 volts (see MG1-12.34.a).

Table 5

SINGLE-PHASE FRACTIONAL HORSEPOWER
LOCKED-ROTOR CURRENT

hp	locked-rotor current, amperes*			
	115 volts		230 volts	
	design O	design N	design O	design N
1/6 and smaller	50	20	25	12
1/4	50	26	25	15
1/3	50	31	25	18
1/2	50	45	25	25
3/4	......	61	......	35

[MG1-12.32]

*For 2-, 4-, 6- and 8-pole, 60 cycle motors, single phase.

Note 1. The locked-rotor currents of single-phase general purpose fractional-horsepower motors shall not exceed the values for Design N motors.

Table 6

SINGLE-PHASE INTERGRAL HORSEPOWER
LOCKED-ROTOR CURRENT

hp	locked-rotor current, amperes		
	design L motors		design M motors
	115 volts	230 volts	230 volts
1	70	35	......
1½	......	50	40
2	......	65	50
3	......	90	70
5	......	135	100
7½	......	200	150
10	......	260	200
15	......	390	300
20	......	520	400

[MG1-12.33]

Note 1. The locked-rotor current of single-phase, 60 cycle design L and M motors of all types, when measured with rated voltage and frequency impressed and with the rotor locked shall not exceed the above values.

Table 7

ALLOWABLE SIZE CONDUCTOR FOR COPPER, 115-120 VOLTS, SINGLE PHASE WITH 2% VOLTAGE DROP.

	Minimum Allowable Size of Conductor			Copper up to 200 Amperes, 115-120 Volts, Single Phase, Based on 2% Voltage Drop																					
	In Cable, Conduit, Earth		Overhead in Air*	Length of Run in Feet — Compare size shown below with size shown to left of double line. Use the larger size.																					
Load in Amps	Types R, T, TW	Types RH, RHW, THW	Bare & Covered Conductors	30	40	50	60	75	100	125	150	175	200	225	250	275	300	350	400	450	500	550	600	650	700
5	12	12	10	12	12	12	12	12	12	12	10	10	10	10	8	8	8	8	6	6	6	6	4	4	4
7	12	12	10	12	12	12	12	12	12	10	10	8	8	8	8	6	6	6	6	4	4	4	4	4	3
10	12	12	10	12	12	12	12	10	10	8	8	8	6	6	6	6	4	4	4	4	3	3	2	2	2
15	12	12	10	12	12	10	10	10	8	6	6	6	4	4	4	4	4	3	2	2	1	1	1	0	0
20	12	12	10	12	10	10	8	8	6	6	4	4	4	4	3	3	2	2	1	1	0	0	00	00	00
25	10	10	10	10	10	8	8	6	6	4	4	4	3	3	2	2	1	1	0	0	00	00	000	000	000
30	10	10	10	10	8	8	8	6	4	4	4	3	2	2	1	1	1	0	00	00	000	000	000	4/0	4/0
35	8	8	10	10	8	8	6	6	4	4	3	2	2	1	1	0	0	00	00	000	000	4/0	4/0	4/0	250
40	8	8	10	8	8	6	6	4	4	3	2	2	1	1	0	0	00	00	000	000	4/0	4/0	250	250	300
45	6	8	10	8	8	6	6	4	4	3	2	1	1	0	0	00	00	000	000	4/0	4/0	250	250	300	300
50	6	6	10	8	6	6	4	4	3	2	1	1	0	0	00	00	000	000	4/0	4/0	250	250	300	300	350
60	4	6	8	8	6	4	4	4	2	1	1	0	00	00	000	000	000	4/0	250	250	300	300	350	400	400
70	4	4	8	6	6	4	4	3	2	1	0	00	00	000	000	4/0	4/0	250	300	300	350	400	400	500	500
80	2	4	6	6	4	4	3	2	1	0	00	00	000	000	4/0	4/0	250	300	300	350	400	400	500	500	600
90	2	3	6	6	4	4	3	2	1	0	00	000	000	4/0	4/0	250	250	300	350	400	500	500	500	600	600
100	1	3	6	4	4	3	2	1	0	00	000	000	4/0	4/0	250	250	300	350	400	500	500	500	600	600	700
115	0	2	4	4	4	3	2	1	0	00	000	4/0	4/0	250	300	300	350	400	500	500	600	600	700	700	750
130	00	1	4	4	3	2	1	0	00	000	4/0	4/0	250	300	300	350	400	500	500	600	600	700	750	800	900
150	000	0	2	4	2	1	1	0	000	4/0	4/0	250	300	350	350	400	500	500	600	700	700	800	900	900	1M
175	4/0	00	2	3	2	1	0	00	000	4/0	250	300	350	400	400	500	500	600	700	750	800	900	1M		
200	250	000	1	2	1	0	00	000	4/0	250	300	350	400	500	500	500	600	700	750	900	1M				

Table 8

ALLOWABLE SIZE CONDUCTOR FOR COPPER 115-120 VOLTS, SINGLE PHASE WITH 3% VOLTAGE DROP.

	Minimum Allowable Size of Conductor			Copper up to 200 Amperes, 115-120 Votts, Single Phase, Based on 3% Voltage Drop																					
	In Cable, Conduit, Earth		Overhead in Air*	Length of Run in Feet — Compare size shown below with size shown to left of double line. Use the larger size.																					
Load in Amps	Types R, T, TW	Types RH, RHW, THW	Bare & Covered Conductors	30	40	50	60	75	100	125	150	175	200	225	250	275	300	350	400	450	500	550	600	650	700
5	12	12	10	12	12	12	12	12	12	12	12	12	12	10	10	10	10	8	8	8	8	8	6	6	6
7	12	12	10	12	12	12	12	12	12	12	12	10	10	10	8	8	8	8	6	6	6	6	6	4	4
10	12	12	10	12	12	12	12	12	12	10	10	8	8	8	8	8	6	6	6	4	4	4	4	4	3
15	12	12	10	12	12	12	12	10	10	8	8	8	6	6	6	6	4	4	4	4	3	3	2	2	2
20	12	12	10	12	12	12	10	10	8	8	6	6	6	4	4	4	4	3	3	2	2	2	1	1	0
25	10	10	10	12	12	10	10	8	8	6	6	6	4	4	4	4	3	2	2	1	1	1	0	0	0
30	10	10	10	12	10	10	8	8	6	6	4	4	4	4	3	3	2	2	1	1	0	0	00	00	00
35	8	8	10	12	10	8	8	8	6	6	4	4	3	3	2	2	2	1	0	0	0	00	00	000	000
40	8	8	10	10	10	8	8	6	6	4	4	3	3	2	2	2	1	0	0	00	00	00	000	000	4/0
45	6	8	10	10	8	8	8	6	4	4	4	3	2	2	1	1	1	0	00	00	000	000	000	4/0	4/0
50	6	6	10	10	8	8	6	6	4	4	3	2	2	1	1	1	0	0	00	000	000	000	4/0	4/0	250
60	4	6	8	8	8	6	6	4	4	3	2	2	1	1	0	0	00	00	000	000	4/0	4/0	250	250	300
70	4	4	8	8	6	6	6	4	3	2	2	1	0	0	0	00	00	000	4/0	4/0	250	250	300	300	300
80	2	4	6	8	6	6	4	4	3	2	1	0	0	00	00	00	000	4/0	4/0	250	250	300	300	350	350
90	2	3	6	8	6	4	4	4	2	1	1	0	00	00	000	000	000	4/0	250	250	300	300	350	400	400
100	1	3	6	6	6	4	4	3	2	1	0	0	00	000	000	000	4/0	250	250	300	350	350	400	400	500
115	0	2	4	6	4	4	4	3	1	0	0	00	000	000	4/0	4/0	4/0	250	300	350	350	400	500	500	500
130	00	1	4	6	4	4	3	2	1	0	00	000	000	4/0	4/0	250	250	300	350	400	400	500	500	600	600
150	000	0	2	4	4	3	2	1	0	00	000	000	4/0	4/0	250	250	300	350	400	500	500	500	600	600	700
175	4/0	00	2	4	3	2	2	1	0	00	000	4/0	250	250	300	300	350	400	500	500	600	600	700	700	750
200	250	000	1	4	3	2	1	0	00	000	4/0	250	250	300	350	350	400	500	500	600	700	700	750	800	900

The above tables were printed courtesy of the Edison Electric Institute and the Food and Energy Council.

Conductors in overhead spans must be at least No. 10 for spans up to 50 feet and No. 8 for longer spans.

Table 9

ALLOWABLE SIZE CONDUCTOR FOR COPPER, 115-120 VOLTS, SINGLE PHASE WITH 4% VOLTAGE DROP

Minimum Allowable Size of Conductor — Copper up to 200 Amperes, 115-120 Volts, Single Phase, Based on 4% Voltage Drop

Length of Run in Feet — Compare size shown below with size shown to left of double line. Use the larger size.

Load in Amps	In Cable, Conduit, Earth: Types R, T, TW	In Cable, Conduit, Earth: Types RH, RHW, THW	Overhead in Air*: Bare & Covered Conductors	30	40	50	60	75	100	125	150	175	200	225	250	275	300	350	400	450	500	550	600	650	700
5	12	12	10	12	12	12	12	12	12	12	12	12	12	12	12	12	10	10	10	10	8	8	8	8	8
7	12	12	10	12	12	12	12	12	12	12	12	12	12	10	10	10	10	8	8	8	8	6	6	6	6
10	12	12	10	12	12	12	12	12	12	12	10	10	10	10	8	8	8	8	6	6	6	6	4	4	4
15	12	12	10	12	12	12	12	12	10	10	10	8	8	8	6	6	6	6	4	4	4	4	4	3	3
20	12	12	10	12	12	12	12	10	10	8	8	8	6	6	6	6	4	4	4	4	3	3	2	2	2
25	10	10	10	12	12	12	10	10	8	8	6	6	6	6	4	4	4	4	3	3	2	2	1	1	1
30	10	10	10	12	12	10	10	10	8	6	6	6	4	4	4	4	4	3	2	2	1	1	1	0	0
35	8	8	10	12	12	10	10	8	8	6	6	4	4	4	4	3	3	2	2	1	1	0	0	0	00
40	8	8	10	12	10	10	8	8	6	6	4	4	4	4	3	3	2	2	1	1	0	0	00	00	00
45	6	8	10	12	10	10	8	8	6	6	4	4	4	3	3	2	2	1	1	0	0	00	00	00	000
50	6	6	10	10	10	8	8	6	6	4	4	4	3	3	2	2	1	1	0	0	00	00	000	000	000
60	4	6	8	10	8	8	8	6	4	4	4	3	2	2	1	1	1	0	00	00	000	000	000	4/0	4/0
70	4	4	8	10	8	8	6	6	4	4	3	2	2	1	1	0	0	00	00	000	000	4/0	4/0	4/0	250
80	2	4	6	8	8	6	6	4	4	3	2	2	1	1	0	0	00	00	000	000	4/0	4/0	250	250	300
90	2	3	6	8	8	6	6	4	4	3	2	1	1	0	0	00	00	000	000	4/0	4/0	250	250	300	300
100	1	3	6	8	6	6	4	4	3	2	1	1	0	0	00	00	000	000	4/0	4/0	250	250	300	300	350
115	0	2	4	8	6	6	4	4	3	2	1	0	0	00	00	000	000	4/0	4/0	250	300	300	350	350	400
130	00	1	4	6	6	4	4	3	2	1	0	0	00	00	000	000	4/0	4/0	250	300	300	350	400	400	500
150	000	0	2	6	4	4	4	3	1	0	0	00	000	000	4/0	4/0	4/0	250	300	350	350	400	500	500	500
175	4/0	00	2	6	4	4	3	2	1	0	00	000	000	4/0	4/0	250	250	300	350	400	400	500	500	600	600
200	250	000	1	4	4	3	2	1	0	00	000	000	4/0	4/0	250	250	300	350	400	500	500	500	600	600	700

Table 10

ALLOWABLE SIZE CONDUCTOR FOR COPPER, 230-240 VOLTS, SINGLE PHASE WITH 2% VOLTAGE DROP

Minimum Allowable Size of Conductor — Copper up to 400 Amperes, 230-240 Volts, Single Phase, Based on 2% Voltage Drop

Length of Run in Feet — Compare size shown below with size shown to left of double line. Use the larger size.

Load in Amps	In Cable, Conduit, Earth: Types R, T, TW	In Cable, Conduit, Earth: Types RH, RHW, THW	Overhead in Air*: Bare & Covered Conductors	50	60	75	100	125	150	175	200	225	250	275	300	350	400	450	500	550	600	650	700	750	800
5	12	12	10	12	12	12	12	12	12	12	12	12	12	12	10	10	10	10	8	8	8	8	8	6	6
7	12	12	10	12	12	12	12	12	12	12	12	10	10	10	10	8	8	8	8	6	6	6	6	6	6
10	12	12	10	12	12	12	12	12	10	10	10	10	8	8	8	8	6	6	6	6	4	4	4	4	4
15	12	12	10	12	12	12	10	10	10	8	8	8	6	6	6	6	4	4	4	4	4	3	3	3	2
20	12	12	10	12	12	10	10	8	8	8	6	6	6	6	4	4	4	4	3	3	2	2	2	1	1
25	10	10	10	12	10	10	8	8	6	6	6	6	4	4	4	4	3	3	2	2	1	1	1	0	0
30	10	10	10	10	10	10	8	6	6	6	4	4	4	4	4	3	2	2	1	1	1	0	0	0	00
35	8	8	10	10	10	8	8	6	6	4	4	4	4	3	3	2	2	1	1	0	0	0	00	00	00
40	8	8	10	10	8	8	6	6	4	4	4	4	3	3	2	2	1	1	0	0	00	00	00	000	000
45	6	8	10	10	8	8	6	6	4	4	4	3	3	2	2	1	1	0	0	00	00	00	000	000	000
50	6	6	10	8	8	6	6	4	4	4	3	3	2	2	1	1	0	0	00	00	000	000	000	4/0	4/0
60	4	6	8	8	8	6	4	4	4	3	2	2	1	1	1	0	00	00	000	000	000	4/0	4/0	4/0	250
70	4	4	8	8	6	6	4	4	3	2	2	1	1	0	0	00	00	000	000	4/0	4/0	4/0	250	250	300
80	2	4	6	6	6	4	4	3	2	2	1	1	0	0	00	00	000	000	4/0	4/0	250	250	300	300	300
90	2	3	6	6	6	4	4	3	2	1	1	0	0	00	00	000	000	4/0	4/0	250	250	300	300	350	350
100	1	3	6	6	4	4	3	2	1	1	0	0	00	00	000	000	4/0	4/0	250	250	300	300	350	350	400
115	0	2	4	6	4	4	3	2	1	0	0	00	00	000	000	4/0	4/0	250	300	300	350	350	400	400	500
130	00	1	4	4	4	3	2	1	0	0	00	00	000	000	4/0	4/0	250	300	300	350	400	400	500	500	500
150	000	0	2	4	4	3	1	0	0	00	000	000	4/0	4/0	4/0	250	300	350	350	400	500	500	500	600	600
175	4/0	00	2	4	3	2	1	0	00	000	000	4/0	4/0	250	250	300	350	400	400	500	500	600	600	600	700
200	250	000	1	3	2	1	0	00	000	000	4/0	4/0	250	250	300	350	400	500	500	500	600	600	700	700	750
225	300	4/0	0	3	2	1	0	00	000	4/0	4/0	250	300	300	350	400	500	500	600	600	700	700	750	800	900
250	350	250	00	2	1	0	00	000	4/0	4/0	250	300	300	350	350	400	500	600	600	700	700	750	800	900	1M
275	400	300	00	2	1	0	00	000	4/0	250	250	300	350	350	400	500	500	600	700	700	800	900	900	1M	
300	500	350	000	1	1	0	000	4/0	4/0	250	300	350	350	400	500	500	600	700	700	800	900	900	1M		
325	600	400	4/0	1	0	00	000	4/0	250	300	300	350	400	500	500	600	600	700	750	900	900	1M			
350	600	500	4/0	1	0	00	000	4/0	250	300	350	400	400	500	500	600	700	750	800	900	1M				
375	700	500	250	0	0	00	4/0	250	300	300	350	400	500	500	600	600	700	800	900	1M					
400	750	600	250	0	00	000	4/0	250	300	350	400	500	500	500	600	700	750	900	1M						

Conductors in overhead spans must be at least No. 10 for spans up to 50 feet and No. 8 for longer spans.

The above tables were printed courtesy of the Edison Electric Institute and the Food and Energy Council.

Table 11. Approximate Power Requirements of Common Electrical Equipment and Small Appliances Found Around the Home.

EQUIPMENT	WATTS
Air conditioner central, electric	3000-9000
Air conditioner, room	800-3000
Blender	200-300
Blanket, electric	150-200
Clock	2-3
Coffee maker	500-1000
Corn popper	450-600
Crock-pot, slow cooker	150-300
Dishwasher	600-1200
Dryer, clothes	4000-5000
Entertainment center no TV	30-100
Fan, 8-12 inch portable	40-80
Fan, kitchen vent	100
Freezer, household	300-500
Fryer, deep fat	1200-1650
Frying pan	1000-1200
Furnace, oil fired (fan and burner)	600-800
Garbage disposal	300-600
Heater, permanent wall type	1000-2300
Heater, portable household	1000-1500
Heater, water	2000-5000
Hot plate, per burner	600-1000
Ironer	1200-1500
Iron, hand	660-1200
Knife, electric	100
Lamps, fluorescent	15-60
Lamps, incandescent	15-1000
Micro-wave oven	700-1500
Mixer, food	120-250
Motors, less than 1 hp	1200/hp
1 hp and above	1000/hp
Polisher, floor	250
Projector, movie or slide	300-1000
Radio	10-80
Range, oven only	4000-5000
Range, top only	4000-6000
Razor	8-12
Refrigerator, household	200-400
Roaster	1200-1650
Sewing machine	60-90
Soldering iron	100-300
Television	200-400
Toaster	600-1200
Vacuum cleaner	250-800
Waffle iron	600-1000
Washer, automatic	600-800
Washing machine	350-550

Table 12. Circular Mil and Resistance Chart For Solid Copper Wire

(American wire gage--B & S)

Gage number	Diameter (mils)	Cross section		Ohms per 1,000 ft.		Ohms per mile 25° C. (=77° F.)	Pounds per 1,000 ft.
		Circular mils	Square inches	25° C. (= 77° F.)	65° C. (= 149° F.)		
0000	460.0	212,000.0	0.166	0.0500	0.0577	0.264	641.0
000	410.0	168,000.0	.132	.0630	.0727	.333	508.0
00	365.0	133,000.0	.105	.0795	.0917	.420	403.0
0	325.0	106,000.0	.0829	.100	.116	.528	319.0
1	289.0	83,700.0	.0657	.126	.146	.665	253.0
2	258.0	66,400.0	.0521	.159	.184	.839	201.0
3	229.0	52,600.0	.0413	.201	.232	1.061	159.0
4	204.0	41,700.0	.0328	.253	.292	1.335	126.0
5	182.0	33,100.0	.0260	.319	.369	1.685	100.0
6	162.0	26,300.0	.0206	.403	.465	2.13	79.5
7	144.0	20,800.0	.0164	.508	.586	2.68	63.0
8	128.0	16,500.0	.0130	.641	.739	3.38	50.0
9	114.0	13,100.0	.0103	.808	.932	4.27	39.6
10	102.0	10,400.0	.00815	1.02	1.18	5.38	31.4
11	91.0	8,230.0	.00647	1.28	1.48	6.75	24.9
12	81.0	6,530.0	.00513	1.62	1.87	8.55	19.8
13	72.0	5,180.0	.00407	2.04	2.36	10.77	15.7
14	64.0	4,110.0	.00323	2.58	2.97	13.62	12.4
15	57.0	3,260.0	.00256	3.25	3.75	17.16	9.86
16	51.0	2,580.0	.00203	4.09	4.73	21.6	7.82
17	45.0	2,050.0	.00161	5.16	5.96	27.2	6.20
18	40.0	1,620.0	.00128	6.51	7.51	34.4	4.92
19	36.0	1,290.0	.00101	8.21	9.48	43.3	3.90
20	32.0	1,020.0	.000802	10.4	11.9	54.9	3.09
21	28.5	810.0	.000636	13.1	15.1	69.1	2.45
22	25.3	642.0	.000505	16.5	19.0	87.1	1.94
23	22.6	509.0	.000400	20.8	24.0	109.8	1.54
24	20.1	404.0	.000317	26.2	30.2	138.3	1.22
25	17.9	320.0	.000252	33.0	38.1	174.1	0.970
26	15.9	254.0	.000200	41.6	48.0	220.0	0.769
27	14.2	202.0	.000158	52.5	60.6	277.0	0.610
28	12.6	160.0	.000126	66.2	76.4	350.0	0.484
29	11.3	127.0	.0000995	83.4	96.3	440.0	0.384
30	10.0	101.0	.0000789	105.0	121.0	554.0	0.304
31	8.9	79.7	.0000626	133.0	153.0	702.0	0.241
32	8.0	63.2	.0000496	167.0	193.0	882.0	0.191
33	7.1	50.1	.0000394	211.0	243.0	1,114.0	0.152
34	6.3	39.8	.0000312	266.0	307.0	1,404.0	0.120
35	5.6	31.5	.0000248	335.0	387.0	1,769.0	0.0954
36	5.0	25.0	.0000196	423.0	488.0	2,230.0	0.0757
37	4.5	19.8	.0000156	533.0	616.0	2,810.0	0.0600
38	4.0	15.7	.0000123	673.0	776.0	3,550.0	0.0476
39	3.5	12.5	.0000098	848.0	979.0	4,480.0	0.0377
40	3.1	9.9	.0000078	1,070.0	1,230.0	5,650.0	0.0299

N/O SPST MAGNETIC RELAY CIRCUIT

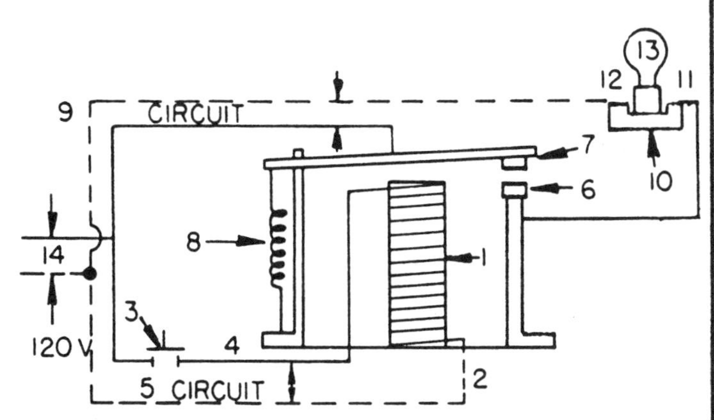

SPST Magnetic Relay

Part Identification:

1.	________	8.	________
2.	________	9.	________
3.	________	10.	________
4.	________	11.	________
5.	________	12.	________
6.	________	13.	________
7.	________	14.	________

Operation Teaches: Ability to ...

1. Identify parts of the N/O SPST magnetic relay circuits.
2. Define specifications on the SPST magnetic relay nameplate.
3. Outline the operating principles of an electromagnet.
4. Connect a SPST magnetic relay in a circuit.
5. Need to control a large ampere electrical load with a low ampere electrical load control circuit.
6. Properly connect the circuit.
7. Identify the two circuits in a SPST magnetic relay lamp circuit and switch.
8. List applications for SPST magnetic relays.

Name ________________

Date ________ Grade ________

Operational Procedure:

1. Complete the part identification.
2. Determine the voltage of the relay load circuit ________
3. Determine the voltage of the relay control circuit ________
4. Determine the largest load that can be controlled ________
5. Connect the SPST magnetic relay switch as diagrammed.
6. What happens when you complete the control circuit?

   ____

   Why does this happen? ________

   ____
7. What happens when you open the switch? ________

   ____

   Why does this happen? ________
8. Trace the current flow in the control and load circuits.
9. List specific applications for the SPST magnetic relay.

   ____

Materials:

1. Single pole single throw (SPST) magnetic relay.
2. Light bulb and socket
3. Push button or two-way switch
4. Conductors: black and white
5. Ground receptacle extension cord with 3 amp fuse
6. Screwdriver
7. Grounded plug cap
8. Electrical and Basic Controls in Ag. Production, Hobar Publications #2177

Evaluation Score Sheet:

Item	Points Possible	Points Earned
1. Part identification	28	________
2. Operational questions	24	________
3. Correct wiring of relay control and load circuit	18	________
4. Correct wiring switch	10	________
5. Correct wiring lamp	10	________
6. Safety and work habits	10	________
Total	100	________

NOTES

SPDT MAGNETIC RELAY CIRCUIT

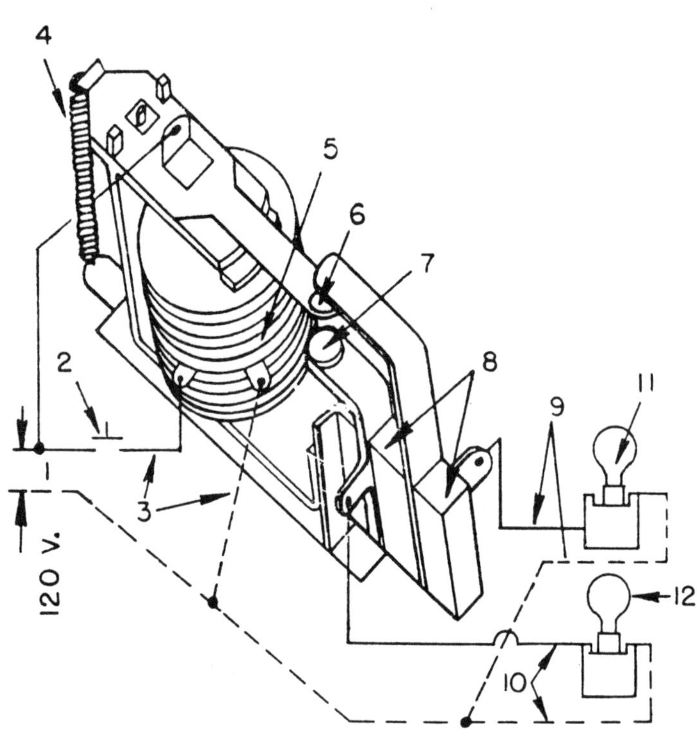

SPDT Magnetic Relay

Part Identification:

1. ______________ 7. ______________

2. ______________ 8. ______________

3. ______________ 9. ______________

4. ______________ 10. ______________

5. ______________ 11. ______________

6. ______________ 12. ______________

Evaluation Score Sheet:

Item	Points Possible	Points Earned
1. Part identification	36	________
2. Operational questions	28	________
3. Proper wiring of the circuit	26	________
4. Safety and work habits	10	________
Total	100	________

Operational Procedure:

1. Complete the part identification.
2. Trace the control circuit in the diagram.
3. Trace the N/O load circuit in the diagram.
4. Trace the N/C load circuit in the diagram.
5. What causes the N/O switch to close? ______________
   The N/C switch to open? ______________
6. What is the largest load that this SPDT switch can control? ______________
7. Wire the SPDT magnetic relay, switch and lamps as diagrammed.
8. What happens to the lamps when you press the push button switch? ______________
9. What happens when you release the push button switch? ______________
10. Which lamp is on the N/C and which is on the N/O circuit? ______________
11. What are some uses for a SPDT magnetic relay? ______________

Operation Teaches: Ability to ..

1. Identify the parts of a SPDT magnetic relay.
2. Trace the control and load circuits.
3. Describe the principles on which the SPDT magnetic relay operates.
4. Read the specifications for this SPDT magnetic relay.
5. Wire the SPDT magnetic relay.
6. List uses of a SPDT magnetic relay.

Materials:

1. Single pole, double throw (SPDT) magnetic relay
2. Conductors: black and white
3. Light bulbs and sockets
4. Push button or two-way switch
5. Screwdriver
6. Grounded receptacle extension cord with 3 amp fuse
7. Grounded plug cap
8. Electrical and Basic Controls in Ag. Production, Hobar Publications #2177

Name ______________________

Date ____________ Grade ______________

NOTES

PUSH BUTTON MAGNETIC STARTER SWITCH CIRCUIT

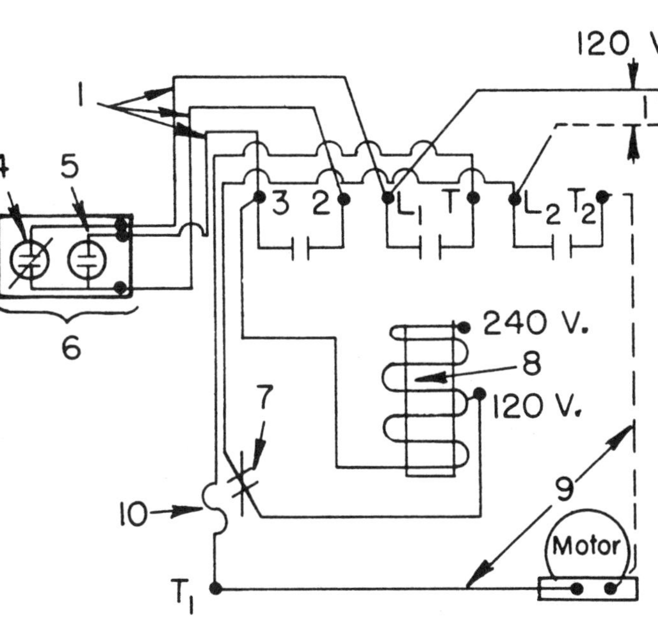

Magnetic Starter Switch

Part Identification:

1.	______	4.	______
2.	______	5.	______
3.	______	6.	______
L1.	______	7.	______
L2.	______	8.	______
T.	______	9.	______
T1.	______	10.	______
T2.	______	11.	______

Evaluation Score Sheet:

Item	Possible	Earned
1. Part identification	32	______
2. Operational questions		
3. Proper wiring of the push button and magnetic starter switch	24	______
	34	______
4. Safety and work habits	10	______
Total	100	______

Operational Procedure:

1. Complete the part indentification.
2. Trace the control circuit in the diagram.
3. Trace the load circuit in the diagram.
4. What causes the switch to close the load circuit?
   ______
5. What causes the switch to open the load circuit?
   ______
6. What is the largest load that this switch can control?
   ______
7. Wire the push button and magnetic starter as diagrammed.
8. After you press the start button, it returns to its N/O position. Why does the motor keep running? (How is the control circuit still closed?)
   ______
9. How do you open the control circuit and stop the motor?
   ______
10. What are some uses for a push button and magnetic starter switch?
   ______

Operation Teaches:

Ability to ...

1. Identify the parts of the push button and magnetic starter switch.
2. Describe the operating principles for the magnetic starter switch.
3. Trace the control and load circuits.
4. Properly wire this push button and magnetic starter switch.
5. List uses of a push button and magnetic starter switch.

Materials:

1. Three pole magnetic starter switch
2. Push button switch
3. Electric motor
4. Conductors: black and white
5. Screwdriver
6. Grounded receptacle extension cord with 3 amp fuse
7. Grounded plug cap
8. Electrical and Basic Controls in Ag. Production, Hobar Publications #2177

Name ______

Date ______ Grade ______

NOTES

THERMOSTAT CIRCUIT

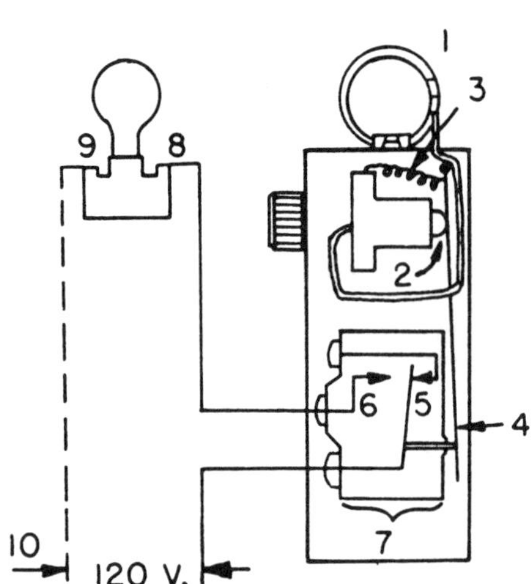

Thermostat

Part Identification:

1. ________________ 6. ________________

2. ________________ 7. ________________

3. ________________ 8. ________________

4. ________________ 9. ________________

5. ________________ 10. ________________

Evaluation Score Sheet: Points

Item	Possible	Earned
1. Part identification	30	______
2. Operational questions	36	______
3. Proper wiring of the thermostat	24	______
4. Safety and work habits	10	______
Total	100	______

Operational Procedure:

1. Complete the part identificaiton.
2. What type of thermostat is this? ________________
   Why? ________________
3. List two other types of thermostats: ________________
   ____
4. Desciribe the principles on which the thermostat in steps 2 and 3 operate:
   ____
   ____
5. What is the maximum load for this thermostat? ________
6. Wire the thermostat so that the switch is N/O. (Wired as diagrammed.)
   Does the light come on when the temperature is higher or lower than the setting? ________________
   What are uses for a thermostat wired like this? ______
7. Wire the thermostat so that the switch is N/C. Does the light come on when the temperature is higher or lower than the setting?
   ____
   What are the uses of a thermostat wired like this?
   ____

Operation Teaches: Ability to ...

1. Identify the parts of a thermostat.
2. Differentiate between different kinds of thermostats.
3. Describe the principles on which thermostats work.
4. Read the specifications for the thermostat.
5. Properly wire the thermostat circuit.
6. Adjust the temperature setting of the thermostat.
7. List uses for a thermostat.

Materials:

1. Thermostat
2. Conductors: black and white
3. Screwdriver
4. Light bulb and socket
5. Grounded receptacle extension cord with 3 amp fuse
6. Ground cap plug
7. Electrical and Basic Controls in Ag. Production, Hobar Publications #2177

Name ________________________

Date ____________ Grade ________________

NOTES

24 HOUR TIME SWITCH CIRCUIT

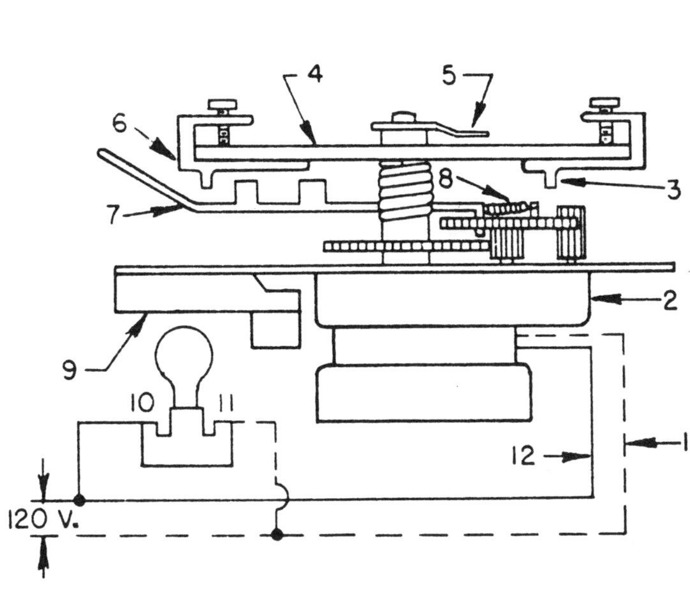

24 HOUR TIME SWITCH

Part Identification:

1.	________	7.	________
2.	________	8.	________
3.	________	9.	________
4.	________	10.	________
5.	________	11.	________
6.	________	12.	________

Operational Procedure:

1. Complete the part identification.
2. Trace the control circuit in the diagram.
3. Trace the load circuit in the diagram.
4. What causes the switch to close the load circuit? ________
5. What causes the switch to open the load circuit? ________
6. How do you set the trippers? ________
7. How do you set the time? ________
8. Can the switch be manually operated? ________
9. What is the largest load that this 24 hour time switch can control? ________
10. What is the recommended voltage for this 24 hour time switch? ________
11. Wire the 24 hour time switch as diagrammed.
12. Set the switch so that the light stays on for 2 hours.
13. Will the switch close the control circuit again automatically? ________
    How long does it take? ________
14. What are some of the farm uses for a 24 hour time switch? ________

Operation Teaches: Ability to ...

1. Identify the parts of the 24 hour time switch.
2. Trace the control and load circuits.
3. Describe the operating principles for the 24 hour time switch.
4. Read the specifications for the 24 hour time switch.
5. Properly wire and adjust the 24 hour time switch.
6. List uses for a 24 hour time switch.

Materials:

1. 24 hour time switch
2. Light bulb and socket
3. Conductors: black and white
4. Screwdriver
5. Grounded receptacle extension cord with 3 amp fuse
6. Grounded cap plug
7. Electrical and Basic Controls in Ag. Production, Hobar Publications #2177

Name ____________________

Date __________ Grade __________

Evaluation Score Sheet:

Item	Points Possible	Points Earned
1. Part identification	36	______
2. Operational questions	20	______
3. Proper wiring of the 24 hour time switch	34	______
4. Safety and work habits	10	______
Total	100	______

NOTES

REPEAT TIMER SWITCH CIRCUIT

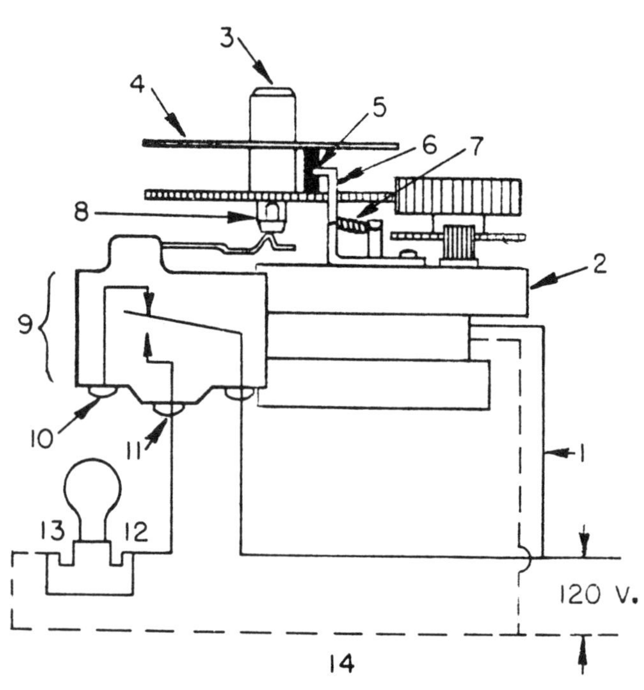

REPEAT TIMER SWITCH

Operational Procedure:

1. Complete the part identification.
2. Trace the control circuit in the diagram.
3. Trace the load circuit in the diagram.
4. Is the load circuit hooked up to the N/O or N/C switch? ______
5. What causes the switch to close the load circuit? ______
6. What causes the switch to open the load circuit? ______
7. What is the largest load that this repeat timer switch can control? ______
8. What is the recommended voltage for this repeat timer switch? ______
9. Wire the repeat timer switch as diagrammed.
10. Set the switch so that the light stays on for 2 minutes.
11. Will the switch close the control circuit again automatically? ______
12. What are some of the uses for a repeat timer switch? ______

Part Identification:

1.	______	8.	______
2.	______	9.	______
3.	______	10.	______
4.	______	11.	______
5.	______	12.	______
6.	______	13.	______
7.	______	14.	______

Operation Teaches: Ability to ...

1. Identify the parts of the repeat timer switch.
2. Trace the control and load circuits.
3. Describe the principles on which the repeat timer switch operates.
4. Read the specifications for the repeat timer switch.
5. Properly wire and adjust the repeat timer switch.
6. List uses for a repeat timer switch.

Materials:

1. Repeat timer switch
2. Light bulb and socket
3. Conductors: black and white
4. Screwdriver
5. Grounded receptacle extension cord with 3 amp fuse
6. Grounded cap plug
7. Electrical and Basic Controls in Ag. Production, Hobar Publications #2177

Evaluation Score Sheet:

Item	Points Possible	Points Earned
1. Part identification	42	______
2. Operational questions	24	______
3. Proper wiring of the repeat timer switch	24	______
4. Safety and work habits	10	______
Total	100	______

Name ______

Date ______ Grade ______

NOTES

Index

Index

Index

Index

Index

NOTES